D1060187

INORGANIC SYNTHESES

Volume XIX

Editor-in Chief
DUWARD F. SHRIVER

Department of Chemistry
Northwestern University
Evanston, Illinois 60201

●●●●●●●●●●●●●●●●●●●●●●●●●⌐●●●●●●●●●●●●●●●●

INORGANIC
SYNTHESES

Volume XIX

A Wiley-Interscience Publication
JOHN WILEY & SONS

New York   Chichester   Brisbane   Toronto

Published by John Wiley & Sons, Inc.

Library of Congress Catalog Number: 39-23015

ISBN 0-471-04542-X

Printed in the United States of America

10 9 8 7 6 5 4 3 2 1

PREFACE

The wide range of synthetic techniques and compounds described here reflects the diversity of current inorganic chemistry. For example, the recent enthusiasm among inorganic chemists and solid-state physicists for studying anisotropic electronic conductors and fast-ion conductors is represented by a collection of syntheses for linear-chain conductors, layered tantalum sulfide intercalation compounds, and derivatives of the ionic conductor β-alumina. A chapter on Werner-type complexes includes the first reliable synthesis of hydrido(triphenyl-phosphine)copper(I), as well as a variety of other useful compounds. The widespread interest in organometallic compounds prompted the inclusion of a large chapter containing subsections devoted to thiocarbonyl and carbene complexes and to improved syntheses for more familiar compounds, such as arenetricarbonylchromium complexes and pentacarbonylmanganese halides. Improved syntheses for important main group compounds, such as hexaborane and dimethylzinc, are given in the final chapter. In a significant departure from previous volumes, one chapter is devoted to syntheses based on a single technique, the codeposition of metal atoms with reactive substrates. Metal atom syntheses are of recent origin, and the techniques are not familiar to many chemists; therefore the decision was made to provide a detailed practical introduction to the method and to locate all of the metal atom syntheses in one chapter, irrespective of compound type.

To ensure reliability, the procedures in *Inorganic Syntheses* are checked in a laboratory that is independent from that of the submitter, and the manuscripts are examined for clarity and completeness by the Editor, with the help of the Editorial Board. I greatly appreciate the efforts of the submitters, checkers, and Editorial Board members who were willing to undertake these tasks. Recent editors who helped me in many ways during the preparation of this book are Fred Basolo, Bodie Douglas, and George Parshall. I also appreciate suggestions made by Steven Ittel. Considerable help in the editing was provided by John Bailar, Conrad Fernelius, Smith Holt, Dennis Lehman, Warren Powell, Therald Moeller, and Thomas Sloan. I thank my secretary Madeleine Ziebka, who monitored the progress of manuscripts and efficiently handled many other chores.

I have personally benefited from earlier volumes of *Inorganic Syntheses*, which stimulated my interest in synthetic inorganic chemistry when I was an undergraduate and later provided reliable syntheses for compounds that I needed in research. I hope that this new volume is equally useful to others.

Duward F. Shriver

Evanston, Illinois
July 1978

NOTICE TO CONTRIBUTORS

The *Inorganic Syntheses* series is published to provide all users of inorganic substances with detailed and foolproof procedures for the preparation of important and timely compounds. Thus the series is the concern of the entire scientific community. The Editorial Board hopes that all chemists will share in the responsibility of producing *Inorganic Syntheses* by offering their advice and assistance in both the formulation of and the laboratory evaluation of outstanding syntheses. Help of this kind will be invaluable in achieving excellence and pertinence to current scientific interests.

There is no rigid definition of what constitutes a suitable synthesis. The major criterion by which syntheses are judged is the potential value to the scientific community. An ideal synthesis is one that presents a new or revised experimental procedure applicable to a variety of related compounds, at least one of which is critically important in current research. However, syntheses of individual compounds that are of interest or importance are also acceptable. Syntheses of compounds that are available commercially at reasonable prices are not acceptable.

The Editorial Board lists the following criteria of content for submitted manuscripts. Style should conform with that of previous volumes of *Inorganic Syntheses*. The introductory section should include a concise and critical summary of the available procedures for synthesis of the product in question. It should also include an estimate of the time required for the synthesis, an indication of the importance and utility of the product, and an admonition if any potential hazards are associated with the procedure. The Procedure should present detailed and unambiguous laboratory directions and be written so that it anticipates possible mistakes and misunderstandings on the part of the person who attempts to duplicate the procedure. Any unusual equipment or procedure should be clearly described. Line drawings should be included when they can be helpful. All safety measures should be stated clearly. Sources of unusual starting materials must be given, and, if possible, minimal standards of purity of reagents and solvents should be stated. The scale should be reasonable for normal laboratory operation, and any problems involved in scaling the procedure either up or down should be discussed. The criteria for judging the purity of the final product should be delineated clearly. The section of Properties should supply and discuss those physical and chemical characteristics that are relevant to judging the purity of the product and to permitting its handling and use in an intelligent manner. Under References, all pertinent literature citations should be listed in order. A style sheet is available from the Secretary of the Editorial Board.

The Editorial Board determines whether submitted syntheses meet the general specifications outlined above. Every synthesis must be satisfactorily reproduced in a laboratory other than the one from which it was submitted.

Each manuscript should be submitted in duplicate to the Secretary of the Editorial Board, Professor Jay H. Worrell, Department of Chemistry, University of South Florida, Tampa, FL 33620. The manuscript should be typewritten in English. Nomenclature should be consistent and should follow the recommendations presented in *Nomenclature of Inorganic Chemistry,* 2nd Ed., Butterworths & Co, London, 1970 and in *Pure Appl. Chem.,* **28**, No. 1 (1971). Abbreviations should conform to those used in publications of the American Chemical Society, particularly *Inorganic Chemistry.*

TOXIC SUBSTANCES AND LABORATORY HAZARDS

The obvious hazards associated with the preparations in this volume have been delineated in each experimental procedure but it is impossible to foresee and discuss all possible sources of danger. Therefore, the synthetic chemist should be familiar with general hazards associated with toxic, flammable, and explosive materials.

Recently it has become apparent that many common laboratory chemicals have subtle biological effects that were not suspected previously. A list of 400 toxic substances is available in the *Federal Register,* Vol. 40, No. 23072, May 28, 1975, and an abbreviated version of this list is presented in *Inorganic Syntheses,* Vol. 18, p. xv, 1978. For a current assessment of the hazards of any particular chemical, see the most recent edition of *Threshold Limit Values for Chemical Substances and Physical Agents in the Workroom Environment* published by the American Conference of Governmental Industrial Hygienists.

In light of the primitive state of our knowledge of the biological effects of chemicals, it is prudent that all the syntheses reported in this and other volumes of *Inorganic Syntheses* be conducted with rigorous care to avoid contact with all reactants, solvents, and products.

Attention also must be given to the explosion and fire hazards presented by combustible organic vapors and combustible gases such as hydrogen and methane. These vapors are readily ignited by static electricity, electrical sparks from most laboratory appliances, open flames, and other highly exothermic reactions. Thus appreciable atmospheric concentrations of combustible vapors should be avoided.

The drying of impure ethers presents a potential explosion hazard. Those unfamiliar with this hazard should consult *Inorganic Syntheses,* Vol. 12, p. 317, 1970.

Duward F. Shriver

CONTENTS

Chapter One   ELECTRICALLY CONDUCTING SOLIDS

Chapter Two   METAL ATOM SYNTHESES

Chapter Three   TRANSITION METAL COMPOUNDS AND COMPLEXES

Chapter Four TRANSITION METAL ORGANOMETALLIC COMPOUNDS

Chapter Five MAIN GROUP COMPOUNDS

Chapter One

ELECTRICALLY CONDUCTING SOLIDS

1. POTASSIUM TETRACYANOPLATINATE BROMIDE (2:1:0.3) TRIHYDRATE, $K_2[Pt(CN)_4]Br_{0.3} \cdot 3H_2O$*

Submitted by J. A. ABYS,[†] N. P. ENRIGHT,[†] H. M. GERDES,[†]
T. L. HALL,[†] and JACK M. WILLIAMS[‡]
Checked by J. ACKERMAN[§] and A. WOLD[§]

The highly anisotropic physical properties of the mixed-valence square planar inorganic compounds, of which $K_2[Pt(CN)_4]Br_{0.3} \cdot 3H_2O$ is the prototype, have attracted great interest because of their so-called one-dimensional metallic character.[1] Bromine nonstoichiometry in $K_2[Pt(CN)_4]Br_{0.3} \cdot 3H_2O$ has been extensively studied and is well understood.[1,2] The Li^+, Na^+, and Rb^+ analogues of the parent compound are not easily prepared because of difficulties encountered during hydrazine reduction of Pt^{IV} (Eq. 3). However, they appear to be accessible by way of the Ba^{2+} salt[3] by addition of the appropriate alkali metal sulfate to $Ba[Pt(CN)_4] \cdot 4H_2O$, which may be prepared from $K_2[Pt(CN)_4] \cdot 3H_2O$ followed by partial bromination and so forth. The following preparation is a

*Research performed under the auspices of the Division of Basic Energy Sciences of the U.S. Department of Energy.

[†]Research participants sponsored by the Argonne Center for Educational Affairs: (a) J. A. Abys from Marist College, Poughkeepsie, NY, (b) N. P. Enright from Middlebury College, Middlebury, VT, (c) H. M. Gerdes from Marist College, Poughkeepsie, NY, and (d) T. L. Hall from University of North Carolina, Chapel Hill, NC 27514.

[‡]Correspondent: Chemistry Division, Argonne National Laboratory, Argonne, IL 60439.

[§]Department of Chemistry, Brown University, Providence, RI 02912.

combination of the previously reported methods, including those used by Saillant et al.,[4] but with a considerable improvement in yield.

Materials

The platinum sponge* used in this preparation should be either 99.99 or 99.999% pure. All other chemicals should be reagent grade. Only KCN, which generally assays 97-98% pure, is recrystallized from absolute methanol.[5] It is dried at 50° for 2 hours before it is weighed, as are the KBr and $[N_2H_6]SO_4$. Distilled water is used throughout the entire preparation.

Procedure

1. *Dihydrogen hexabromoplatinate(IV)*. Pt + 2Br$_2$ + 2HBr → $H_2[PtBr_6]$. (■**Caution.** *Bromine and hydrobromic acid are extremely corrosive. It is recommended that all steps involving these chemicals be carried out in a hood and that gloves and face shields be worn.*) Hydrobromic acid (48%) (32 mL) is added to platinum[6] (4.5 g, 0.023 mole) in a 250-mL, round-bottomed flask equipped with a reflux condenser and heating mantle. Bromine (15 mL, 0.275 mole) is added, and the mixture is refluxed for approximately 5 hours or until all the platinum has reacted. The solution of hexabromoplatinic(IV) acid is allowed to cool and is then suction filtered using a medium-porosity sintered-glass filter. (■**Caution.** *A cold trap should be used to prevent unreacted bromine from escaping during filtration.*)

2. *Dipotassium hexabromoplatinate(IV)*. $H_2[PtBr_6]$ + 2KBr → $K_2[PtBr_6]$ + 2HBr. The hexabromoplatinic(IV) acid filtrate (reddish-purple) is placed in a 250-mL beaker, and potassium bromide (8.2 g, 0.069 mole) is added with vigorous mechanical stirring.[7,8] Approximately 20 mL of distilled water is added to the resulting reddish-orange slurry, and stirring is continued for 1 hour. The resulting precipitate of dipotassium hexabromoplatinate(IV) is isolated on a medium-porosity, sintered-glass filter in a hood and allowed to air dry for 20 minutes. It is then oven dried for 1 hour at 110°, at which temperature the salt turns dark red. Upon cooling to room temperature the salt assumes the original orange-red color. The yield for this step is 16.4-17.3 g (0.0218-0.0230 mole), representing 94-99% based on the original platinum.

3. *Dipotassium tetrabromoplatinate(II).*** 2K$_2$[PtBr$_6$] + $[N_2H_6]SO_4$ → 2K$_2$[PtBr$_4$] + 4HBr + H_2SO_4 + N$_2$. The potassium hexabromoplatinate(IV) obtained in reaction 2 is placed in a 250-mL beaker, and approximately 75 mL

---

*Available from United Mineral and Chemical Corp., 129 Hudson Street, NY 10013. Foil should be avoided because of the excessive reaction time required.

**Available from Research Organic/Inorganic Chemical Co., 11686 Sheldon St., Sun Valley, CA 91352.

of distilled water is added, producing a reddish-orange slurry. Vigorous stirring is begun and one-half of the stoichiometric amount of hydrazine sulfate[9] [based on the yield of dipotassium hexabromoplatinate(IV) obtained in reaction 2] is added slowly. The temperature is raised to 40-50° by means of a steam bath, and the mixture is stirred for 1 hour. During this period the color of the mixture changes from reddish-orange to a dark reddish-brown. Hydrazine sulfate is added in 0.1-0.3 g amounts, allowing 10 minutes between additions until only a trace (no more than enough to cover the bottom of the beaker) of unreacted dipotassium hexabromoplatinate(IV) remains. Excess hydrazine sulfate reduces the tetrabromoplatinate(IV) complex to platinum metal. Stirring is continued for 30 minutes after the last addition of hydrazine sulfate and then the mixture is filtered. The brown filtrate is placed in a porcelain dish and evaporated on a steam bath until a thick dark-brown slush remains. Crystals of dipotassium tetrabromoplatinate(II) form during evaporation and are broken up with a porcelain spatula while the crystals are kept wet at all times. The brown dipotassium tetrabromoplatinate(II) is filtered and air dried for 20 minutes on a sintered-glass filter and then oven dried at 50° for 1 hour before it is weighed. The yield for this step is 11.2-13.0 g (0.020-0.022 mole), representing 86-95% based on original platinum metal.

4. *Dipotassium tetracyanoplatinate(II) trihydrate.** $K_2[PtBr_4]$ + 4KCN +

$3H_2O \xrightarrow{H_2O} K_2[Pt(CN)_4] \cdot 3H_2O$ + 4KBr. The dipotassium tetrabromoplatinate(II) from reaction 3 is placed in a 100-mL beaker, 20 mL of distilled water is added with stirring, and the temperature is raised to 60°. The required stoichiometric amount of potassium cyanide is added in 0.2-0.4 g portions, allowing approximately 3 minutes between additions. (■**Caution.** *Because of the extremely poisonous nature of cyanide these steps should be carried out in a well-ventilated hood using protective gloves, clothing, and a face shield.*) When the brownish color begins to fade, smaller portions of potassium cyanide are added until full stoichiometry is achieved. Stirring is continued at about 60° for 20-30 minutes. Color changes occur during the reaction, with the initial cloudy and dark-brown solutions forming a lighter brown suspension early in the reaction. Near the termination of the reaction the color changes vary depending on two factors: (1) apparent differences in the hydration states of the products and (2) the amount of platinum black (metal) present in solution. Solution color variations from a yellow-green to a dark grey or grey-green sometimes occur. It is absolutely necessary to filter the solution *hot* to remove any insoluble compounds formed during the reaction. The resulting solution is transfered to a 100-mL beaker and cooled in a NaCl-ice bath for 20-30 minutes. Needlelike crystals are obtained and variations in color due to the difference in hydration state of

*Available from ICN-K & K Laboratories, Inc., 121 Express St. Plainview, NY 11803. Recrystallization may be necessary.

the product are observed. For a better yield it is possible to collect a second crop by dissolving the insolubles obtained from the filtration of $K_2[Pt(CN)_4] \cdot xH_2O$ in 25 mL of H_2O, filtering the resulting solution, evaporating to 10 mL, and then cooling as described above. If the filtrate is clear and colorless, the trihydrate crystals obtained are light green with a yellow tint. A clear yellow filtrate yields, upon crystallization, a yellow product, which is presumably the monohydrate. The yield of the first crop ranges from 7.0 to 7.4 g, which represents approximately 70-75% based on the original platinum.

5. *Dipotassium dibromotetracyanoplatinate(IV) dihydrate.* $K_2[Pt(CN)_4] \cdot 3H_2O + Br_2 \xrightarrow{H_2O} K_2[Pt(CN)_4Br_2] \cdot 2H_2O + H_2O$. At this point, one-sixth of the product obtained in reaction 4 is placed in a 100-mL beaker and stirred with 15-20 mL of distilled water and 2 mL of bromine.[10,11] The clear, colorless solution becomes reddish-brown upon the addition of bromine, which is added in an approximate tenfold excess of the stoichiometric requirement. The solution temperature is raised to 50-60° with stirring for 20-30 minutes, and excess bromine is removed (in a fume hood), yielding a yellow solution of dipotassium dibromotetracyanoplatinate(IV). This salt is generally not isolated, but if it is desired to do so only 5-10 mL of water is used at the start of this step, and the final product, which has a bright canary-yellow color, is isolated by crystallizing by cooling with a NaCl-ice bath.

6. *Potassium tetracyanoplatinate bromide (2:1:0.3) trihydrate.* $5K_2[Pt(CN)_4] \cdot 3H_2O + K_2[Pt(CN)_4Br_2] \cdot 2H_2O \rightarrow 6K_2[Pt(CN)_4]Br_{0.3} \cdot 3H_2O$. The remainder of the unbrominated product (five-sixths of the yield of reaction 4) is added, at room temperature, to the final yellow solution obtained in reaction 5. The resulting clear yellow solution is placed in an ice bath for 30 minutes to precipitate the final product. The resulting pH is 2-3. The yield of $K_2[Pt(CN)_4]Br_{0.3} \cdot 3H_2O$ is approximately 5.7-6.3 g, representing 54-60% based on original platinum.

Properties

The compound $K_2[Pt(CN)_4]Br_{0.3} \cdot 3H_2O$ (KCP(Br)) forms lustrous, coppery-colored tetragonal[12] crystals [a = 9.907(3) and c = 5.780(2) Å, space group *P4mm*]. The structure has been established by neutron diffraction analysis.[12]

Large single crystals may be grown by slow cooling of seeded aqueous solutions or by the addition of KBr and urea to aqueous solutions.[4] The crystals gain or lose water easily and may be stored indefinitely over a saturated solution of NH_4Cl and KNO_3 that has a relative humidity of 72% at 23°. Thermogravimetric analysis of the H_2O and D_2O forms of KCP(Br) has established that the former[13] contains 3.0 H_2O and that the latter[14] contains 3.2 D_2O.

The salt may be best described as a mixed-valence compound containing platinum in the 2.3+ oxidation state. The combination of the mixed-valence state, the very short Pt—Pt separation of 2.89 Å (only approximately 0.1 Å longer than in platinum metal), and the formation of linear Pt—Pt chains along the c axis gives rise to the high, anisotropic metallic conductivity in this material.[1]

Excluding the broad absorptions due to H_2O (>3000 cm^{-1}) the observed infrared absorption frequencies are:[15] 2156 (vs); 2148 (sh); 509 (s); 498 (sh); 475 (vw); 411 (s); 296 (w); 288 (w) and 244 (m) cm^{-1}. The first 10 reflections in the X-ray powder diffraction pattern occur at the following d spacings: 9.69 (vs); 6.89 (vs); 4.90 (vs); 4.38 (vs); 3.46 (m); 3.27 (w); 3.10 (s); 2.97 (w); 2.85 (m); and 2.75 (s) Å.

References

1. J. S. Miller and A. J. Epstein, *Prog. Inorg. Chem.*, **20**, 1 (1975).
2. H. R. Zeller, in *Festkorperprobleme*, Vol. 13, H. J. Queisser (ed.), Pergamon Press Ltd., New York, 1973, p. 31.
3. (a) G. Brauer, *Handbuch der Praparativen Anorganischen Chemie*, Vol. 2, Ferdinand Enke Verlag, Stuttgart, 1962, p. 1372; (b) R. L. Maffly and J. M. Williams, *Inorg. Synth.*, **19**, 112 (1979).
4. R. B. Saillant, R. C. Jaklevic, and C. D. Bedford, *Mater. Res. Bull.*, 9, 289 (1974).
5. T. J. Neubert and S. Susman, *J. Chem. Phys.*, **41**, 722 (1964); see Reference 3.
6. G. Brauer, *Handbuch der Praparativen Anorganischen Chemie*, Vol. 2, Ferdinand Enke Verlag, Stuttgart, 1962, p. 1365.
7. G. Brauer, *Handbuch der Praparativen Anorganischen Chemie*, Vol. 2, Ferdinand Enke Verlag, Stuttgart, 1962, p. 1367.
8. R. N. Keller, *Inorg. Synth.*, **2**, 247 (1946).
9. G. A. Shagisultanova, *Russ. J. Inorg. Chem.*, **6**, 8, 904 (1961).
10. Gmelins *Handbuch der Anorganischen Chemie*, Vol. 68C, 8th ed., Verlag Chemie, Berlin, Germany, 1940, p. 207.
11. I. I. Chernyaev, A. V. Babkov, and N. N. Zheligoyskaya, *J. Inorg. Chem. (USSR)* (Engl. Ed.), 8, 1279 (1963).
12. J. M. Williams, J. L. Petersen, H. M. Gerdes, and S. W. Peterson, *Phys. Rev. Lett.*, **33**, 1079 (1974).
13. J. M. Williams, Abstracts of the 170th National Meeting of the American Chemical Society, August 1975, Chicago, Illinois, Paper No. 127.
14. C. Peters and C. F. Eagen, *Phys. Rev. Lett.*, **34**, 1132 (1975). Notes that although the title of this article suggests that the material under study was $K_2[Pt(CN)_4]Br_{0.3} \cdot 3.0 H_2O$ the thermogravimetric analysis (Fig. 3 this reference) was conducted on $K_2[Pt(CN)_4]Br_{0.3} \cdot 3.2D_2O$.
15. J. R. Ferraro, L. J. Basile, and J. M. Williams, *J. Chem. Phys.*, **64**, 732 (1976).

2. CATION-DEFICIENT, PARTIALLY OXIDIZED TETRACYANOPLATINATES*

The metallic character of the cation-deficient partially oxidized tetracyano-platinate complexes is of considerable interest because of their unusual aniso-tropic electrical properties. The method used in the preparation of these com-pounds is a modification of the procedure used by Levy[1] and gives a higher yield of pure product than the syntheses previously described in the literature[2-5] (only the potassium compound has been reported to date).

■**Caution.** *Because of the extremely poisonous nature of cyanide, all steps of these syntheses should be carried out in a well-ventilated hood using protective gloves, clothing, and a face shield. Also, hydrogen peroxide is a very strong oxidizing agent. Care should be taken to not allow contact with the skin or with clothing.*

A. CESIUM TETRACYANOPLATINATE (2:1) MONOHYDRATE AND CESIUM TETRACYANOPLATINATE (1.75:1) DIHYDRATE

Submitted by R. L. MAFFLY,** J. A. ABYS,** and J. M. WILLIAMS†
Checked by R. N. RHODA‡

Procedure

The $Ba[Pt(CN)_4] \cdot 4H_2O$ used as the starting material for this preparation is synthesized and purified as described by Maffly et al.[6] The Cs_2SO_4 used should be 99.9% pure*** to minimize any possible cation contamination, especially by potassium. All other chemicals are reagent grade. Distilled water is used through-out the procedure.

1. *Cesium tetracyanoplatinate(II) monohydrate.* $Ba[Pt(CN)_4] \cdot 4H_2O$ + $Cs_2SO_4 \xrightarrow{H_2O} Cs_2[Pt(CN)_4] \cdot H_2O + BaSO_4 + 3H_2O$. Initially 7.0 g (1.38 ×

*Research performed under the auspices of the Division of Basic Energy Sciences of the U.S. Department of Energy.

**Research participants sponsored by the Argonne Center for Educational Affairs: (a) R. L. Maffly from Whitman College, Walla Walla, WA and (b) J. A. Abys from Marist College, Poughkeepsie, NY.

†Correspondent: Chemistry Division, Argonne National Laboratory, Argonne, IL 60439.

‡Research Laboratory, International Nickel Co., Inc., Sterling Forest, Suffern, NY 10901.

***Available from Alfa Products, Ventron Corp., P.O. Box 299, Danvers, MA 01923.

10^{-2} mole) of $Ba[Pt(CN)_4] \cdot 4H_2O$ is dissolved, with stirring, in 25 mL of hot (70-80°) water. A solution of 5.0 g (1.38×10^{-2} mole) Cs_2SO_4 in 10 mL of warm H_2O is added to the stirred solution, and an immediate precipitation of $BaSO_4$ occurs. The temperature is maintained and the stirring is continued for 30 minutes. The solution is then filtered through a fritted filter (fine porosity). The $BaSO_4$ is rinsed with two 3-mL portions of hot water and then discarded. The filtrate is transferred to a beaker and placed in a hot water bath. The solution is evaporated until the volume is 20 mL. The beaker is then placed in an ice bath for 1 hour to allow $Cs_2[Pt(CN)_4] \cdot H_2O$ to crystallize. The blue-white crystals are isolated on a fritted filter (medium porosity) and air dried for 15 minutes (yielding 6.0-6.3 g of product). The filtrate is returned to the hot water bath and the volume is reduced to 5 mL. It is then cooled and a second crop of crystals is isolated in a similar manner (yielding 0.4-0.7 g of product). Total yield 6.4-7.0 g (1.1×10^{-2} to 1.2×10^{-2} mole), 80-87% based on $Ba[Pt(CN)_4] \cdot 4H_2O$.

2. *Cesium tetracyanoplatinate (1.75:1) dihydrate,* $Cs_{1.75}[Pt(CN)_4] \cdot 2H_2O$.

$$8Cs_2[Pt(CN)_4] \cdot H_2O + H_2O_2 + H_2SO_4 + 6H_2O \xrightarrow{H_2O} 8Cs_{1.75}[Pt(CN)_4] \cdot 2H_2O + Cs_2SO_4.$$ The 7.0 g (1.20×10^{-2} mole) of $Cs_2[Pt(CN)_4] \cdot H_2O$ obtained in reaction 1 is dissolved, with stirring, in 25 mL of hot water. Then $1 M H_2SO_4$ is added slowly to acidify the solution (pH = 1) followed by 0.65 mL of 20% H_2O_2[7] (4×10^{-3} mole). The heating and stirring are continued for 1 hour. The solution is then covered and allowed to cool, and the product is allowed to crystallize for 24 hours.

The fine needle-shaped crystals (coppery brown in appearance) are isolated by filtering through a fritted filter (fine porosity) and air dried for 20 minutes. Care must be taken to prevent the crystals from dehydrating. Because of the necessity of keeping the crystals moist, the determination of yield is approximate. Yield: 6.0-7.0 g (1.1×10^{-2} to 1.2×10^{-2} mole) or 92-100% based on original $Ba[Pt(CN)_4] \cdot 4H_2O$. *Anal.* Calcd. for $C_4H_4N_4O_2Cs_{1.75}Pt$: C, 8.45; H, 0.71; N, 9.87; O, 5.64; Cs, 40.96; Pt, 34.36. Found: C, 8.68; H, 0.69; N, 10.21; O, 5.69; Cs, 40.83; Pt, 33.40.

Properties

Since the crystals of $Cs_{1.75}[Pt(CN)_4] \cdot 2H_2O$ lose water easily, they must be stored over a saturated solution of NH_4Cl and KNO_3 that has a relative humidity of 72% at 23°.[8] Recrystallization of $Cs_{1.75}[Pt(CN)_4] \cdot 2H_2O$ have not been successful.

The first 10 most consistent and easily recognized reflections on the X-ray powder pattern* occur at the following *d* spacings: 9.30 (s); 4.65 (m); 4.50 (m);

---

*We wish to thank Ms. E. Sherry for obtaining all X-ray powder patterns.

4.25 (m); 4.04 (m); 3.92 (s); 3.63 (w); 3.29 (w); 3.08 (w) and 2.85 (m) Å. It should be noted that the 2.85-Å spacing corresponds to the Pt–Pt separation in other partially oxidized tetracyanoplatinates of known structure.

B. POTASSIUM TETRACYANOPLATINATE (1.75:1) SESQUIHYDRATE, $K_{1.75}[Pt(CN)_4] \cdot 1.5H_2O$

$$8K_2[Pt(CN)_4] \cdot 3H_2O + H_2SO_4 + H_2O_2 \xrightarrow{H_2O}$$

$$8K_{1.75}[Pt(CN)_4] \cdot 1.5H_2O + 14H_2O + K_2SO_4$$

Submitted by T. R. KOCH,* N. P. ENRIGHT,* and J. M. WILLIAMS†
Checked by J. ACKERMAN‡ and A. WOLD‡

Procedure

The $K_2[Pt(CN)_4] \cdot 3H_2O$ starting material used in this synthesis is prepared as in Section A., by substituting an equal molar amount of K_2SO_4 for the Cs_2SO_4 specified. All other chemicals are reagent grade. Distilled water is used throughout the procedure.

Dipotassium tetracyanoplatinate(II) trihydrate (4.0 g, 9.3×10^{-3} mole) is dissolved in 8 mL of hot water (70°) in a polypropylene beaker and acidified with 1 M H_2SO_4[7] (pH 1-2). Only 0.5 mL (3.1×10^{-3} mole) of 20% H_2O_2 is required for oxidation. The rest of the procedure is identical to that in Section A for cesium tetracyanoplatinate (1.75:1) dihydrate, with the exception that after cooling for about 24 hours, and prior to collection of the crystals, the beaker is placed in a desiccator under reduced pressure over magnesium perchlorate and allowed to stand until about 5 mL of solution remains.

The yield of $K_{1.75}[Pt(CN)_4] \cdot 1.5H_2O$ is approximately 1.7 g (4.3×10^{-3} mole), representing 46% based on the initial $K_2[Pt(CN)_4] \cdot 3H_2O$. *Anal.* Calcd. for $C_4H_3N_4O_{1.5}K_{1.75}Pt$: C, 12.18; H, 0.77; N, 14.20; O, 6.08. Found: C, 12.39; H, 0.75; N, 14.28; O, 6.50.

Properties

The compound $K_{1.75}[Pt(CN)_4] \cdot 1.5H_2O$ forms bronze-colored triclinic crystals

*Research participants sponsored by the Argonne Center for Educational Affairs: (a) T. R. Koch from Coe College, Cedar Rapids, IA and (b) N. P. Enright from Middlebury College, Middlebury, VT.
†Correspondent: Chemistry Division, Argonne National Laboratory, Argonne, IL 60439.
‡Department of Chemistry, Brown University, Providence, RI 02912.

2. *Cation-Deficient, Partially Oxidized Tetracyanoplatinates* 9

with Delauney-reduced cell constants $a = 10.360(17)$ Å, $b = 11.832(19)$ Å, $c = 9.303(15)$ Å, $\alpha = 102.43(8)°$, $\beta = 106.36(7)°$, and $\gamma = 114.74(5)°$, space group $C_i\text{-}P\bar{1}$.[9] Single-crystal neutron diffraction analysis[9] has established the existence of the first known zigzag platinum-atom chain in this type of compound. The Pt—Pt separations are all equal at 2.96 Å.

Crystals of $K_{1.75}[Pt(CN)_4] \cdot 1.5H_2O$ lose water easily and may be stored over a saturated solution of NH_4Cl and KNO_3, which has a relative humidity of 72% at $23°$[8] (H_2O vapor pressure; 15.15 torr). The compound may be recrystallized from water.

The infrared spectrum of $K_{1.75}[Pt(CN)_4] \cdot 1.5H_2O$ is essentially the same as that of $K_2[Pt(CN)_4]Br_{0.3} \cdot 3H_2O$. Excluding the broad absorptions due to H_2O (> 3000 cm^{-1}), the observed frequencies are: 2156 (vs); 2148 (sh); 509 (s); 500 (sh); 482 (m); 413 (s); 303 (w); 293 (w); 281 (w); and 248 (m) cm^{-1}.[10]

The first 10 reflections in the X-ray powder pattern occur at the following d spacings: 8.54 (vs); 8.17 (vs); 7.64 (vs); 4.94 (s); 4.64 (s); 4.49 (s); 4.23 (m); 4.08 (m); 3.86 (w); and 3.70 (vw) Å. The pattern is identical to the X-ray powder pattern originally reported by Krogmann.[11]

C. RUBIDIUM TETRACYANOPLATINATE (1.6:1) DIHYDRATE, $Rb_{1.6}[Pt(CN)_4] \cdot 2H_2O$

$$10Rb_2[Pt(CN)_4] \cdot 1.5H_2O + 2H_2O_2 + 2H_2SO_4 + H_2O \rightarrow$$

$$10Rb_{1.6}[Pt(CN)_4] \cdot 2H_2O + 2Rb_2SO_4$$

Submitted by T. R. KOCH,* J. A. ABYS,* and J. M. WILLIAMS†
Checked by R. N. RHODA‡

Procedure

The $Rb_2[Pt(CN)_4] \cdot 1.5H_2O$ used in this synthesis is prepared as in Section A., by using an equal molar amount of Rb_2SO_4 in place of the Cs_2SO_4 specified. All other chemicals are reagent grade. Distilled water is used throughout the procedure. This procedure must be followed without deviation.

Rubidium tetracyanoplatinate(II) sesquihydrate (8.8 g, 1.8×10^{-2} mole) is dissolved in 10 mL of hot water (70°) and acidified by slow addition of 1 M sul-

*Research Participants sponsored by the Argonne Center for Educational Affairs: (a) T. R. Koch from Coe College, Cedar Rapids, IA and (b) J. A. Abys from Marist College, Poughkeepsie, NY.
†Correspondent: Chemistry Division, Argonne National Laboratory, Argonne, IL 60439.
‡Research Laboratory, International Nickel Co., Inc., Sterling Forest, Suffern, NY 10401.

furic acid (pH $\cong$ 1-2). Partial oxidation is achieved by the addition of a 20% solution of hydrogen peroxide[7] (1.0 mL, 6.0 $\times$ 10^{-3} mole). The remainder of the procedure is identical to that in Section B. The yield of approximately 5.7 g (1.2 $\times$ 10^{-2} mole) of $Rb_{1.6}[Pt(CN)_4] \cdot 2H_2O$ represents 52% based on the initial $Rb_2[Pt(CN)_4] \cdot 3H_2O$. *Anal.* Calcd. for $C_4H_4N_4O_2Rb_{1.6}Pt$: C, 10.18; H, 0.85; N, 11.87; O, 6.78; Rb, 28.97. Found: C, 10.13; H, 0.41; N, 11.76; O, 6.51; Rb, 29.0.

Properties

Crystals of $Rb_{1.6}[Pt(CN)_4] \cdot 2H_2O$ lose water easily and must be stored over a saturated solution of NH_4Cl and KNO_3, which has a relative humidity of 72% at 23° (H_2O vapor pressure: 15.15 torr).[8] The compound may be recrystallized from water. This platinum complex has a density of 3.31 g/cm^3.

The first eight strong reflections in the X-ray powder pattern occur at the following d spacings: 8.31 (vs); 4.79 (vs); 4.40 (m); 4.15 (s); 3.90 (m); 3.81 (m); 3.14 (m); and 2.89 (s) Å.

References

1. L. A. Levy, *J. Chem. Soc. (Lond.)*, **101**, 1081 (1912).
2. E. A. Hadow, *Q. J. Chem. Soc. (Lond.)*, **14**, 106 (1861).
3. T. Wilm, *Chem. Ber.*, **21**, 1434 (1888).
4. H. Terrey, *J. Chem. Soc. (Lond.)*, 202 (1928).
5. P. Weselsky, *J. Prakt. Chem.*, **69**, 276 (1856).
6. R. L. Maffly and J. M. Williams, *Inorg. Synth.*, **19**, 112 (1979).
7. L. A. Levy, *J. Chem. Soc. (Lond.)*, **101**, 1097 (1912).
8. G. Edgar and W. O. Swan, *J. Am. Chem. Soc.*, **44**, 574 (1922).
9. K. D. Keefer, D. M. Washecheck, N. P. Enright and J. M. Williams, *J. Am. Chem. Soc.*, **98**, 233 (1976); J. M. Williams, K. D. Keefer, D. M. Washecheck and N. P. Enright, *Inorg. Chem.*, **15**, 2446 (1976).
10. J. R. Ferraro, L. J. Basile and J. M. Williams, *J. Chem. Phys.*, **65**, 3025 (1976).
11. K. Krogmann and H. D. Hausen, *Z. Naturforsch.*, **23B**, 1111 (1968).

3. GUANIDINIUM TETRACYANOPLATINATE BROMIDE (2:1:0.25) HYDRATE, $[C(NH_2)_3]_2[Pt(CN)_4]Br_{0.25} \cdot H_2O$*

Submitted by T. F. CORNISH† and JACK M. WILLIAMS‡
Checked by R. N. RHODA§

*Research performed under the auspices of the Division of Basic Energy Sciences of the U.S. Department of Energy.

†Research participant sponsored by the Argonne Center for Educational Affairs from Marist College, Poughkeepsie, NY.

‡Correspondent: Chemistry Division, Argonne National Laboratory, Argonne, IL 60439.

§Paul D. Merica Research Laboratory, International Nickel Co., Inc., Sterling Forest-Suffern, NY 10901.

Partially oxidized tetracyanoplatinates have attracted much attention in recent years because of their highly anisotropic physical properties.[1] Although partially oxidized salts of K^+, Na^+, Rb^+, and Mg^{+2} have been reported, the analogous bromide-deficient guanidinium salt is the only such compound containing an organic cation.[2] A procedure for the preparation of this guanidinium salt has not been previously reported.

■**Caution.** *Because of the extremely poisonous nature of cyanide, these steps should be carried out in a well-ventilated hood using protective gloves, clothing, and a face shield.*

Procedure

The $Ba[Pt(CN)_4] \cdot 4H_2O$ used in this preparation is synthesized and purified as described by Koch et al.[3] This material is air dried for approximately 15 minutes before being used in the preparation. The guanidine sulfate is purchased from Eastman Kodak and is used without further purification. The bromine is reagent grade. Distilled water is used throughout.

1. *Guanidinium tetracyanoplatinate(II).* $Ba[Pt(CN)_4] \cdot 4H_2O + [C(NH_2)_3]_2$-$SO_4 \xrightarrow{H_2O} [C(NH_2)_3]_2[Pt(CN)_4] + BaSO_4 + 4H_2O$. Barium tetracyanoplatinate(II) tetrahydrate (9.00 g,* ~ 0.018 mole) is partially dissolved in about 100 mL of hot (70-90°) water forming a milky white suspension. The suspension is stirred vigorously on a steam bath as 4.14 g (0.0191 mole) of guanidine sulfate is added, producing an immediate white precipitate of $BaSO_4$. After stirring for 45 minutes on a steam bath, the mixture is cooled to room temperature and then suction filtered using a fine-porosity sintered-glass filter. (Alternatively, the suspension may be left on the steam bath for about 2 hours to allow the $BaSO_4$, precipitate to coarsen. The precipitate is then allowed to settle and as much liquid as possible is decanted.) The clear colorless filtrate is then evaporated to about 10 mL on a steam bath. Some of the desired white product precipitates before this volume is reached. The mixture is cooled in a NaCl ice bath for ½ hour and the product is collected and air dried on a fine-porosity, sintered-glass filter for about 10 minutes. Approximately 6.2-7.0 g of white crystals is recovered. This represents a 84-94% yield based on the initial $Ba[Pt(CN)_4] \cdot 4H_2O$.

2. *Bis(guanidinium) dibromotetracyanoplatinate(IV) hydrate.* $[C(NH_2)_3]_2$-$[Pt(CN)_4] + Br_2 \xrightarrow{H_2O} [C(NH_2)_3]_2[Pt(CN)_4Br_2] \cdot xH_2O$. At this point, one-sixth of the product obtained in step 1 is dissolved in 30 mL of water. Then,

---

*A portion of this weight is due to water, since the material is only air dried for 15 minutes after preparation. If the synthesis is performed with completely dried barium salt, difficulty is encountered in this step since all $Ba[Pt(CN)_4] \cdot 4H_2O$ is not consumed by 4.14 g of guanidine sulfate.

1 mL of bromine is added, and the mixture is stirred gently on a steam bath. The clear, colorless solution becomes reddish-brown upon the addition of bromine, which is added in a greater than tenfold excess of the stoichiometric requirement. The mixture is stirred for ½ hour, after which time a clear yellow solution of bis(guanidium) dibromotetracyanoplatinate(IV) forms, indicating that the excess bromine has been removed. This salt is not isolated.

3. *Guanidinium tetracyanoplatinate bromide* (2:1:0.25) *monohydrate.*
$5[C(NH_2)_3]_2[Pt(CN)_4] + [C(NH_2)_3]_2[Pt(CN)_4Br_2] \cdot xH_2O \rightarrow 6[C(NH_2)_3]_2$-$[Pt(CN)_4]Br_{0.25} \cdot H_2O$. The remainder of the unbrominated product (five-sixths of the yield from procedure 1) is added to the clear yellow solution of bis(guanidinium) dibromotetracyanoplatinate(IV) from step 2. The resulting mixture is heated on a steam bath with gentle stirring until a clear yellow solution results. Upon dissolution the stirring is stopped. Heating is continued until about 20 mL of solution remains. The solution is then cooled for about ½ hour in a sodium chloride-ice bath before filtering (fine-porosity sintered-glass filter) and isolating the red-brown precipitate. The yield is 3.5-3.9 g of $[C(NH_2)_3]_2[Pt(CN)_4]Br_{0.25} \cdot H_2O$, which is collected after air drying for about 10 minutes (43-48% based on the original $Ba[Pt(CN)_4] \cdot 4H_2O$). The yield may be increased approximately 30% by evaporating the filtrate to a volume of about 5 mL and repeating the procedure for isolation described above. However, the product recovered from this second isolation often contains starting material and must therefore be recrystallized. *Anal.* Calcd. for $C_6H_{14}N_{10}OBr_{0.25}Pt$: C, 15.76; H, 3.09; N, 30.63; O, 3.50; Br, 4.37; Pt, 42.66. Found: C, 15.49;.H, 2.86; N, 30.61; O, 3.00; Br, 4.37;* Pt, 42.50.

Properties

The compound $[C(NH_2)_3]_2[Pt(CN)_4]Br_{0.25} \cdot H_2O$ forms reddish-bronze tetragonal crystals. The cell constant as determined from neutron diffraction are as follows: $a = 15.621$ Å, $c = 5.816$ Å. The structural elucidation of $[C(NH_2)_3]_2$-$[Pt(CN)_4]Br_{0.25} \cdot H_2O$ has been completed.[4] The $[C(NH_2)_3]_2[Pt(CN)_4]Br_{0.25} \cdot H_2O$ appears to be relatively stable toward hydration and dehydration and may be safely stored over a saturated solution of NaCl, which has a relative humidity of 75.1% at 25°.[5] The compound may be recrystallized from water. The density of $[C(NH_2)_3]_2[Pt(CN)_4]Br_{0.25} \cdot H_2O$ is 2.16 g/cm³.

The $[C(NH_2)_3]_2[Pt(CN)_4]Br_{0.25} \cdot H_2O$ may be characterized by its X-ray powder pattern. The first 12 reflections occur at the following d spacings: 10.99

---

*The analyses for bromine actually varied from 3.0 to 6.03%. However, the majority of results agreed quite well with 4.37% bromine, which yields the formula $[C(NH_2)_3]_2$-$[Pt(CN)_4]Br_{0.25} \cdot H_2O$.

(vs); 7.73 (vs); 5.50 (vs); 4.90 (vs); 4.48 (w); 3.89 (m); 3.66 (w); 3.47 (s); 3.17 (w); 3.04 (m); 2.89 (s); 2.78 (s) Å*.

References

1. J. S. Miller and A. J. Epstein, *Prog. Inorg. Chem.*, **20**, 1 (1976).
2. J. S. Miller and A. J. Epstein, *Prog. Inorg. Chem.*, **20**, 48 (1976).
3. T. R. Koch, J. A. Abys, and J. M. Williams, *Inorg. Synth.*, **19**, 9 (1979).
4. J. M. Williams, T. F. Cornish, D. M. Washecheck, and P. L. Johnson, *Ann. N. Y. Acad. Sci.*, in press, 1978.
5. J. F. Young, *J. Appl. Chem.*, **17**, 241 (1967).

4. ELECTROCHEMICAL SYNTHESES OF PARTIALLY OXIDIZED PLATINUM COMPLEXES

Submitted by JOEL S. MILLER
Checked by A. L. BALCH[†] and C. HARTMAN[†]

In recent years there has been a keen interest in chemical and physical properties of highly conducting linear-chain inorganic,[1a] organic,[1b] and polymeric[1a] materials as a result of the observation of a number of unusual phenomena in such materials.[1] As a consequence of the pseudo-one dimensionality of the linear-chain conductors, impurities and/or defects that interrupt the chain may strongly influence physical properties such as electrical conductivity. Thus preparative routes to high-quality single crystals of highly conducting materials are important. Judging from electrical properties, electrochemical syntheses that use the crystal as an extension of the anode provide the highest quality materials in a reproducible manner. Herein is described the electrochemical growth for the nonstoichiometric highly conducting platinum chain complexes, $K_{1.75}Pt(CN)_4 \cdot 1.5H_2O$, $K_2Pt(CN)_4Cl_{0.3} \cdot 3H_2O$, $K_2Pt(CN)_4Br_{0.3} \cdot 3H_2O$, and $K_{1.64}Pt(C_2O_4)_2 \cdot 2H_2O$.[2]

*The authors wish to thank Ms. E. Sherry for providing us with the X-ray powder photographs.

**Xerox Corporation, Webster Research Center, Xerox Square-114, Rochester, NY 14644. Current address, Rockwell International Science Center, Box 1085, Thousand Oaks, CA 91360.

†Department of Chemistry, University of California, Davis, CA 95616.

A. PARTIALLY OXIDIZED POTASSIUM TETRACYANOPLATINATES

A variety of highly conducting complexes containing platinum have been synthesized by chemical oxidation[1a] of $[Pt(CN)_4]^{2-}$ and $[Pt(C_2O_4)_2]^{2-}$. However, the electrical properties of such materials can be utilized to synthesize single crystals by means of electrolysis and growth at the anode.[2] The electrolysis may be carried out in a single cell compartment, since the cathode reaction does not interfere with the synthesis and/or isolation of the desired product. A variety of cell designs are suitable for the syntheses; however, we have routinely used the rectangular glass cell, for example, $10 \times 5 \times 5$ mm (Fig. 1), fitted with platinum wire electrodes.

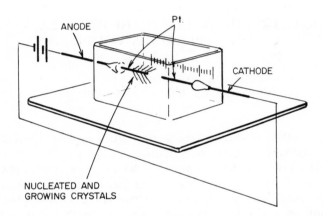

Fig. 1. Electrolytic cell used for the electrochemical syntheses of highly conducting inorganic complexes.

1. Potassium Tetracyanoplatinate (1.75:1) Sequihydrate, $K_{1.75}Pt(CN)_4 \cdot 1.5H_2O$

$$1.75K^+ + Pt(CN)_4^{2-} \xrightarrow{H_2O} 0.25e^- + K_{1.75}Pt(CN)_4 \cdot 1.5H_2O$$

Procedure

Potassium tetracyanoplatinate may be prepared as described elsewhere in this volume,[3] or commercial material* may be used after chloride impurities are eliminated by two recrystallizations from distilled water. An aqueous solution at

---

*Available from Platinum Chemicals, Inc., P.O. Box 565, Asbury Park, NJ, 07712.

room temperature (4 mL) containing 500 mg of potassium tetracyanoplatinate-(II)[3] (1.325 mmole; 0.3 M) is placed in the electrolysis cell (Fig. 1), and upon application of an 1.5-V dc potential (1½-V dry cell), nucleation and growth of metallic-appearing needle crystals of $K_{1.75}Pt(CN)_4 \cdot 1.5H_2O$ are effected at the anode, with gas evolution (H_2) occurring at the cathode. After continuous application of the voltage for 24 hours, the crystals are gently broken away from the anode with a fine spatula and collected by filtration. The resultant crystals are checked under illumination with an ultraviolet lamp* to determine the presence of the highly fluorescent $K_2Pt(CN)_4 \cdot xH_2O$ starting material as an impurity. The impurities may either be removed with tweezers or quickly washed away with a small amount of cold distilled water. (Under similar conditions $Na_2Pt(CN)_4$ solutions do not yield a partially oxidized product.) The yield depends on the size and shape of the electrodes, as well as current density, concentration of $[Pt(CN)_4]^{2-}$, and the duration of electrolysis. Long well-formed highly reflecting brass-appearing needle crystals (30% yield) can be grown in this manner. The ν_{CN} is given in Table I. The X-ray powder diffraction pattern is identical to that of authentic $K_{1.75}Pt(CN)_4 \cdot 1.5H_2O$.[4] Use of D_2O as solvent leads to isolation of the deuterated product.

TABLE I. Characteristic ν_{CN} Absorptions for Tetracyanoplatinates[6]

Complex	$\nu_{C\equiv N}$ (cm^{-1}), Nujol mull
$K_2Pt(CN)_4 \cdot 3H_2O$	2148
$K_2Pt(CN)_4Cl_{0.3} \cdot 3H_2O$	2154
$K_2Pt(CN)_4Br_{0.3} \cdot 3H_2O$	2153
$K_2Pt(CN)_4Cl_2$	2181
$K_2Pt(CN)_4Br_2$	2170
$K_{1.75}Pt(CN)_4 \cdot 1.5H_2O$	2142

2. Potassium Tetracyanoplatinate Halide (2:1:0.3) Trihydrate, $K_2Pt(CN)_4$-$X_{0.3} \cdot 3H_2O$ (X = Cl, Br)

$$2\,K^+ + Pt(CN)_4^{2-} + 0.3X^- \xrightarrow{H_2O} 0.3e^- + K_2Pt(CN)_4X_{0.3} \cdot 3H_2O$$

Procedure

When the above procedure is repeated with approximately 1 M KX (X = Cl, Br)

---

*Suitable lamps include models UVS-12 and UVL-22 manufactured by UltraViolet Products, Inc., San Gabrial, CA 91778.

instead of distilled water (or D_2O), copper-colored crystals of $K_2Pt(CN)_4X_{0.3} \cdot 3H_2O$ (X = Cl, Br) (30% yield) appear at the anode. It is imperative that the synthesis of $K_2Pt(CN)_4Br_{0.3} \cdot 3H_2O$ be carried out in a chloride-free environment, for traces of Cl^- tend to be preferentially incorporated into the structure.[5] The values for ν_{CN} are given in Table 1. The X-ray powder diffraction patterns were identical to those of authentic $K_2Pt(CN)_4X_{0.3} \cdot 3H_2O$ samples prepared by chemical oxidation.[6,7]

Properties

The partially oxidized complexes $K_{1.75}Pt(CN)_4 \cdot 1.5H_2O$, $K_2Pt(CN)_4X_{0.3} \cdot 3H_2O$ (X = Cl, Br) all form dichroic needlelike crystals that exhibit high reflectivity of visible light and a metallic luster. These hydrated complexes are quite soluble in water and tend to lose water easily under ambient conditions; this water loss may significantly alter their physical properties. Thus it is desirable to store these crystals in an environment of high humidity, for example, in a desiccator containing a saturated solution of ammonium chloride and potassium nitrate (72% relative humidity).[5] The ν_{CN} value is given in Table I. The physical properties of these complexes have been studied by many chemists and physicists. The results of their ongoing studies point toward a band metallic state at room temperature undergoing a metal-insulator transition at lower temperature. The room-temperature conductivities for $K_{1.75}Pt(CN)_4 \cdot 1.5H_2O$, $K_2Pt(CN)_4X_{0.3} \cdot 3H_2O$ (X = Cl, Br) are approximately 25[8] and 300 ohms^{-1} cm^{-1}.[1a]

B. PARTIALLY OXIDIZED POTASSIUM BIS(OXALATO)PLATINATE

In addition to the partially oxidized tetracyanoplatinates, bis(oxalato)platinate-(II) can be nonstoichiometrically oxidized by chemical oxidants[1a] to form highly lustrous needlelike crystals containing platinum in the 2.36 oxidation state. These complexes have not been characterized to the extent of the tetracyanoplatinate complexes; however, the oxalato complexes are reported to be highly conducting.[1a] The starting material is bis(oxalato)platinate(II), which can be prepared in 30% yields from hexachloroplatinate(IV) and potassium oxalate.[9]

1. Potassium Bis(oxalato)platinate(II) (2:1) Dihydrate, $K_2Pt(C_2O_4)_2 \cdot 2H_2O$

$$K_2PtCl_6 + 3K_2C_2O_4 \xrightarrow{H_2O} K_2Pt(C_2O_4)_2 \cdot 2H_2O + 2CO_2 + 6KCl$$

Procedure

One hundred twenty milliliters of an aqueous slurry containing 2.686 g K_2PtCl_6

(5.527 mmole) and 3.049 g $K_2C_2O_4$ (16.55 mmole) is heated under reflux for 18 hours. After filtration with a fine-porosity sintered-glass funnel to remove small amounts of platinum, the warm filtrate is reduced in volume to 40 mL and is then cooled in a refrigerator $(-8°)$ for 2 hours. The product is recrystallized from a minimum amount of water (approximately 25 mL). Yield 0.75 g (30%). *Anal.* Calcd.[10] for $C_4H_4K_2O_{10}Pt$: C, 9.90; H, 0.83; Pt, 40.19. Found: C, 9.75; H, 0.75; and Pt, 40.10.

2. Potassium Bis(oxalato)platinate (1.64:1) Dihydrate, $K_{1.64}Pt(C_2O_4)_2 \cdot 2H_2O$

$$1.64\ K^+ + Pt(C_2O_4)_2^{2-} \xrightarrow{H_2O} 0.36e^- + K_{1.64}Pt(C_2O_4)_2 \cdot 2H_2O$$

Procedure

The procedure described for $K_{1.75}Pt(CN)_4 \cdot 1.5H_2O$ is utilized; however, because of the lower solubility of $K_2Pt(C_2O_4)_2$ in water at ambient temperature a 0.02 *M* solution of $Pt(C_2O_4)_2^{2-}$ is employed.[9] *Anal.* Calcd. for $C_4H_4K_{1.64}O_{10}Pt$: K, 13.61; C, 10.18; H, 0.43. Found: K, 13.70; C, 10.42; H, 0.30.

Properties

The potassium salts of bis(oxalato)platinate(II) and potassium-deficient bis-(oxalato)platinate complexes are stable in the solid state. However, prolonged heating of an aqueous solution of $[Pt(C_2O_4)_2]^{2-}$ tends to deposit platinum metal. Thus care must be taken in handling solutions of $[Pt(C_2O_4)_2]^{2-}$. The tetra-*n*-butylammonium salt of $[Pt(C_2O_4)_2]^{2-}$ can be prepared from a dichloro-methane extraction of an aqueous solution of 1:2::$K_2Pt(C_2O_4)_2$:$(n$-$C_4H_9)_4NBr$. Removal of dichloromethane solvent leaves analytically pure $[(n$-$C_4H_9)_4N]_2$-$Pt(C_2O_4)_2$. The infrared spectra exhibit characteristic bands at 1709, 1674, 1388, 1236, 900, 825, 575, 559, 469, 405, 370, and 328 cm^{-1}.[11]

Crystals of the partially oxidized $K_{1.64}Pt(C_2O_4)_2 \cdot 2H_2O$ are similar in appearance to the partially oxidized tetracyanoplatinates and tend to lose water under ambient conditions. Thus it is advisable to store these partially oxidized complexes in a high-humidity environment, for example, saturated solutions of NH_4Cl and KNO_3 (72% relative humidity). The physical properties of $K_{1.64}Pt(C_2O_4)_2 \cdot 2H_2O$ can be found in a recent review.[1a]

References

1. (a) J. S. Miller and A. J. Epstein, *Prog. Inorg. Chem.*, **20**, 1 (1976); (b) A. F. Garito and A. J. Heeger, *Acc. Chem. Res.*, **7**, 232 (1974).
2. J. S. Miller, *Science*, **193**, 189 (1976).

3. J. A. Abys, N. P. Enright, H. M. Gerdes, T. L. Hall, and J. M. Williams, *Inorg. Synth.*, **19**, 1 (1979).
4. K. Krogmann and H. D. Hausen, *Z. Naturforsch.*, **23**, 111 (1968); A. H. Reis, Jr., S. W. Peterson, D. M. Washecheck, and J. S. Miller, *J. Am. Chem. Soc.*, **98**, 234 (1976); K. D. Keefer, D. M. Washecheck, N. P. Enight, and J. M. Williams, *J. Am. Chem. Soc.*, **98**, 233 (1976).
5. R. B. Saillant, R. C. Jaklevic, and C. D. Bedford, *Mater. Res. Bull.*, **9**, 289 (1974).
6. K. Krogmann and H. D. Hausen, *Z. Anorg. Allg. Chem.*, **258**, 67 (1968).
7. J. S. Miller and S. Z. Goldberg, *J. Chem. Educ.*, **54**, 54 (1977).
8. A. J. Epstein and J. S. Miller, in press.
9. Gmelin Handbuch, **68D**, 199 (1957).
10. J. S. Miller and A. O. Goedde, *J. Chem. Educ.*, **50**, 431 (1973); J. S. Miller, S. H. Kravitz, and S. Kirschner, P. Ostrowski, and P. Nigray, *J. Chem. Educ.*, **55**, 181 (1978).
11. J. Fujita, A. E. Martell, and K. Nakamura, *J. Chem. Phys.*, **36**, 324, 331 (1962).

5. LINEAR-CHAIN IRIDIUM CARBONYL HALIDES

Submitted by A. P. GINSBEF G,* J. W. KOEPKE,* and C. R. SPRINKLE*
Checked by VICKIE HAGLEY[†] and A. H. REIS, JR.[†]

Conducting linear chains of partially oxidized metal-metal bonded iridium atoms are present in the compounds $Ir(CO)_3Cl_{1.10}$[1,2**] and $K_{0.60}Ir(CO)_2Cl_2 \cdot 0.5H_2O$.[3,4] In $Ir(CO)_3Cl_{1.10}$ the chains are made up of planar $[Ir(CO)_3Cl]^{0.1+}$ units stacked along the axis perpendicular to the ligand plane; the extra chloride appears to occur interstitially in random positions between the chains.[2] In $K_{0.60}Ir(CO)_2Cl_2 \cdot 0.5H_2O$ stacks of planar *cis*-$[Ir(CO)_2Cl_2]^{0.6-}$ units make up the chains.[3] Preliminary X-ray measurements indicate that both compounds have essentially the same intrachain Ir—Ir distance [2.85 Å in $Ir(CO)_3Cl_{1.10}$ and 2.86 Å in $K_{0.60}Ir(CO)_2Cl_2 \cdot 0.5H_2O$],[1,4] although they have quite different degrees of partial oxidation. Both compounds are of considerable interest as candidates for detailed physical studies, provided that methods for growing single crystals of high quality can be developed.

Hieber et al.[5] originally isolated $Ir(CO)_3Cl_{1.10}$ from the products of the reaction of finely divided $IrCl_3 \cdot H_2O$ with CO (1 atm) at 150°C, but they formulated the compound as $Ir(CO)_3Cl$. The preparation given below is due to Fischer and Brenner,[6] who also assigned the integral stoichiometry. The method for prepar-

*Bell Laboratories, Murray Hill, NJ 07974.
†Chemistry Division, Argonne National Laboratory, Argonne, IL 60439.

**As noted in reference 8, new X-ray results on this compound indicate that it is in fact stoichiometric $Ir(CO)_3Cl$.

ing $K_{0.60}Ir(CO)_2Cl_2 \cdot 0.5H_2O$ is a modification of a procedure due to Cleare and Griffith,[7] who formulated the compound as a salt of the cluster anion $[Ir_4(CO)_8Cl_8]^{2-}$. $K_{0.60}Ir(CO)_2Cl_2 \cdot 0.5H_2O$ can also be obtained by high-pressure carbonylation of K_3IrCl_6.[3]

A. $Ir(CO)_3Cl_{1.10}$

$$IrCl_3 \cdot xH_2O \xrightarrow[180°]{CO} Ir(CO)_3Cl_{1.10}$$

Procedure

■**Caution.** *Carbon monoxide and chlorine are highly toxic gases. The reaction should be performed only in an efficient fume hood.*

Chromatographic-grade silica gel (18 g) is mixed with a solution of iridium trichloride trihydrate (3.4 g, 9.6 mmole) in distilled water (200 mL). The mixture is stripped dry on a rotary evaporator (60°/aspirator vacuum) and the black solid residue is transferred to a crystallizing dish and dried overnight in a vacuum oven ($120°/10^{-2}$ torr, liquid nitrogen trap). A Pyrex reaction tube, prepared as shown in Fig. 1, is charged with the dry $IrCl_3$-silica gel. Inlet and outlet connections to the reaction tube are made with Tygon tubing. C.P. grade CO and Cl_2 are dried by passing them through a sulfuric acid gas washing bottle followed by a phosphorus pentoxide drying tower (separate drying systems should be used for the CO and Cl_2); prepurified N_2 is used without further treatment.

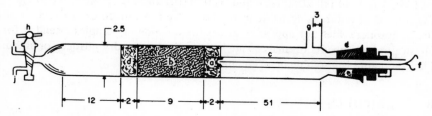

Fig. 1. All dimensions in centimeters. (*a*) 2-cm-thick glass wool plugs; (*b*) $IrCl_3$ on silica gel; (*c*) Pyrex jacket for thermocouple; (*d*) 24/40 ⊤ joint; (*e*) Kontes 24/25 Teflon adapter with O-ring seals; (*f*) thermocouple leads to indicating controller; (*g*) outlet to H_2SO_4 bubbler by way of trap; (*h*) 2-mm double oblique Teflon stopcock; (*i*) N_2 inlet; (*j*) CO or Cl_2 inlet.

The region of the charge plus about 5 cm beyond the glass-wool plugs is wrapped with heating tape and heated at 150° in a slow Cl_2 stream for 2.5 hours. After cooling to room temperature and flushing out the Cl_2 with N_2, CO is passed through the system and the charge is reheated to 180°. The rate of flow of CO must be sufficiently rapid so that the $Ir(CO)_3Cl_{1.10}$ formed is promptly carried out of the hot zone, but not so rapid that large amounts are carried out of the reaction tube; this is easily judged visually. After about 12 hours, the initially black charge appears almost colorless and a fairly large amount of brown-coppery material has condensed as a plug outside of the hot zone. A small amount of yellow solid $[Ir_4(CO)_{12}]$ is found at the end of the charge before the plug of $Ir(CO)_3Cl_{1.10}$. The product is resublimed by adjusting the heating tape to cover it and then heating to 180° with CO flowing. Again, the flow rate must be regulated to carry the sublimate well clear of the hot zone but not out of the tube. Several hours are required for the sublimation. The product is most easily removed from the tube by sliding the thermocouple jacket out of the way and then cutting out the section of the tube containing the sublimate. Yields between 0.5 and 1.3 g (15-43%) may be expected. A small amount of yellow impurity may be present in the end of the sublimate plug nearest the charge, but microscopic examination in reflected light shows the bulk of the product to be free of this impurity. *Anal.* Calcd. for $Ir(CO)_3Cl_{1.10}$: C, 11.43; Cl, 12.37. Found: C, 11.37; Cl, 12.44.[2]

Properties

$Ir(CO)_3Cl_{1.10}$ prepared by the above procedure is a copper-brown fluffy cottonlike mass of interlaced fibrous needles. Under a microscope the needles are dichroic, transmitting brown for light vibrating parallel to the long direction and pale yellow for the transverse direction; they have a metallic copper luster in reflected light. The infrared spectrum (Nujol or Fluorolube mulls) has ν_{CO} at ~2135 (sh), 2080 (s), 2050 (sh), and ~2020 (sh) and ν_{IR-Cl} at 320 (m) cm^{-1}. The crystal structure is available in the literature.[8] Although brief exposure to the atmosphere does not appear to affect the compound, samples intended for physical measurements or analysis are best handled under dry nitrogen or argon. The compound is insoluble in all common solvents.

B. $K_{0.60}Ir(CO)_2Cl_2 \cdot 0.5H_2O$

$$K_2IrCl_6 + HCOOH \xrightarrow[\text{reflux}]{\text{conc. HCl}} K_{0.60}Ir(CO)_2Cl_2 \cdot 0.5H_2O$$

The following operations are all carried out under oxygen-free nitrogen using standard Schlenk techniques.[9,10] Solvents and liquid reagents are degassed before use.

A 100-mL, two-necked flask with 24/40 ⦗ joints is charged with potassium hexachloroiridate(IV) (2.0 g, 4.1 mmole), concentrated hydrochloric acid (10 mL), and 90% formic acid (10 mL). The side neck of the flask is stoppered, and the center neck is connected to a reflux condenser, which in turn is connected, by way of its upper joint, to a vacuum/nitrogen line.[9,10] All joints should be covered with Teflon sleeves. The system is degassed by repeatedly evacuating and backfilling with nitrogen, and it is then heated to vigorous reflux (bp 112°) under a positive pressure (50-100 torr); excess nitrogen escapes through a mercury valve or pressure release bubbler on the nitrogen line. It is essential that the reaction be brought to full reflux. The product is not formed below the boiling point. After the mixture is refluxed for 12 hours it is cooled to room temperature. Degassed water (24 mL) is added to the copper-colored mass, and a fine-porosity Schlenk filter and receiver are attached to the side neck of the flask. The solid is redissolved by reheating to reflux, and the hot solution is filtered through the Schlenk frit. The filtrate is a clear yellow solution that rapidly turns green. Potassium chloride (2 g) is added to the filtrate, which is then heated with stirring to about 90°. After the solution has cooled to room temperature it is allowed to stand in an ice bath for 8 hours. The product is collected by Schlenk filtration, washed with cold water, and dried at $56°/10^{-3}$ torr; yield 1.25 g (85%). *Anal.* Calcd. for $K_{0.60}Ir(CO)_2Cl_2 \cdot 0.5H_2O$: K, 6.67; Ir, 54.66; Cl, 20.17; C, 6.83; H, 0.29. Found: K. 6.72; Ir, 54.75; Cl, 19.94; C, 7.02; H, 0.22.[3] The bromide analogue may be prepared by a similar procedure, but the yield is very erratic.[3]

Properties

The compound $K_{0.60}Ir(CO)_2Cl_2 \cdot 0.5H_2O$ forms as clumps of interlaced fibers comprised of many parallel needles. The needles are dichroic, transmitting red-brown for light vibrating parallel to the long dimension and pale yellow for light vibrating parallel to the transverse direction; they have a metallic copper luster in reflected light. The infrared spectrum (Nujol or Fluorolube mulls) has ν_{CO} at 2040 (s), 2080 (m), and 2115 (w) and ν_{IR-Cl} at 317 (m), 300 (m), and 280 (sh) cm^{-1}. Exposure to air for short periods does not affect the dry solid, but for long-term storage argon or nitrogen should be used. Recrystallization has been carried out from dilute aqueous HCl and from acetone.[3] A salt with the tetrathiafulvalene cation, (2,2'-bi-1,3-dithiolylidene cation radical), $(TTF)_{0.61} Ir(CO)_2Cl_2$, may be prepared by metathesis between $(TTF)_3(BF_4)_2$ and $K_{0.60}Ir(CO)_2Cl_2 \cdot 0.5H_2O$ in acetone solution.[3] The synthesis of $(TTF)_3(BF_4)_2$ is given in this volume.[11]

References

1. K. Krogmann, W. Binder, and H. D. Hausen, *Angew. Chem. (Int. Ed. Engl.),* 7, 812 (1968).
2. A. P. Ginsberg, R. L. Cohen, F. J. Di Salvo, and K. W. West, *J. Chem. Phys.,* 60, 2657 (1974).

3. A. P. Ginsberg, J. W. Koepke, J. J. Hauser, K. W. West, F. J. Di Salvo, C. R. Sprinkle, and R. L. Cohen, *Inorg. Chem.*, **15**, 514 (1976).
4. K. Krogmann and H. P. Geserich, *Am. Chem. Soc. Symp. Ser.*, **5**, 350 (1974).
5. W. Hieber, H. Lagally, and A. Mayr, *Z. Anorg. Allg. Chem.*, **246**, 138 (1941).
6. E. O. Fischer and K. S. Brenner, *Z. Naturforsch.*, **B17**, 774 (1962).
7. M. J. Cleare and W. P. Griffith, *J. Chem. Soc. A*, **1970**, 2788.
8. A. H. Reis, Jr., V. S. Hagley, and S. W. Peterson, *J. Am. Chem. Soc.*, **99**, 4184 (1977).
9. S. Herzog and J. Dehnert, *Z. Chem.*, **4**, 1 (1964).
10. D. F. Shriver, *The Manipulation of Air-Sensitive Compounds*, McGraw-Hill Book Co., Inc., New York, 1969, Chap. 7.
11. F. Wudl and M. L. Kaplan, *Inorg. Synth.*, **19**, 27 (1979).

6. POLYMERCURY CATIONS

Submitted by B. D. CUTFORTH* and R. J. GILLESPIE*
Checked by A. G. MACDIARMID[†] and N. D. MIRO[†]

Compounds that display highly anisotropic metallic conductivity have recently generated considerable interest. The preparations of two such compounds of mercury, $Hg_{2.86}(AsF_6)$ and $Hg_{2.91}(SbF_6)$, are described below. A determination of the crystal structure of $Hg_{2.86}(AsF_6)$[1] has shown it to contain linear nonintersecting infinite chains of mercury atoms in two mutually perpendicular directions. The compound $Hg_{2.91}(SbF_6)$ has been shown to be isostructural with $Hg_{2.86}(AsF_6)$. Both compounds are anisotropic metallic conductors[2] and as such they constitute a new class of anisotropic materials. The syntheses of the compounds $Hg_3(AsF_6)_2$ and $Hg_3(Sb_2F_{11})_2$, which are used in the preparation of large single crystals of the compounds $Hg_{2.86}(AsF_6)$ and $Hg_{2.91}(SbF_6)$, have been described briefly previously.[3] The synthesis of both of these compounds, which contain the linear Hg_3^{2+} cation, are now described in detail.

General Procedure

The apparatus used in the preparations described below (Fig. 1) is similar to that used for the polychalcogen cations.[4] It consists of two 18-mm od Pyrex glass tubes *A* and *B*, terminating at the top in 3-cm long 6-mm od glass tubes *E* and *F*. The tubes are joined by a tube with a 10-mm medium-porosity glass sinter *C* and a sealing constriction *D*. Side-arm *A* contains a magnetic Teflon-covered

*Department of Chemistry, McMaster University, Hamilton, Ontario, L8S 4M1 Canada.
[†]Department of Chemistry, University of Pennsylvania, Philadelphia, PA 19104.

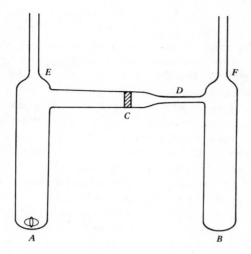

Fig. 1. Reaction vessel.

stirring bar. Tubes *E* and *F* can be closed using a glass plug in a Swagelok union containing Teflon Ferrules.

A. TRIMERCURY BIS-[μ-FLUORO-BIS(PENTAFLUOROANTIMONATE-(V))], $Hg_3(Sb_2F_{11})_2$, AND TRIMERCURY BIS[HEXAFLUOROAR-SENATE(V)], $Hg_3(AsF_6)_2$

$$3Hg + 5SbF_5 \rightarrow Hg_3(Sb_2F_{11})_2 + SbF_3$$

$$3Hg + 3AsF_5 \rightarrow Hg_3(AsF_6)_2 + AsF_3$$

Procedure for $Hg_3(Sb_2F_{11})_2$

Antimony pentafluoride* (4.94 g, 22.8 mmole) is added, in a good dry box, to side *B* of the reaction flask using an all-glass syringe. Elemental mercury (2.74 g, 13.7 mmole) is added to side *A*. The reaction vessel is fitted at *E* and *F* with Nupro or similar metal valves, removed from the dry box, and attached to a vacuum line that has been thoroughly flamed before use. Approximately 10 mL of sulfur dioxide[†] is transferred to side *B* by cooling with liquid N_2, and the

---

*SbF$_5$ (Ozark Mahoning) is distilled twice in a thoroughly dried all-glass apparatus under dry air and stored in glass flasks.

[†]Anhydrous sulfur dioxide is dried by storing it over phosphorus(V) oxide before use.

reaction flask is flame-sealed at E and F. The flask is then allowed to warm to room temperature, at which time the adduct $SbF_5 \cdot SO_2$ forms in B (■**Caution.** *Liquid sulfur dioxide solutions can generate up to 3 atm of pressure at room temperature; therefore the handling of such solutions in sealed glass equipment necessitates suitable shields and the use of face shields and heavy gloves.*) The adduct is completely soluble in SO_2 and is transferred to side A by pouring the solution through the glass frit while cooling A in a Dry Ice-acetone slush bath. The SbF_5 is quantitatively transferred to side arm A by redistilling some of the solvent back to side arm B and then retransferring. The reaction proceeds quite rapidly, initially giving a light-yellow solution with unreacted mercury and then an insoluble orange solid. After 48 hours a very pale-yellow solution containing a yellow solid is obtained. The solution is poured into B and SO_2 is distilled back onto the yellow solid in side A. After repeated extraction of the yellow solid with SO_2 the solid in A is white and consists mainly of SbF_3. (Three or four extractions are usually adequate, but more may be desirable if a high yield of product is desired.) Flask A is then reimmersed in liquid N_2 to remove all the SO_2 from the product and after 12 hours the flask is flame sealed at D. The following analysis is obtained for the light yellow semicrystalline solid remaining in B after recrystallization from SO_2. *Anal.* Calcd. for $Hg_3(Sb_2F_{11})_2$: Hg, 39.94; Sb, 32.32; F. 27.74. Found: Hg, 39.80; Sb, 32.23; F, 27.92. The yield is essentially quantitative if a sufficient number of extractions are carried out. The product is transferred to a glass storage vessel in the dry box.

Procedure for $Hg_3(AsF_6)_2$

Tube B is flame sealed at F and elemental mercury (2.7 g, 13 mmole) is added to A. The flask is closed at E with a Nupro or similar valve and is attached to a calibrated vacuum line. Sulfur dioxide is condensed onto the mercury by cooling A in liquid N_2. While the flask is immersed in liquid nitrogen, arsenic pentafluoride (2.7 g, 16 mmole) is condensed into the flask, which is then sealed at E. (■**Caution.** *Arsenic pentafluoride is a highly toxic gas.*) When the gas is warm the pressure in the reaction vessel rises several atmospheres. The vessel is allowed to warm to room temperature (behind a shield) and stirred for a few hours. The reaction proceeds vigorously, initially giving a mass of golden insoluble crystals, $Hg_{2.86}(AsF_6)$, then a red solution, and finally a light-yellow solution, leaving no insoluble material. The SO_2 is removed from the product by immersing side arm B in liquid nitrogen. After several hours the flask is sealed at D. The product is a light-yellow semicrystalline solid. *Anal.* Calcd. for $Hg_3(AsF_6)_2$: Hg, 61.43; As, 15.30; F, 23.27. Found: Hg, 61.39; As, 15.28; F, 22.92. The yield is essentially quantitative. The product is transferred to a glass storage flask in a dry box.

Properties

The compounds $Hg_3(Sb_2F_{11})_2$ and $Hg_3(AsF_6)_2$ are both light-yellow solids that disproportionate immediately in moist air to give elemental mercury. Both compounds are soluble in SO_2 and HSO_3F, and the solution Raman spectrum shows the characteristic ν_1 mode of the linear Hg_3^{2+} cation at $113\ cm^{-1}$. The solution in HSO_3F is not stable, and oxidation to Hg_2^{2+} occurs in a few hours at room temperature. The X-ray crystal structure of $Hg_3(AsF_6)_2$ has been determined[3] and shows that the Hg_3^{2+} cation is linear in the solid state.

B. THE INFINITE-CHAIN COMPOUNDS $Hg_{2.86}(AsF_6)$ AND $Hg_{2.91}(SbF_6)$

$$5.72Hg + 3AsF_5 \rightarrow 2Hg_{2.86}(AsF_6) + AsF_3$$

$$2.72Hg + Hg_3(AsF_6)_2 \rightarrow 2Hg_{2.86}(AsF_6)$$

$$2.82Hg + Hg_3(Sb_2F_{11})_2 \rightarrow 2Hg_{2.91}(SbF_6) + 2SbF_5$$

Procedure for $Hg_{2.86}(AsF_6)$

The same procedure is followed as in the preparation of $Hg_3(AsF_6)_2$. Elemental mercury (2.47, 12.3 mmole) is allowed to react with a very slight excess of AsF_5 over the amount required by the above equation. When the reaction is complete, golden metallic crystals are present together with a red solution. The crystals are filtered by transferring the SO_2 to side arm B, washed with SO_2, and pumped dry by freezing B in liquid nitrogen. Care must be taken not to cool the crystals of $Hg_{2.86}(AsF_6)$ in SO_2 below $-30°$ because they disproportionate in SO_2 at low temperature:

$$2Hg_{2.86}(AsF_6) \rightarrow Hg_3(AsF_6)_2 + 2.72Hg$$

Alternate Procedure for $Hg_{2.86}(AsF_6)$

Large single crystals of $Hg_{2.86}(AsF_6)$ suitable for conductivity measurements are best grown by oxidizing elemental mercury with a solution of $Hg_3(AsF_6)_2$ in SO_2 at low temperature $(-20°)$. Elemental mercury (1.07 g, 5.3 mmole) is added to side arm A of a typical reaction vessel containing a magnetic stirring bar in side B rather than side A. The compound $Hg_3(AsF_6)_2$ (2.46 g, 2.15 mmole) is added to side arm B. It is not necessary to use exactly stoichiometric amounts of reactants as the reaction does not go to completion and unreacted Hg_3^{2+} and Hg^0 always remain in equilibrium with the product. The vessel is transferred to the vacuum line, and approximately 10 mL of SO_2 is condensed into sidearm B. The $Hg_3(AsF_6)_2$ is then dissolved in the SO_2, and the entire apparatus is cooled to $-20°$. Without significantly warming the apparatus, the solution of $Hg_3(AsF_6)_2$ is transferred onto the mercury and left to stand at $-20°$. After several days a reaction usually begins on the surface of the mercury. If no reaction is visible after this time the temperature should be raised to $-10°$. If left undisturbed for upwards of a week, crystals of $Hg_{2.86}(AsF_6)$ with large flat faces are usually obtained. The crystals are recovered by carefully transferring the SO_2 to side arm B and washing them very carefully with fresh SO_2, which is distilled back into side arm A by gently cooling the side of the tube with a cotton swab soaked in liquid nitrogen. If the crystals are cooled in the presence of SO_2, they disproportionate and this damages the crystal faces. A certain amount of unreacted mercury is present with the crystals of $Hg_{2.86}(AsF_6)$, but it is easily removed by simply pouring it off. The tube is then sealed at D after keeping side arm B immersed in liquid nitrogen to remove traces of the solvent from the crystal. *Anal.* Calcd. for $Hg_{2.86}(AsF_6)$: Hg, 75.23; As, 9.82; F, 14.95. Found: Hg, 75.83; As, 9.52; F, 14.61. The exact composition $Hg_{2.86}(AsF_6)$ cannot be obtained by standard analytical procedure but was established from the X-ray crystallographic determination of the structure.

Large crystals can be obtained by use of a slightly modified apparatus. A small bulb is attached to the lower end of side arm A by means of a 1-mm capillary. It is helpful to flatten the capillary slightly by squeezing with a pair of tweezers after heating in a flame. The bulb is filled with mercury up to the top of the capillary. Using the procedure described above a single large crystal of dimensions up to $10 \times 10 \times 2$ mm often grows from the top of the capillary of mercury. This method is an adaption of that used by MacDiarmid.[5]

Procedure for $Hg_{2.91}(SbF_6)$

Crystals of $Hg_{2.91}(SbF_6)$ are prepared in a manner analogous to that used for $Hg_{2.86}(AsF_6)$. Elemental mercury (1.50 g, 7.5 mmole) is oxidized with a solution of $Hg_3(Sb_2F_{11})_2$ (4.01 g, 21.7 mmole) in SO_2 at $-20°$. Crystals are obtained within 3 days. If crystal growth is allowed to continue for a longer time a white insoluble material is slowly formed, contaminating the product. The crystals are washed several times with small amounts of fresh SO_2 at $-20°$ and isolated in the same manner as for $Hg_{2.86}(AsF_6)$. It is necessary to keep the solution at $-20°$ during the washing. If it is allowed to warm up to room temperature an insoluble white material, probably SbF_3, is formed. The composition has been established from the X-ray crystal structure analysis.[6]

Properties

The compounds $Hg_{2.86}(AsF_6)$ and $Hg_{2.91}(SbF_6)$ are both shiny golden crystals with a distinct metallic luster. Both immediately disproportionate in the presence of moisture and they must be handled in a very good dry box. The crystal surfaces appear to be affected by oxygen. Both materials are anisotropic metallic conductors with a room-temperature conductivity for directions in the *ab* plane of 8.7×10^3 ohm^{-1} cm^{-1} in $Hg_{2.86}(AsF_6)$ and 10^4 ohm^{-1} cm^{-1} in $Hg_{2.91}(SbF_6)$.[7] The anisotropy ratio of the conductivity σ_a/σ_c is 100 for $Hg_{2.86}(AsF_6)$ and 40 for $Hg_{2.91}(SbF_6)$. Anisotropy in the electrical properties has also been observed in optical reflectivity experiments.[2,7] There is a metallic-like plasma edge with a large reflectivity in the infrared with light polarized parallel to the mercury chains. The reflectance for light polarized perpendicular to the mercury chains is an order of magnitude smaller. The electrical conductivity of both compounds increases as the temperature is lowered from room temperature to 4.2°K.

References

1. I. D. Brown, B. D. Cutforth, C. G. Davies, R. J. Gillespie, P. R. Ireland, and J. E. Vekris, *Can. J. Chem.*, **52**, 791 (1974).
2. B. D. Cutforth, W. R. Datars, R. J. Gillespie, and A. van Schyndel, *Solid State Commun.*, **21**, 377 (1977).
3. B. D. Cutforth, C. G. Davies, P. A. W. Dean, R. J. Gillespie, P. R. Ireland, and P. K. Ummat, *Inorg. Chem.*, **12**, 1343 (1973).
4. P. A. W. Dean, R. J. Gillespie, and P. K. Ummat, *Inorg. Synth.*, **15**, 213 (1974).
5. N. D. Miro, A. G. MacDiarmid, A. J. Heeger, A. F. Garito and C. K. Chang, *J. Inorg. and Nucl. Chem.*, **40**, 1351 (1978).
6. B. D. Cutforth, Ph.D. Thesis, McMaster University, Hamilton, Ontario, 1975.
7. E. S. Koteles, W. R. Datars, B. D. Cutforth, and R. J. Gillespie, *Solid State Commun.*, **20**, 1129 (1976).

7. 2,2′-BI-1,3-DITHIOLYLIDENE (TETRATHIAFULVALENE, TTF) AND ITS RADICAL CATION SALTS

Submitted by F. WUDL* and M. L. KAPLAN*
Checked by E. M. ENGLER[†] and V. V. PATEL[†]

In the past decade or two, physical scientists have become intrigued by the limitless possibilities afforded them by synthetic chemists who have produced materials that are not metals, but have "metallic" properties.[1-4] Chief among

*Bell Laboratories, Murray Hill, NJ 07974.
[†]IBM Thomas J. Watson Research Center, Yorktown Heights, NY 10598.

this class of compounds, from the point of view of interest shown, are the charge-transfer salts and other salts of 7,7,8,8-tetracyanoquinodimethane[5] (TCNQ) [2,2'-(2,5-cyclohexadiene-1,4-diylidene)bis(propanedinitrile)] .

Shortly after TTF was first synthesized[6-9] it was found to possess electrical properties superior to other comparable organic donors.[10] Within a year, the donor properties of TTF and the acceptor properties of TCNQ were combined and together they were shown to form a highly conducting donor-acceptor complex,[11] the first example of an organic "metal." Subsequent work with TTF has shown that a wide variety of its salts can be prepared[12] and thus are amenable to study. The first salts available were prepared by direct oxidation of TTF by an acceptor (e.g., Cl_2,[6] TCNQ,[11] 1,3-butadiene-1,1,2,3,4,4-hexacarbonitrile,[13] 2,2'-(2,6-naphthalenediylidene)bis[propanedinitrile],[14a] $Ni[S_2C_2(CF_3)_2]_2$,[14b] $Ni[S_2C_2H_2]_2$,[14a] $Ni[S_2C_2(CN)_2]_2$[14c]). The more general approach presented here is founded on a metathetical reaction and provides access to TTF salts based on anions not solely derived from strong oxidizing agents.

In the past, TTF has been prepared from 1,3-dithiole-2-thione, which, by an oxidative step, was converted to a 1,3-dithiolylium salt followed by coupling with base.[6-9,15] The present method uses a reductive sequence, thereby permitting milder conditions and better yields.[16]

A. 2,2'-BI-1,3-DITHIOLYLIDENE (TETRATHIAFULVALENE) (TTF)

Materials

The 2-(methylthio)-1,3-dithiolylium iodide can be prepared by a published procedure[17] or purchased commercially (Strem Chemical Co., Danvers, MA).

Procedure

2-(Methylthio)-1,3-dithiolylium iodide (50.0 g, 0.18 mole) and 500 mL of absolute, reagent grade, methanol are added to a 2-L Erlenmeyer flask that contains a magnetic stirring bar and thermometer. The slurry is stirred and cooled in an ice bath. The temperature is maintained at or below 15° and $NaBH_4$ is added in small portions (34.5 g, 0.93 mole). (**■Caution.** *Vigorous hydrogen evolution and foaming occur when $NaBH_4$ is added.*) Upon completion of the addition, stirring is continued for 2 hours. One liter of reagent grade, anhydrous diethyl ether is added to precipitate the NaI. After the mixture has been stored at $-15°$ overnight (16 hr) the solution is decanted from the solid into a 2-L separatory funnel, washed three times with water (200 mL each time), dried over anhydrous $MgSO_4$, filtered, and concentrated on a rotary evaporator.

The resulting orange-yellow oil [(2-methylthio)-1,3-dithiole] [16] need not be further purified. Fluoroboric acid reagent is prepared by adding fluoroboric acid (hydrogen tetrafluoroborate) (33.3 g of 48% HBF_4, 0.18 mole) dropwise to stirred, cooled (0°) acetic anhydride (100 mL). (**■Caution.** *The fluoroboric acid reagent preparation is very exothermic.*) This reagent is added dropwise to a stirred, cold (0°) solution of the 2-(methylthio)-1,3-dithiole in acetic anhydride (100 mL) in a 500-mL Erlenmeyer flask. (**■Caution.** *Stench, use fume hood.*)

When the addition is complete, anhydrous diethyl ether (about 200 mL) is added. The near-white salt, 1,3-dithiolylium tetrafluoroborate,[12] is collected on a Büchner funnel, washed with anhydrous diethyl ether, and dried under nitrogen. The 1,3-dithiolylium tetrafluoroborate (28 g, 0.15 mole) from the precious step is dissolved in acetonitrile (100 mL) in a 500-mL Erlenmeyer flask. While the solution is stirred magnetically at 0°, triethylamine is added until the formation of yellow crystals becomes obvious (about 50 mL). Another quantity of triethylamine (about 10 mL) is added and stirring is continued for 10 minutes. Water (approximately 300-400 mL) is added to precipitate the product, which separates as yellow-orange crystals. The solid is collected on a Büchner funnel, washed with water, and dried under nitrogen. The yield of crude product is 15.0 g (0.0736 mole), which corresponds to an 82% overall yield based on the starting 2-(methylthio)-1,3-dithiolylium iodide.

Recrystallization from cyclohexane-hexane (500 mL:300 mL), including clarification with Norit and drying with magnesium sulfate, affords 75% recovery of TTF as large orange-yellow needles with a melting point of 119.1-119.3° (corrected).

Properties

Tetrathiafulvalene crystallizes as long orange-yellow needles and melts at 119.1-119.3° (corrected) (lit. mp 119-119.5°,[15] 118.5-119°,[6] 120°[9]). The ^{1}H NMR spectrum in CCl_4 (TMS) is a sharp singlet at τ 3.75. (The checkers report τ 3.62.) The infrared spectrum in KBr exhibits bands at 1530 (w), 1250 (w), 1075 (w), 870 (w), 797 (m), 782 (m) and 734 (m) cm^{-1}. The optical spectrum consists of the following: [hexane, λ_{max} (nm) (ϵ)] 303 (12,800), 316 (11,400), 368 (2000), 450 (230). No change is observed in TTF when it is stored for many months in amber bottles in a refrigerator under nitrogen. However, in solution, TTF is readily photooxidized in the presence of oxygen, to a violet, water-soluble substance.[6] In the solid-state photooxidation occurs more slowly. Two reversible oxidation steps are observed in acetonitrile with tetraethylammonium tetrafluoroborate as the supporting electrolyte. These occur[8] at E_1^0 = +0.36 (the checkers report +0.33) and E_2^0 = + 0.70 versus aqueous saturated calomel. The ESR spectrum of the radical cation of TTF, in acetonitrile at room temperature, is a quintet, a_H = 1.26, g = 2.00838.

B. [TRIS(2,2'-BI-1,3-DITHIOLYLIDENE) RADICAL CATION(2+)] BIS[TETRAFLUOROBORATE(1-)] (TETRATHIAFULVALENIUM TETRAFLUOROBORATE), (TTF)₃(BF₄)₂

$$3TTF + H_2O_2 + 2HBF_4 \xrightarrow{CH_3CN} (TTF)_3(BF_4)_2 + 2H_2O$$

Procedure

To a solution of TTF (3.50 g, 0.0172 mole) in 70 mL of dried (P_2O_5) distilled (P_2O_5) acetonitrile is added a solution of 30% hydrogen peroxide (0.648 g, 0.00572 mole) in 48% HBF_4 (2.168 g, 0.0130 mole). After it is stirred (magnetic) at room temperature for 10 minutes the dark solution is refrigerated for 1 hour. The dark-purple crystalline solid is removed by Buchner filtration, washed with cold acetonitrile and then diethyl ether, and dried under a nitrogen stream. The first solid weighs 2.71 g (60% yield). Concentration of the mother liquors to half volume and cooling provides an additional 0.42 g (10%) of product. *Anal.* Calcd. for $C_{18}H_{12}S_{12}B_2F_8$: C, 27.48; H, 1.54. Found: C, 27.53; H, 1.51.

Properties

The tetrafluoborate salt of the TTF radical cation forms deep-purple needles; it is soluble in warm acetonitrile and sparingly soluble in cold acetonitrile, acetone, and methyl acetate. The infrared spectrum in Nujol consists of lines at 1450 (m), 1280 (w), 1270 (w), 1250 (w), 1120 (m), 1030 (s, broad), 840 (w), 827 (m), 802 (m), 752 (m), 745 (m), 732 (m), 702 (m), 673 (m) cm⁻¹. The electronic spectrum [CH₃CN, λ_{max} (ϵ)] data are: 250 (sh), 290 (sh), 305 (sh), 316 (19,800), 335 (sh), 375 (sh), 397 (sh), 432 (34,100), 576 (9600) nm.

C. 2,2'-BI-1,3-DITHIOLYLIDENE RADICAL CATION (TETRATHIAFUL-VALENIUM) SALTS[12]

$$[TTF] X + MZ \rightarrow [TTF] Z \downarrow + MX$$

Procedure

This general preparation, by metathesis, presupposes that [TTF] X, MZ, and MX are soluble and that [TTF] Z is not. Most commonly triethylammonium, tetra-

TABLE I[a]

Compound	Yield (%)	Resistance (ohms) (Ω)[b]
$(TTF)_3(BF_4)_2$	60	$>10^6$
$(TTF)_{14}(NCS)_8$ [c]	70	2-6
$(TTF)_{14}(NCSe)_8$	45	7
$(TTF)_{11}I_8$ [c,d]	28	0.5
$(TTF)_{24}I_{63}$ [c,e]	37	4.5
$(TTF)_8I_{15}$ [c,f]	75	55
$(TTF)_2[Pt(CN)_4]$	92	$>10^6$
$(TTF)_2[Cu(mnt)_2]$ [g]	60	$>10^6$
$(TTF)_2[Ni(mnt)_2]$ [g]	70	$>10^6$
$(TTF)_2[Pt(mnt)_2]$ [g]	70	$>10^6$
$(TTF)[Pt(mnt)_2]$ [g,h]	94	$>10^6$

[a] All elemental analyses based on C, H, and N were correct for formulae shown.

[b] Determined at room temperature on about 0.1 mg compressed in a glass capillary between two steel pistons of 2-mm diameter.

[c] Corresponds to "$(TTF)_2(NCS)$." This and other compounds below exhibit "nonstoichiometric" elemental analyses that are best described (within 0.1-0.05% of experimental) by the formulae given in the table.

[d] This compound is also known as "TTF_3I_2." It is prepared from $(TTF)_3(BF_4)_2$ and $Bu_4N^+I_3$ and forms hollow needles.

[e] Corresponds to $TTF_{1.0}I_{2.63}$. Prepared from $(TTF)_3(BF_4)_2$ and $[Bu_4N]^+[I_3]^-$; has the appearance of silver wool.

[f] Corresponds exactly to TTF_8I_{15}. Also prepared from $[Bu_4N]^+$-$[I_3]^-$ and $(TTF)_3(BF_4)_2$.

[g] Prepared from $(TTF)_3(BF_4)_2$ and $[(Bu_4N)_2][M(mnt)_2]$, where mnt = 2,3-dimercapto-2-butenedinitrilato(2−) (maleonitriledithiolato), $[N\equiv C-C(S^-)=C(S^-)-C\equiv N]$, and M = Cu, Ni, Pt, etc.

[h] Prepared from $[Bu_4N][Pt(Mnt)_2]$ and $(TTF)_3(BF_4)_2$.

butylammonium, and tetraethylammonium salts of Z have been utilized. The preparation of the thiocyanate salt of the TTF radical cation is presented as a typical example of how this method can be used.

To a warm (about 60°C) filtered solution of $(TTF)_3(BF_4)_2$ (157 mg, 0.2 mmole) in 200 mL of freshly distilled (from P_2O_5) acetonitrile is added a warm, filtered solution of tetrabutylammonium thiocyanate (120 mg, 0.4 mmole) in 25 mL of acetonitrile. The flask containing the warm mixture is then purged with argon, sealed, and wrapped with insulating material. After it has come to room temperature (about 2 hr) the solution is stored in a refrigerator for about 16 hours. The dark-purple needlelike crystals are collected by suction filtration, washed first with cold acetonitrile and then diethyl ether, and finally dried in a nitrogen stream. The yield of product is 100 mg. *Anal.* Calcd. for $(TTF)_{14}(NCS)_8$, $(C_{92}H_{56}S_{64}N_8)$: C, 33.23; H, 1.70; N, 3.37. Found: C, 33.09; H, 1.71; N, 3.31. Examples of other salts of the TTF radical cations prepared by metathesis are shown in Table I.

Properties

The thiocyanate salt of the TTF radical cations exists as dark-purple needlelike crystals that are insoluble in nonpolar solvents and moderately soluble in warm polar solvents (e.g., acetonitrile, dichloromethane). Stoichiometries derived from elemental analysis for this and most salts of the TTF radical cations are not unequivocal since several calculated values come within the acceptable limits (i.e., ±0.3% per element) of the percentages actually determined. For identification purposes we have called the thiocyanate $(TTF)_{14}(NCS)_8$, although, for example, $(TTF)_{11}(NCS)_6$, $(TTF)_{12}(NCS)_7$, and $(TTF)_{15}(NCS)_8$ are also consistent with our results. The compressed pellet resistance is 2-6 ohms (see Table I).

D. [2,2'-BI-1,3-DITHIOLYLIDENE RADICAL CATION(1+)][2,2'-(2,5-CYCLOHEXADIENE-1,4-DIYLIDENE)BIS[PROPANEDINITRILE] RADICAL ANION(1−)] (TETRATHIAFULVALENIUM-TETRACYANOQUINODIMETHANIDE), [(TTF)(TCNQ)]

Procedure

The preparation of (TTF)(TCNQ) is straightforward. Equimolar quantities of the two starting materials are dissolved in an appropriate solvent (usually acetonitrile) at convenient concentrations (0.5-1.0 mg/mL). When the two yellow solutions are mixed, an immediate dark-green color develops and small black needles separate within 30 minutes at room temperature. The product can be collected, washed, and then dried to give a material of analytical purity.

No special precautions are required to prepare material of this quality. However, since the loss of conductivity in samples of (TTF)(TCNQ) has been attributed to inclusion of impurities[18] it is appropriate to use purified starting materials and distilled solvents and to perform the reaction in the absence of oxygen. Although the product is easy to prepare, the crystals are small and therefore not suitable for single-crystal measurements.

The growth of relatively large single crystals has been facilitated by the development of a three-chamber apparatus (Fig. 1) that uses a diffusion-growth method.[19,20] The chambers each have a capacity of about 50 mL and are separ-

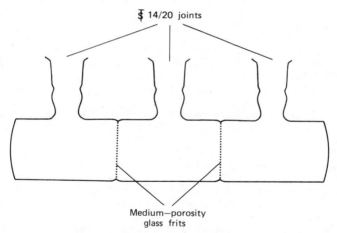

$\overline{\underline{\underline{\mathbf{S}}}}$ 14/20 joints

Medium—porosity
glass frits

Fig. 1. Three-chamber crystal-growing apparatus.

ated by medium-porosity glass frits. The apparatus is carefully washed and dried prior to use. The acetonitrile used should be freshly distilled from P_2O_5. A solution of TTF (35 mg/50 mL) is filtered into one side chamber, and a solution of TCNQ (35 mg/50 mL) is filtered into the opposite chamber. Pure solvent is filtered into the middle chamber. The chambers are closed with Teflon stoppers and the whole apparatus is thermostated for 7 days at 30°. Single crystals as large as 1-2 cm long and 1 mm in width can then be harvested from the center chamber by decantation. Yield 20 mg (28%). *Anal.* Calcd. for (TTF)(TCNQ) ($C_{18}H_8$-S_4N_4): C, 52.92; H, 1.97; N, 13.71. Found: C, 53.14, H. 1.81; N, 13.97.

Properties

The characteristic shiny black needles of (TTF)(TCNQ) are insoluble in hydrocarbon solvents and of limited solubility in acetonitrile and dichloromethane. The resistance of a compressed pellet of (TTF)(TCNQ) is less than 1 ohm.[21] The optical spectrum [CH_3CN, λ_{max} (ϵ)] consists of lines at 305 (sh), 316 (14,600), 375 (sh), and 393 (64,700) nm.

References

1. I. F. Shchegolev, *Phys. Status Solidi (B)*, **12**, 9 (1972).
2. A. F. Garito and A. J. Heeger, *Acc. Chem. Res.*, **7**, 218 (1974).
3. Z. Soos, *Ann. Rev. Phys. Chem.*, **25**, 121 (1974).
4. A. N. Block, in *Charge and Energy Transfer in Organic Semiconductors*, K. Masuda and M. Silver (eds.), Plenum Press, New York, 1974.
5. D. S. Acker and W. R. Hertler, *J. Am. Chem. Soc.*, **84**, 3370 (1962).
6. F. Wudl, G. M. Smith, and E. J. Hufnagel, *Chem. Commun.*, **1970**, 1453.
7. D. L. Coffen, *Tetrahedron Lett.*, **1970**, 2633.
8. D. L. Coffen, J. Q. Chambers, D. R. Williams, P. E. Garrett, and N. D. Chambers, *J. Am. Chem. Soc.*, **93**, 2258 (1971).
9. S. Hünig, G. Kiesslich, H. Quast, and D. Scheutzow, *Justus Liebig's Ann. Chem.*, **1973**, 310.
10. F. Wudl, D. Wobschall, and E. J. Hufnagel., *J. Am. Chem. Soc.*, **94**, 670 (1972).
11. J. Ferraris, D. O. Cowan, V. Walatka, Jr., and J. Perlstein, *J. Am. Chem. Soc.*, **95**, 948 (1973).
12. F. Wudl, *J. Am. Chem. Soc.*, **97**, 1962 (1975).
13. F. Wudl and E. W. Southwick, *Chem. Commun.*, **1974**, 254.
14. (a) D. J. Sandman and A. F. Garito, *J. Org. Chem.*, **39**, 1165 (1974); (b) L. V. Interrante, K. W. Browall, H. R. Hart, Jr., I. S. Jacobs, G. D. Watkins, S. H. Wee, *J. Am. Chem. Soc.*, **97**, 890 (1975); (c) F. Wudl, C. H. Ho, and A. Nagel, *J. Chem. Soc. Chem. Commun.*, **1973**, 923.
15. L. R. Melby, H. D. Hartzler, and W. H. Sheppard, *J. Org. Chem.*, **39**, 2456 (1974); for large-scale preparations of TTF this procedure is probably more suitable.
16. F. Wudl and M. L. Kaplan, *J. Org. Chem.*, **39**, 3608 (1974).
17. F. Challenger, E. A. Mason, E. C. Holdsworth, and R. Emmott, *J. Chem. Soc.*, **1953**, 292.
18. R. V. Gemmer, D. O. Cowan, T. O. Poehler, A. N. Bloch, R. E. Pyle, and R. H. Bauks, *J. Org. Chem.*, **40**, 3544 (1975).
19. M. L. Kaplan, *J. Cryst. Growth*, **33**, 161 (1976).
20. H. Angai, *J. Cryst. Growth*, **33**, 185 (1976); this reference provides another technique.
21. F. Wudl, M. L. Kaplan, and J. J. Hauser, unpublished observations.

8.   TANTALUM DISULFIDE (TaS$_2$)
AND ITS INTERCALATION COMPOUNDS

Submitted by J. F. REVELLI*
Checked by F. J. DISALVO†

Over the past 5 years, a considerable amount of research has been devoted to the study of layered transition metal dichalcogenides and their so-called intercalation complexes.[1-5] In particular, the group IV, V, and VI transition metal dichalcogenides form layered structures with hexagonal, rhombohedral, or trigonal symmetry. The basic layer is composed of a covalently bound X—M—X sandwich (M = transition metal, X = chalcogen), and successive layers are bound together by relatively weak chalcogen-chalcogen van der Waals bonds. Hence, under appropriate conditions, these layers can be "pried" apart, and other chemical species can be inserted between them—much in the same fashion as two decks of cards are mixed together by shuffling. From the point of view of chemical bonding, intercalation can be regarded as a charge-transfer process in which the orbitals of the guest species are mixed together with those of the layered "host."[2-4] In the case of the group VI layered transition metal dichalcogenides the energy gained in charge transfer is presumably small. Hence, in these materials stable intercalation complexes have been observed only for highly electropositive guest species, such as the alkali metals. For this case a net donation of charge occurs from the alkali metal to the transition metal dichalcogenide.[4-8] The group V layered dichalcogenides, on the other hand, exhibit the phenomenon of intercalation for a wide range of organic (Lewis base) materials,[2-5] transition and post-transition metals,[9] and the alkali metals.[7] Synthesis techniques of the group V layered compounds and their intercalation complexes are described in the following sections. In particular, TaS$_2$ and the tantalum sulfide intercalates are used as the primary examples.

Systematic studies of the structural properties of TaS$_2$ by Jellinek[10] revealed the presence of several polymorphic forms of the compound as a function of temperature. Within a layer, the tantalum atom sits in the holes formed between two layers of sulfur atoms in the S—Ta—S sandwich. The coordination of the tantalum is trigonal prismatic or octahedral, depending on whether the two sulfur layers lie one on top of the other or are rotated by 60°. Thus the various poly-

*Electrical Engineering and Materials Science Departments, Northwestern University, Evanston, Ill. Current address: Xerox Corp., Webster Research Center, Rochester, NY 14600.
†Bell Laboratories, Murry Hill, NJ 07974.

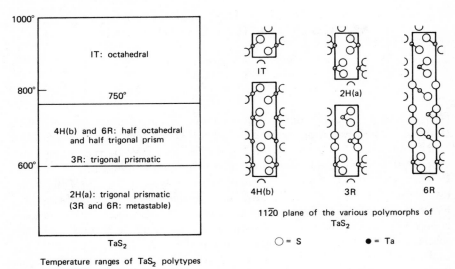

Temperature ranges of TaS_2 polytypes

1120 plane of the various polymorphs of
TaS_2

$\bigcirc$ = S $\bullet$ = Ta

Fig. 1. The a notation indicates that the Tantalum atoms are aligned along the crystallographic c-axis. R indicates rhombohedral symmetry. The 3R phase has three trigonal prismatic slabs per unit cell, whereas the 6R phase has six slabs that are alternately trigonal prismatic and octahedral. F. Jellinek, *J. Less-Common Met.,* **4,** 9-15 (1962); F. J. DiSalvo et al., *J. Phys. Chem. Solids,* **34,** 1357 (1973).

morphs observed in TaS_2 result from the variety of stacking sequences of the basic S–Ta–S slabs. Figure 1 shows the various phases that are formed as a functions of temperature. The high-temperature "1T" (T = trigonal symmetry) phase has octahedral coordination of the tantalum atom and *one* S–Ta–S slab in the unit cell. The room-temperature 2H(a) phase (H = hexagonal symmetry) has trigonal prismatic coordination of the tantalum atoms and *two* S–Ta–S slabs per unit cell (see 1120 section, Fig. 1). For reasons as yet not fully understood, only those phases that have trigonal prismatic character tend to form intercalation compounds readily. Of these, the 2H(a) polytype of TaS_2 seems to be the most favorable host material.

A. POLYCRYSTALLINE 2H(a) PHASE OF TaS_2

$$Ta + 2S \xrightarrow{850^\circ} 1T\text{-}TaS_2$$

$$1T\text{-}TaS_2 \xrightarrow{750^\circ} 6R\text{-}TaS_2 \xrightarrow{550^\circ} 2H(a)\text{-}TaS_2$$

Procedure

The techniques used in the preparation of polycrystalline and single-crystal TaS_2 are much the same as those described for TiS_2, ZrS_2, and SnS_2 and have been described in this series.[12,13] The latter compounds, however, exist in the 1T phase structure throughout the entire range from the crystal growth temperature down to room temperature. The compound TaS_2, on the other hand, must undergo two first-order phase transitions as it is cooled from above 750° to room temperature. To assure complete transformation to the 2H(a) phase, the TaS_2 must be cooled very slowly through these transition temperatures.

The synthesis of 2H(a)-TaS_2 powder is carried out in fused quartz or Vycor ampules—typically 10 cm long and 1.7 cm in diameter, with a wall thickness of 1.0-2.0 mm (Fig. 2a). About 1.41 g (0.008 mole) of 0.020-in. diameter tantalum

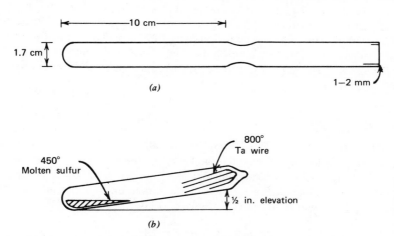

Fig. 2. (a) Typical sample tube (Vycor or quartz). (b) Loaded sample tube.

wire (99.8% purity)* is cut in the form of 3.8 cm lengths, washed in hot dilute HCl to remove iron contamination introduced in the cutting process, rinsed in distilled water, and dried. Sulfur powder, 0.500 g (0.016 mole, 99.9999% purity),[†] is then added, and the ampule is sealed under a vacuum of about 10^{-3} torr. The sealed ampule is placed into a cold laboratory tube furnace with the tantalum wire charge placed in the end of the ampule towards the center of the oven and elevated about 1 cm (see Fig. 2b). Thus, as the sample is heated, the

*Available from Materials Research Corp., Route 303, Orangeburg, NY 10962.
[†]Available from United Mineral and Chemical Corp., 129 Hudson St., New York, NY 10013.

sulfur melts and remains in the lower, cooler end of the ampule. A thermocouple is placed near this end to ensure that the temperature is at or below $450°$ (the pressure of sulfur vapor in equilibrium with liquid sulfur is 1 atm at this temperature). The end of the ampule containing the tantalum may be heated to 800 or $900°$.

■**Caution.** *Because the reaction of tantalum and sulfur is exothermic and because of the high vapor pressure of sulfur, it is essential, in order to avoid explosions, that the liquid sulfur be prevented from coming into direct contact with the hot tantalum wire.*

The use of tantalum wire along with the "two-zone" technique ensures a safe reaction rate. When the sulfur has reacted completely with the tantalum wire (usually within 7-10 days) the ampule is removed from the furnace, shaken vigorously to break up clumps of unreacted material, and placed in the center of tne furnace. It may be necessary to repeat this procedure several times to ensure that all the wire has reacted.* After 1 or 2 days at $850-900°$, the cooling process may be started. The sample is cooled initially to $750°$ and annealed for 1 day. It is then annealed at $650°$ for about a day and at $550°$ for another 2 to 3 days. The furnace is then shut off and the sample is allowed to cool for 5-6 hours to room temperature.

Properties

The polycrystalline $2H(a)$-TaS_2 should be free flowing and in the form of black platelets. Any gold-colored material indicates the presence of the 1T phase and results from improper annealing. Further, the presence of a fibrous or needlelike material indicates incomplete reaction of the tantalum. This fibrous material is most likely TaS_3, which decomposes above $650°$.[14]

$$Ta + 2TaS_3 \xrightarrow{650°} 3TaS_2$$

This material may be removed by reheating the sample to $850°$, followed by the same annealing procedure outlined above. The X-ray diffraction pattern for $2H(a)$-TaS_2 may be used for identification. The following d values have been obtained for major low angle X-ray diffraction lines (and intensities): 6.05 (1); 3.025 (0.06); 2.8709 (0.32); 2.7933 (0.07); 2.3937 (0.80); and 2.3389 (0.04) Å. Note that it is difficult to obtain the ideal intensities because of preferred orientation of the crystallites. This material is a superconducting metal with $T_c = 0.8 \pm 0.05°K$.[2]

---

*An alternate method for quickly obtaining smaller (1-g) batches of TaS_2 powder is to place tantalum powder (−60 mesh) in a 10-12 mm (inside diameter fused) quartz tube, 10 cm long, with a stoichiometric quantity of sulfur. This mixture will react completely in a relatively short while (2-3 days) at $450°$. The sample may then be heated uniformly to 850-$900°$ as described for the larger batch.

B. SINGLE-CRYSTAL 2H(a) PHASE OF TaS$_2$

$$TaS_2 + 2I_2 \underset{750°}{\overset{850°}{\rightleftharpoons}} TaI_4 + 2S$$

Procedure

Single crystals of 2H(a)-Tas$_2$ may be obtained by chemical transport either from prereacted TaS$_2$ powder or directly from the elements. If the crystals are to be prepared from the elements, care must be taken to heat the ampule slowly in the manner described previously to avoid explosions. The reactants are loaded into a 20-cm long, 1.7-cm diameter fused-quartz ampule. About 5 mg of iodine is added per cubic centimeter of volume of the reaction ampule. This serves as a transport agent according to the equation given above.[15] Both the sulfur and TaI$_4$ are volatile components. The equilibrium constant is such that at the higher temperature the reaction proceeds to the right, as described, while at the cooler end of the ampule the equilibrium favors TaS$_2$ and I$_2$. Thus the iodine is regenerated constantly as the TaS$_2$ crystals grow in the cooler zone.

Several techniques exist for loading iodine into the reaction vessel, and some are described in Reference 13. The simplest method is to load iodine quickly in air followed by pumping down to a few microns pressure and sealing. The ampule is then heated to 900° (with appropriate precautions taken if the elements are undergoing reaction for the first time). The ampule must then be placed in a thermal gradient of 850-750° such that the charge is at the hot end and the 750° zone extends over a region of 50-75 mm. This may be accomplished by using two tubular furnaces joined together end to end or by a single furnace with a continuously wound filament having taps every 3.8 cm or so. These taps are then shunted with external resistances to achieve the desired temperature profile. One week is usually sufficient to achieve 100% transport of the charge. This transport period must then be followed by the annealing procedure described above; however, the annealing times should be prolonged somewhat to ensure that the large crystals transform properly; 1½ days at 650° and 3 days at 550° are usually sufficient. Upon cooling, the ampule may be cracked open and the crystals are removed and rinsed in CCl$_4$ and then CS$_2$ to remove I$_2$ and S, respectively.

■**Caution.** *The tube should be wrapped in several layers of cloth before it is opened.*

As mentioned in Reference 13, larger crystals can be obtained by using larger ampule diameter (2.5 cm) and/or heating the growth zone above 900° for a few hours (with the charge end at 800° or so) before beginning the transport. This results in "back-transporting" multiple TaS$_2$ nucleation sites that are in the growth zone.

Properties

The crystals obtained in this fashion have hexagonal symmetry (space group $P6_3/mmc$) with a = 3.314 Å and $c/2$ = 6.04 Å ($c/2$ is the basic S–Ta–S slab thickness). The d values given above for the polycrystalline material may be used to check the identity of a crushed crystal.

C. INTERCALATION COMPOUNDS OF TaS$_2$

Three main categories of intercalation compounds can be formed with layered transition metal dichalcogenides: Lewis base complexes, alkali metal complexes, and transition metal or post-transition metal "complexes." A representative synthesis for each category is included. For the Lewis bases the tendency to form intercalation compounds increases as the Lewis basicity of the molecule increases and as the molecular size of the base decreases.[2-4] Hence, NH$_3$ (p$K_\alpha \approx 9$) readily intercalates a wide variety of layered transition metal dichalcognides (including TiS$_2$ and ZrS$_2$), whereas pyridine (p$K_\alpha \approx 5.3$) has been shown to intercalate only 2H(a)-TaS$_2$ or NbS$_2$. Under certain circumstances, the smaller NH$_3$ molecule may be used to "pry" the layered dichalcogenide apart, making possible subsequent intercalation of a larger molecule.[2-4] This "double intercalation" is carried out by preintercalating with NH$_3$.

1a. Pyridine Intercalate of Tantalum Disulfide

$$2TaS_2 + C_5H_5N \text{ (excess)} \xrightarrow{200°C} 2TaS_2 \cdot C_5H_5N$$

Procedure

In the compound 2TaS$_2$·C$_5$H$_5$N the pyridine rings are normal to the crystallographic c-axis.[2-4] Two such rings inserted between the layers give the stoichiometric composition 2TaS$_2$·C$_5$H$_5$N (see Fig. 3). Under certain conditions, the complex 4TaS$_2$·C$_5$H$_5$N also may be obtained.[4]

2H(a)-TaS$_2$ powder (1.91 g about 0.008 mole) is placed in a Pyrex tube about 20 cm long and 1 cm in diameter, with a 2-mm wall thickness. An excess of redistilled pyridine is then added. The volume of pyridine should be three or four times that of the TaS$_2$ powder. (■**Caution.** *Direct contact with pyridine or pyridine vapor should be avoided.*) The tube is connected to a vacuum system and is quickly pumped down to 15 torr pressure (vapor pressure of liquid pyridine at room temperature). The pyridine is then frozen with liquid nitrogen, and the evacuation is continued to a pressure of 10^{-3} torr. To remove dissolved air,

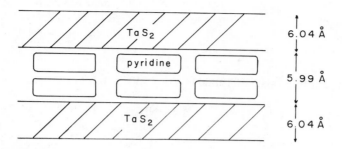

Fig. 3. Schematic representation of tantalum disulhide-pyridine intercalute.

the sample is carried through two or more additional cycles of freezing, pumping on the frozen material, and thawing. The final cycle is followed by sealing the tube under vacuum with the contents frozen at liquid nitrogen temperature. The sample is then heated gradually to 200° with stirring (an oil bath with magnetic stirring is adequate). After 15 minutes to 1 hour at this temperature (if the TaS_2 powder is of good quality), the powder swells to nearly twice its original volume as the pyridine intercalate forms.

Intercalation of single crystals of 2H(a)-TaS_2 may be carried out in a similar manner, although it is possible that, if the crystals are too large in surface area, intercalation may proceed only around the edges. It should be noted that the time for full intercalation of crystals increases with crystal dimensions.

Properties

The pyridine intercalate is blue-black. Examination of the individual platelets under a microscope reveals a characteristic exfoliated appearance. Hexagonal symmetry is retained with $a = 3.325$ Å and $c/2 = 12.03$ Å. Comparison with the slab thickness for the parent 2H(a)-TaS_2 shows a c-axis expansion, δ, of 5.99 Å. The following d values have been obtained for low-angle X-ray diffraction lines: 12.03, 6.015, 4.010, 3.008, 2.880, and 2.859 Å. $2TaS_2 \cdot C_5H_5N$ is also super conducting with a transition temperature of $3.5 \pm 0.3°K$[2,3]

To check the composition of the complex thus synthesized, a weight-gain analysis may be carried out. The Pyrex tube is scribed carefully and broken open after being wrapped in a cloth. The contents are filtered through a tared fritted-glass filter funnel (medium- or fine-porosity filter). The Pyrex tube is rinsed with CH_2Cl_2 to remove any material left clinging to its walls. The excess pyridine liquid may take on a brownish color. This coloration is due to a small amount of sulfur being extracted from the TaS_2 by the pyridine and can be avoided by the addition of some sulfur to the pyridine before intercalation.[16] After suction filtration of its contents, the funnel is stoppered and the $2TaS_2 \cdot C_5H_5N$ powder is *vacuum* dried ($10\text{-}50\mu$) at room temperature and carefully weighed. The weight of intercalated pyridine is determined by the difference between the original TaS_2 charge and the $2TaS_2 \cdot C_5H_5N$.

1b. Ammonia Intercalate of Tantalum Disulfide as an Intermediate

$$TaS_2 \ + \ NH_3 \ \rightarrow \ TaS_2 \cdot NH_3$$

$$2(TaS_2 \cdot NH_3) \ + \ B \ \rightarrow \ 2TaS_2 \cdot B \ + \ 2NH_3$$

Dried, gaseous NH_3 is condensed into a Pyrex combustion tube (2-mm wall thickness, 10-mm id) to which the host material has already been added. The NH_3 is frozen with liquid nitrogen and the tube is evacuated and sealed.

■**Caution.** *The vapor pressure of liquid NH_3 at room temperature is 8 atm. The seal on the Pyrex tube should be carefully annealed so that the tube will not explode. The sealed tube always should be kept behind a protective shield or in a protective metal pipe.*

After a short while (½ to several hours) at room temperature the layered host material reacts fully to yield the stoichiometry $MX_2 \cdot NH_3$. The excess NH_3 may turn light blue because of the extraction of a slight amount of sulfur from the TaS_2, but here again, as in the case of pyridine, a small amount of excess sulfur may be added to the Pyrex tube (i.e., a few milligrams) before it is sealed. The Pyrex tube is then removed from its bomb, cooled to refreeze the NH_3, wrapped in a protective cloth, and cracked open. The intercalated powder is quickly transferred to a flask containing refluxing liquid of the molecules to be intercalated. (■**Caution.** *The Pyrex tube should be opened in a hood to avoid inhalation of NH_3.*) Care should be taken to avoid prolonged exposure of the $TaS_2 \cdot NH_3$ powder to water vapor in the air during transfer.

Other intercalation techniques involve melting or dissolving solid organic materials in solvents such as benzene to obtain a mobile species of the intercalate. Table I gives other organic materials that have been intercalated in $2H(a)\text{-}TaS_2$ along with the reaction times and temperatures. The intercalation complexes retain hexagonal symmetry and the crystallographic a and c parameters are listed with the expansion of the c-axis because of the inclusion of the organic molecule (δ). Also included in the table are the onset temperatures of superconductivity.

2. Sodium Intercalate of Tantalum Disulfide

$$TaS_2 \ + \ xNa \ \rightarrow \ Na_xTaS_2$$

$$(0.4 \leqslant x \leqslant 0.7)$$

TABLE I   Other Molecules that Intercalate TaS_2[a]

Intercalate	Time (days)	Temp (°C)	a(Å)	c(Å)	δ(Å)	T_{onset} (°K)
Amides						
Butyramide	21	150	—	11.0	5.0	3.1
Hexanamide	21	150	—	11.2	5.2	3.1
Stearamide(I)	10	150	—	57.0	51.0	3.1
Thiobenzamide	8	160	—	11.9	5.9	3.3
Phenylamines						
Aniline	16	150	—	18.15	12.11	3.1
N,N-Dimethylaniline	3	170	—	12.45	6.41	4.3
N,N,N',N'-Tetramethyl-p-phenylenediamine	13	200	3.335	2 × 9.66	3.62	2.9
Cyclic amines						
4,4′-Bipyridyl	21	160	3.316	2 × 12.08	6.04	2.5
Quinoline	6	160	—	12.08	6.04	2.8
Pyridine N-oxide	8 hr	100	3.335	11.97	5.93	2.5
Pyridinium chloride	4	170	3.329	9.30	3.26	3.1
Hydroxides						
CsOH	30 min	25	3.330	2 × 9.28	3.24	3.8
LiOH	30 min	25	3.330	2 × 8.92	2.88	4.5
NaOH	30 min	25	3.326	2 × 11.86	2.82	4.8
Triton B[b]	30 min	25	3.331	2 × 11.98	5.94	5.0
Alkylamines						
Ammonia	3	25	3.328	2 × 9.22	3.17	4.2
Methylamine	30	25	3.329	2 × 9.37	3.32	4.2
Ethylamine	30	25	3.334	2 × 9.58	3.53	3.3
Propylamine	30	25	3.330	2 × 9.66	3.61	3.0
Butylamine	30	25	—	2 × 9.73	3.68	2.5
Decylamine	30	25	—	14.6	8.5	—
Dodecylamine	30	25	3.325	34.4	28.3	—
Tridecylamine	30	25	3.322	40.5	36.4	2.5
Tetradecylamine	30	25	3.325	46.2	40.1	2.4
Pentadecylamine	30	25	3.328	45.1	39.0	2.8
Hexadecylamine	30	25	—	39.7	33.6	—
Heptadecylamine	30	25	—	48.5	43.4	2.7
Octadecylamine	30	25	—	55.8	49.7	3.0
Tributylamine	7	200	—	2 × 10.28	4.23	3.0
Miscellaneous						
Ammonium acetate	1 hr	150	3.330	9.08	3.04	2.0
Hydrazine	10 min	10	3.334	9.16	3.12	4.7
Potassium formate	1 hr	200	3.334	9.05	3.01	4.7
Guanidine	3	25	—	—	—	—

[a] From F. J. DiSalvo, Ph.D. Dissertation, Stanford University 1971.
[b] Tetrasodium (ethylenedinitrilo)tetraacetate.

43

Omloo and Jellinek[7] have described the synthesis and characterization of intercalation compounds of alkali metals with the group V layered transition metal dichalcogenides. Typically, these types of intercalation complexes are sensitive to moisture and must be handled in dry argon or nitrogen atmospheres. The alkali metal atoms occupy either octahedral or trigonal prismatic holes between X—M--X slabs. There are two principal means by which these compounds may be prepared.

Procedure 2a

A 1.911-g quantity (0.008 mole) of 2H(a)-TaS$_2$ powder (or an appropriate mixture of the elements; see Sec. A) is loaded into a quartz ampule such as the one described in Fig. 2a. The ampule is then transferred to a dry box with an argon or nitrogen atmosphere, and 74-129 mg (0.0032-0.0056 mole) of freshly cut sodium metal is added. Very low moisture and oxygen concentrations must be maintained in the dry box to prevent attack of the sodium. The techniques for maintaining pure inert atmospheres are described elsewhere.[17] The specified amount of sodium produces a product within the range of $0.4 \leqslant x \leqslant 0.7$[7] over which the intercalation compound exists. The ampule is stoppered, removed from the dry box, attached to a vacuum pump, quickly evacuated to 10^{-3} torr, and sealed under vacuum. It is then heated to 800° for 1 day (again, caution must be exercised if the sample is prepared from the elements; see part Sec. 8.A) and cooled slowly to room temperature. The resulting powder is black or grey and, as mentioned earlier, is very sensitive towards moisture. The crystal structure exhibits hexagonal symmetry with $a = 3.337 \pm 1$ and $c/2 = 7.30 \pm 1$ Å (corresponding to $\delta \approx 1.2$ Å) for $x = 0.7$. When the amount of sodium is decreased below $x = 0.7$, δ increases and the a-axis decreases slightly as x decreases toward 0.4.[7] If the amount of sodium added to the ampule initially corresponds to less than 0.4 mole per mole of TaS$_2$, a phase separation into unreacted TaS$_2$ and Na$_{0.4}$TaS$_2$ occurs. On the other hand, if more than 0.7 mole of sodium is added per mole of TaS$_2$, free sodium remains to attack the quartz reaction vessel. For identification purposes, the prominent low-angle powder X-ray diffraction lines of Na$_{0.7}$TaS$_2$ are: 7.295, 3.648, 2.890, 2.835, 2.687, and 2.484 Å. Two polymorphic forms of Na$_x$TaS$_2$ are reported: η phase and δ phase. The η phase is obtained by heating 1T-TaS$_2$ and sodium at 500°, while the δ phase is obtained when 2H(a)-TaS$_2$ is used as the starting material.[7]

Alternate Procedure 2b

In this method the metallic sodium is dissolved in a solvent such as liquid NH$_3$ or tetrahydrofuran and the resulting solution is used to intercalate 2H(a)-TaS$_2$. This technique was used by J. Cousseau[18] and Trichet et al.[19] in preparation of group

IV layered transition metal dichalcogenide-alkali metal intercalation compounds. The advantage of this method is that it is carried out at room temperature and, consequently, there is less likelihood of reaction between sodium and the reaction vessel. On the other hand, this method is more difficult in that it involves the use of liquid NH_3. Furthermore, undesirable side reactions may occur if the NH_3 is not dried thoroughly or if the reaction vessel is not clean. For example,

$$Na + 2NH_3 \rightarrow NaNH_2\downarrow + H_2(g)$$

is a competing reaction that can occur under "dirty" conditions and is evidenced by the formation of a white precipitate (sodium amide) in the Na-NH_3 solution. Cousseau[18] describes a technique that employs a Pyrex reaction vessel such as the one shown in Fig. 4. A 1.911-g quantity (0.008 mole) of 2H(a)-TaS_2 is

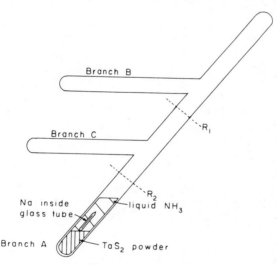

Fig. 4. Adapted from J. Cousseau, Ph.D. Dissertation, Université de Nantes, France, 1973.

placed in branch A of the vessel along with a small sealed glass ampule containing a known weight of distilled sodium (between 74 and 129 mg, as described earlier). Dried ammonia gas is condensed (using liquid nitrogen) into this branch, and the vessel is sealed off under vacuum. The sodium ampule is then broken by briskly shaking the vessel (■**Caution.** *This should be done behind a protective barrier*), and a characteristic blue solution results as the sodium is dissolved. The blue coloration of the liquid NH_3 solution disappears as the TaS_2 is intercalated

TABLE II. Summary of Intercalation Data for $A_x MX_2$ Complexes (*maximum* alkali metal concentration)

Group IVb[a]			Group Vb[b]			Group VIb[a]		
Compound	c-Axis (Å)	δ (Å)	Compound	c-Axis (Å)	δ (Å)	Compound	c-Axis (Å)	δ (Å)
TiS_2	5.67	–	TaS_2	2 × 6.05	–	MoS_2	2 × 6.15	–
$Li_{0.6} TiS_2$	6.16	0.49	$Li_{0.7} TaS_2$	2 × 6.45	0.40	$Li_{0.8}(NH_3)_{0.8} MoS_2$	2 × 9.5	3.35
$Na_{0.8} TiS_2$	3 × 6.68	1.01	$Na_{0.7} TaS_2$	2 × 7.27	1.22	$Na_{0.6} MoS_2$	2 × 7.5	1.35
$K_{0.8} TiS_2$	3 × 7.56	1.89	$K_{0.7} TaS_2$	2 × 8.10	2.05	$K_{0.6} MoS_2$	2 × 8.1	1.95
$Cs_{0.6} TiS_2$	3 × 8.36	2.69				$Cs_{0.5} MoS_2$	2 × 8.89	2.74
$TiSe_2$	6.01	–	$TaSe_2$	2 × 6.36	–	$MoSe_2$	2 × 6.46	–
$Na_{0.95} TiSe_2$	3 × 6.99	0.98	$Na_{0.7} TaSe_2$	2 × 7.70	1.34		–	–
			$K_{0.7} TaSe_2$	2 × 8.52	2.16	$K_{0.5} MoSe_2$	2 × 8.57	2.21

[a] Alkali metal-NH_3 solution intercalation. Data from W. Rudorff, *Chimia*, **19**, 496 (1965).
[b] Direct heating intercalation. Data from W. P. F. A. M. Omloo and F. Jellinek, *J. Less-Common Met.*, **20**, 121 (1970).

by the sodium. The liquid NH_3 is then poured off into branch B and cooled with liquid nitrogen. The temperature gradient causes any residual NH_3 to condense in branch B. This branch is removed by sealing at point R_1. Residual NH_3 in the Na_xTaS_2 may be removed by heating branch A gently (about $200°$), while simultaneously cooling branch C to liquid nitrogen temperature. This branch is removed by sealing at point R_2 after the NH_3 has been frozen in C. The final product should be removed from the Pyrex tube only in an argon- or nitrogen-filled dry box as described in the preceding procedure.

Other layered materials intercalated with alkali metals have been prepared and are given in Table II. Only the maximum alkali metal concentration is listed for the various A_xMX_2.

Properties

Table II gives the c-axes and slab expansions δ. These materials react with air and moisture. No superconductivity has been found in the group V-alkali metal complexes.

3. Tin Intercalate of Tantalum Disulfide

$$TaS_2 + Sn \xrightarrow{850°} SnTaS_2$$

DiSalvo et al.[9] have carried out a systematic survey of intercalation compounds of $2H(a)$-TaS_2 with post-transition metals. In particular, the system Sn_xTaS_2 was found to exist in two composition domains, $0 < x \leqslant \frac{1}{3}$ and $x = 1$. The following discussion briefly describes the techniques used by DiSalvo to synthesize the compound $SnTaS_2$. Syntheses of other transition and post-transition metal intercalation complexes with the layered transition metal dichalcogenides are discussed in References 9 and 20-24.

Procedure

$2H(a)$-TaS_2 is synthesized according to the procedure outlined in Section A. A 1.91-g sample of TaS_2 powder (0.008 mole) is loaded into a fused quartz ampule such as the one shown in Fig. 2a, along with 1.19 g of tin powder (about 0.010 mole; -50 mesh powder; 99.5% purity)* The tube is then evacuated to 10^{-3} torr, sealed, and heated in a small temperature gradient ($\Delta T \sim 20°$) with TaS_2 at the hot end ($850°$). The excess tin not intercalated sublimes to the cooler end of the tube over a period of several weeks. An alternate technique involves pressing a

---

*Available from Alfa Products, Ventron Corp., P.O. Box 299, Danvers, MA 01923.

pellet of a mixture of TaS_2 and Sn powders (1.91 g TaS_2 and 0.950 g Sn) at 40,000 psi. This pellet is then sealed in a fused quartz ampule under vacuum, fired to 600° for a week or so, cooled, and then removed from the ampule. Regrinding, pressing, and refiring several times will ensure a homogeneous sample.

Properties

The $SnTaS_2$ has hexagonal symmetry with lattice parameters $a = 3.28$ Å and $c/2 = 8.7$ Å.[9] For identification purposes, the prominent low-angle X-ray powder d spacings and (relative intensities) are: 8.6 (mw), 4.33 (ms), 2.85 (m), 2.81 (s), 2.71 (s), and 2.56 (m). $SnTaS_2$ undergoes a superconducting transition at $T_c = 2.95°K$.[9]

References

1. A. Weiss and R. Ruthardt, *Z. Naturforsch.*, **24**, 256, 355 and 1066 (1969).
2. F. R. Gamble, F. J. DiSalvo, R. A. Klemm, and T. H. Geballe, *Science,* **168**, 568 (1970).
3. F. R. Gamble, J. H. Osiecki, and F. J. DiSalvo, *J. Chem. Phys.,* **55**, 3525 (1971).
4. F. R. Gamble, J. H. Osiecki, M. Cais, R. Pisharody, and F. J. DiSalvo, *Science,* **174**, 493 (1971).
5. F. R. Gamble and T. H. Geballe, *Treatise on Solid State Chemistry,* Vol. 3, N. B. Hannay (ed.), Plenum Press Publishing Co. 1976. This is a comprehensive survey of intercalation in a wide variety of host materials.
6. J. V. Acrivos, W. Y. Liang, J. A. Wilson, and A. D. Yoffe, *J. Phys. Chem.,* **4**, 6-18 (1971).
7. W.P.F.A.M. Omloo and F. Jellinek, *J. Less-Common Met.,* **20**, 121 (1970).
8. W. Rudolf and H. H. Sich, *Angew. Chem.,* **71**, 724 (1959).
9. F. J. DiSalvo, G. W. Hull, Jr., L. H. Schwartz, J. M. Voorhoeve, and J. V. Waszczak, *J. Chem. Phys.,* **59**, No. 4, 1922 (1973).
10. F. Jellinek, *J. Less-Common Met.,* **4**, 9 (1962).
11. J. A. Wilson and A. D. Yoffe, *Adv. Phys.,* **18**, No. 73, 193-335 (1969). This article offers a complete discussion of the polymorphic phases found in the various layered transition metal dichalcogenides.
12. R. C. Hall and J. P. Mickel, *Inorgan. Synth.,* **5**, 82 (1957).
13. L. E. Conroy and R. J. Bouchard, *Inorg. Synth.,* **12**, 158 (1970).
14. G. Brauer, *Handbook of Preparative Inorganic Chemistry,* Vol. 1, Academic Press Inc., New York, 1963, p. 1328.
15. H. Shaefer, *Chemical Transport Reactions,* Academic Press, New York, 1964.
16. A. H. Thompson, *Nature,* **251**, 492 (1974).
17. D. F. Shriver. *The Manipulation of Air-Sensitive Compounds,* McGraw Hill Book Co., New York, 1969, pp. 164 and 193.
18. J. Cousseau, Thèse de Doctorat de Spécialité, Université de Nantes France, 1973.
19. L. Trichet, J. Cousseau, and J. Rouxel, *C. R. Acad. Sci.,* **273**, 243 (1971).
20. F. Hulliger, *Struct. Bonding,* **4**, 421 (1968).

21. J. M. Voorhoeve, nee van den Berg, and M. Robbins, *J. Solid State Chem.*, **1**, 132 (1970).
22. L. Trichet and J. Rouxel, *C. R. Acad. Sci.*, **267**, 1322 (1969); 269, 1040 (1969).
23. M. S. Whittingham, *Chem. Commun.*, **1974**, 328.
24. G. V. Subba Rao and J. C. Tsang, *Mater. Res. Bull.*, **9**, 921 (1974).

9. PLATINUM DISULFIDE AND PLATINUM DITELLURIDE

$$Pt + 2S \xrightarrow[\text{Cl}_2\ (75\ \text{torr})]{P} PtS_2$$

$$T_1\ (800°) \qquad\qquad T_2\ (740°)$$

$$Pt + 2Te \xrightarrow[\text{Cl}_2\ (75\ \text{torr})]{P} PtTe_2$$

$$T_1\ (875°) \qquad\qquad T_2\ (690°C)$$

Submitted by S. SOLED* AND A. WOLD*
Checked by C. R. SYMON†

Crystals of transition metal chalcogenides often can be grown by chemical vapor transport.[1,2] A charge of the desired stoichiometry is sealed in an evacuated silica tube together with a transport agent, and the tube is placed in one end of a furnace with a thermal gradient. The most commonly used transport agents are the halogens, although several other carriers, such as hydrogen halides, oxygen, and aluminum trichloride, have been tried.[2] Attempts to prepare crystals of platinum disulfide and platinum ditelluride by chemical vapor transport under a variety of thermal conditions with chlorine, bromine, or iodine as the transport agent failed. However, it was found that these platinum dichalcogenides could be transported in the presence of the appropriate mixture of phosphorus and sulfur.

Procedure

Large, well-formed plates of platinum disulfide can be grown when elemental

*Department of Chemistry, Brown University, Providence, RI, 02912.
†Corporate Research Laboratories, Exxon Researchand Engineering Co., P.O. Box 45, Linden, NJ, 07036.

charges of platinum (typically 0.002 mole and in the form of powder or sponge), sulfur, and phosphorus in the molar ratios 1:8:1 are placed in an 11-mm id silica tube approximately 30 cm long. The tube is evacuated to approximately 10^{-3} mm, backfilled with chlorine to a pressure of 75 torr (as monitored by a brass bourdon-type gauge*), and sealed off at a length of 25 cm. The sealed tube is placed in a modified three-zone furnace,[†] with the temperature in the two end zones maintained at 800° and the temperature in the central zone slowly lowered to 740°.

Crystals of platinum ditelluride can be grown when charges of platinum, sulfur, phosphorus, and tellurium in the molar ratio of 1:3:1:2 are held at 875° and the growth zone is cooled to 690°. In the preparation of both the sulfide and the telluride, adding chlorine to a pressure of 75-100 torr enhances the transport, but its presence is not essential. However, the presence of both phosphorus and sulfur is necessary.

Both compounds crystallize with the cadmium diiodide structure (space group $P\bar{3}m1$) as previously reported on polycrystalline samples.[3] For platinum disulfide, $a_0 = 3.542(1)$ Å and $c_0 = 5.043(1)$ Å, and for platinum ditelluride, $a_0 = 4.023(1)$ Å and $c_0 = 5.220(3)$ Å. Direct chemical analysis for the component elements was not carried out. Instead, precision density and unit-cell determinations were performed to characterize the samples. The densities of both compounds as determined by a hydrostatic technique with heptadecafluorodecahydro-1-(trifluoromethyl)naphthalene as the density fluid[4] indicated that they are slightly deficient in platinum. For platinum disulfide, $\rho_{calc} = 7.86$ g/cm^3 and $\rho_{meas} = 7.7(1)$ gm/cm^3, and for platinum ditelluride, $\rho_{calc} = 10.2$ gm/cm^3 and $\rho_{meas} = 9.8(1)$ gm/cm^3. In a typical experiment an emission spectrum of the platinum disulfide showed that phosphorus was present in less than 5 ppm. A mass spectroscopic examination of the platinum ditelluride revealed a small doping by sulfur (less than 0.4%) and traces of chlorine and phosphorus (less than 100 ppm).

Properties

The first 10 powder diffraction lines for each phase are given in Table I. Bulk magnetic susceptibility measurements (77-300°K) show that both platinum disulfide and platinum ditelluride are diamagnetic, with susceptibilities of −31(2) and −12(1) emu/mole respectively, as expected for low-spin d^6 octahedral ions. Platinum disulfide shows semiconducting behavior between 77 and 300°K, with an activation energy of 0.10(1) eV, whereas platinum ditelluride is metallic.

---

*U.S. Gauge (No60S6-A vacuum gauge) available from A. H. Thomas Co., P.O. Box 779, Philadelphia, PA 19105.

†A commercially available furnace can be obtained from Lindberg Hevi-Duty Company, 3709 Westchester Pike, Newton Square, PA 19073.

TABLE I   Platinum Disulfide[a]

d (Å)	hkl	I/I_{100}
5.04	001	100
3.07	100	70
2.621	101	50
2.521	002	70
1.948	102	90
1.771	110	10
1.671	111	90
1.535	200	5
1.468	201	20
1.449	112	5
5.22	001	100
3.49	100	5
2.899	101	65
2.608	002	15
2.089	102	35
2.013	110	20
1.878	111	5
1.740	003	25
1.653	201	5
1.556	103	20

[a]There is always some preferred orientation on grinding either of these crystals, so the intensity ratios may be variable.

References

1. R. J. Bouchard, *Inorg. Synth.,* **14**, 157 (1973).
2. H. Schafer, *Chemical Transport Reactions,* Academic Press Inc., New York, 1964.
3. S. Furuseth, K. Selte, and A. Kjekshus, *Acta Chem. Scand.,* **19** 257 (1965).
4. R. L. Adams, Ph.D. Thesis, Brown University, 1973.

10.   SUBSTITUTED β-ALUMINAS

Submitted by J. T. KUMMER*
Checked by M. STANLEY WHITTINGHAM[†]

β-Alumina has the empirical formula $Na_2O \cdot 11Al_2O_3$. In reality the compound is massively defective and contains considerably more (~25%) Na_2O than indicated

*Ford Motor Company, Research Staff, Dearborn, MI 48121.
[†]Corporate Research Laboratories, Exxon Research and Engineering Co., P.O. Box 45, Linden, NJ 07036.

by the empirical formula. β-Alumina has a hexagonal layer structure[1] of the space group $P6_3/mmc$ with the lattice constants $a = 5.59$ Å and $c = 22.53$ Å. The sodium ions are situated exclusively in planes perpendicular to the c-axis that contain, in a loose packing, an equal number of sodium and oxygen ions. This unusual structure results in the sodium ion possessing a high mobility in this plane, and the resulting high ionic conductivity is the prime reason for the recent interest in this compound.[1]

Substituted β-alumina can be made by an ion exchange procedure in a molten salt using single crystals of sodium β-alumina as starting material.[2] Small single crystals of β-alumina, if not available, can be obtained from fusion-cast bricks of β-alumina (Monofrax H, 14 kg each).[‡] These bricks fracture easily, yielding single crystals of β-alumina that are very thin in the c direction ($\leqslant 0.03$ cm) and up to 1 cm in diameter. If there is a difference in size between the ion introduced into the β-alumina and the ion removed in the ion exchange process there is a very much larger change in c-axis dimension of the crystal than in the a-axis dimension. In general, for crystals 0.03 cm thick and 1 cm in diameter, the physical integrity of sodium β-alumina crystals is preserved during ion exchange with other monovalent cations. The author does not know if very large single crystals will preserve their physical integrity during ion exchange. Polycrystalline ceramic material will, in general, fracture during the ion exchange process as a result of the unequal change in the a and c dimensions. An exception is the exchange of polycrystalline Na$^+$ β-alumina by Ag$^+$ ion where the change along the c axis is small.

The ion exchange can be done either by direct exchange of the Na$^+$ ion in β-alumina with the desired cation in a molten salt medium using an appropriate anion to improve the exchange equilibria[2] or by first exchanging the Na$^+$ in β-alumina with Ag$^+$ ion in molten silver nitrate and then exchanging the Ag$^+$ ion in this material with the desired cation in a molten salt or other media. In general, the latter procedure is preferred. The use of silver β-alumina as an intermediate has two advantages: (1) By employing a metal chloride melt, the exchange reaction with silver β-alumina can be driven to completion as a result of the formation of silver chloride in the melt. (2) Complete exchange is determined by the absence of silver ion in the exchanged material; there should be no fear of contamination error in the analytical procedure since silver is not a common contaminant in the laboratory. Because of the large atomic weight of silver the silver β-alumina contains approximately 19 wt % silver and the exchange can be monitored by weight changes in many cases.

The analysis of β-aluminas for stable cations can be made by a fusion process. The β-alumina can be dissolved in molten Li$_2$CO$_3$ (or K$_2$CO$_3$), the resulting glass dissolved in dilute nitric acid, and the solution analyzed by atomic adsorption. Activation analysis or X-ray fluorescence analysis also can be used.

[‡]Available from Monofrax Div., Carborundum Co., P.O. Box A, Falconer, NY 14733.

A. SILVER β-ALUMINA (ALUMINUM SILVER(I) OXIDE)

$$Na_2O \cdot 11Al_2O_3 + 2AgNO_{3(l)} \rightarrow Ag_2O \cdot 11Al_2O_3 + 2NaNO_{3(l)}$$

Procedure

Sodium β-alumina single crystals are dried at 500° for 3 hours and cooled in a desiccator. The exchange of the sodium ion by the silver ion is carried out using molten silver nitrate at 300° contained in a Vycor or fused-quartz vessel. Pyrex should not be used because of the presence of potassium and sodium in the glass. One gram of dried crystals of sodium β-alumina are placed in a Vycor test tube approximately 2 cm in diameter and about 14 cm long. Ten grams of reagent grade silver nitrate is added and the mixture is heated to 300° in a furnace. The time required for exchange equilibrium increases as the square of the diameter of the largest crystals used. For crystals of 2-mm diameter, the time to reach 99% equilibrium is 3 hours. For crystals of 1-cm diameter the time is 75 hours. The crystals float at first and then sink to the bottom of the test tube. Stirring is not necessary as sodium nitrate is less dense than silver nitrate. At the end of the time allowed for attainment of 99% equilibrium, the molten silver nitrate is decanted from the crystals into a porcelain crucible and the Vycor test tube containing the exchanged crystals is cooled to room temperature. The crystals are washed with water to dissolve the residual silver nitrate, and then with alcohol or acetone, and are then dried at 200°. The crystals contain less than 0.1% sodium.

Properties

Silver β-alumina crystals are colorless. They stay clear and transparent upon heating to 1000° and do not react when contacted with chlorine for several hours at 560°. The lattice constants are $a = 5.594$ Å, $c = 22.498$ Å. The crystals are not hygroscopic.

B. THALLIUM(I) β-ALUMINA [ALUMINUM THALLIUM(I) OXIDE]

$$Na_2O \cdot 11Al_2O_3 + 2TlNO_{3(l)} \rightarrow Tl_2O \cdot 11Al_2O_3 + 2NaNO_{3(l)}$$

Procedure

■**Caution.** *Thallium salts are extremely toxic, particularly their vapor. Melts should be prepared in a hood.*

The procedure for producing thallium β-alumina from sodium β-alumina is the same as that used above for silver β-alumina. Thallium(I) nitrate is used in place

of silver nitrate. For crystals, 1-mm in diameter the time to reach 99% equilibrium is about 75 hours. The exchanged crystals contain less than 0.1% sodium.

Properties

Thallium(I) β-alumina crystals are colorless and do not react with chlorine at
700°. The lattice constants are $a = 5.597$ Å, $c = 22.883$ Å. The crystals are not
hygroscopic.

C. LITHIUM β-ALUMINA (ALUMINUM LITHIUM OXIDE)

$$Ag_2O \cdot 11Al_2O_3 + 2LiCl_{(l)} \rightarrow Li_2O \cdot 11Al_2O_3 + 2AgCl_{(l)}$$

Procedure

One gram of silver β-alumina (see above) is placed into a fused quartz test tube
about 2 cm in diameter and about 14 cm long. Five grams of lithium chloride is
added. It is important that the lithium chloride used have a very low content of
other alkali metal impurities, except Cs, since the ion exchange equilibria greatly
favor the presence of the other alkali metals in the β-alumina crystals over lithium.
Essentially all of the impurity ends up in the crystals. The fused-quartz test tube
is heated to 650° in a furnace. For crystals 1-cm in diameter the time to reach
99% equilibrium is approximately 16 hours. The molten salt is decanted and
the crystals are allowed to cool to room temperature. Methyl alcohol containing
about 10% propylamine or ethylenediamine is used to wash the product and
thereby remove the silver chloride and residual lithium salts. The sample is dried
at 400° and stored in a dessicator. The lithium β-alumina crystals contain less
than 0.05% Ag. If the lithium chloride used contains a trace of sodium or
potassium, it can be prepurified by treatment with silver β-alumina at 650°.
Each gram of silver β-alumina will remove about 30 mg of sodium from the melt.
The molten lithium chloride, after decantation from the pretreatment silver
β-alumina, can be used to prepare the product, lithium β-alumina.

Properties

Lithium β-alumina crystals are colorless and hygroscopic. They should be kept in
a desiccator and dried at 400° before use. The lattice constants are $a = 5.596$ Å,
$c = 22.570$ Å.

D. POTASSIUM β-ALUMINA (ALUMINUM POTASSIUM OXIDE)

$$Ag_2O \cdot 11Al_2O_3 + 2KCl_{(l)} \rightarrow K_2O \cdot 11Al_2O_3 + 2AgCl_{(l)}$$

The preparation of this compound from silver β-alumina is similar to the preparation of lithium β-alumina. The melt consists of 10 g of potassium chloride. The exchange temperature is 800°. For crystals with diameters of 1 cm it takes about 16 hours to reach 99% of equilibrium. The potassium salts used should contain less than 0.1 wt % sodium. After decantation of the melt the crystals are washed with water containing 2% propylamine or ethylenediamine to remove residual potassium salts and silver chloride. The sample is dried at 200°. The potassium β-alumina contains less than 0.05% silver.

Properties

Potassium β-alumina crystals are colorless and are not hygroscopic. The lattice constants are $a = 5.596$ Å, $c = 22.729$ Å.

E. RUBIDIUM β-ALUMINA (ALUMINUM RUBIDIUM OXIDE)

$$Ag_2O \cdot 11Al_2O_3 + 2RbCl_{(l)} \rightarrow Rb_2O \cdot 11Al_2O_3 + 2AgCl_{(l)}$$

The preparation of this substance from silver β-alumina is similar to the preparation of lithium β-alumina. The melt consists of 10 g of rubidium chloride. The exchange temperature is 800°. For crystals 2 mm in diameter it takes about 16 hours to reach 99% of equilibrium. The rubidium salts used should contain less than 0.02% potassium and less than 0.1% sodium. After decantation of the melt the crystals are washed with water containing 2% propylamine or ethylene-diamine to remove residual potassium salts and silver chloride. They are dried at 200°. The rubidium β-alumina crystals contain less than 0.05 wt % silver.

Properties

Rubidium β-alumina crystals are colorless and are not hygroscopic. The lattice constants are $a = 5.597$ Å, $c = 22.877$ Å.

F. AMMONIUM β-ALUMINA (ALUMINUM AMMONIUM OXIDE)

$$Na_2O \cdot 11Al_2O_3 + 2NH_4NO_{3\,(l)} \rightarrow (NH_4)_2O \cdot 11Al_2O_3 + 2NaNO_{3\,(l)}$$

Preparation

■**Caution**. *Ammonium nitrate is hazardous in large amounts and can be detonated with shock waves. There have been no problems when using 10 g of material, but it should be handled with care.*

Because ammonium salts decompose at high temperature, the exchange must be carried out at temperatures below 180°. At this temperature the rate of exchange is low and the time for complete exchange is high. One gram of sodium β-alumina is placed in a Vycor test tube about 2 cm in diameter and 14 cm long along with 10 g of ammonium nitrate. This test tube is placed in a furnace at 170-180° as close to 170° (the melting point) as is feasible. The time to reach 99% equilibrium is approximately 140 hours for crystals ½ mm in diamter. At the end of the exchange process the molten salt is decanted, and the test tube with the crystals is cooled to room temperature. The crystals are washed with water and dried at 200°. These dried crystals are placed in the Vycor test tube along with a fresh 10 g of ammonium nitrate, and the above procedure is repeated. The crystals are dried at 200°. The ammonium β-alumina crystals contain less than 0.3% sodium.

Properties

Ammonium β-alumina crystals are colorless and are not hygroscopic. The lattice parameters are $a = 5.596$ Å, $c = 22.888$ Å. The IR spectra show characteristic absorption bands at 3180, 3070, and 1430 cm^{-1}.

G. GALLIUM(I) β-ALUMINA [ALUMINUM GALLIUM(I) OXIDE]

$$Ag_2O \cdot 11Al_2O_3 + 2Ga \xrightarrow[\text{GaI}_{(l)}]{} Ga_2O \cdot 11Al_2O_3 + 2Ag$$

Preparation

One gram of silver β-alumina crystals is placed in a Vycor tube with a 1-cm id and 20 cm long and closed at one end. Three grams of gallium and 4 g of iodine are added, and the Vycor tube is necked down near the middle in preparation for sealing off. The tube is evacuated with a mechanical pump to <1 torr pressure and sealed. The sealed tube is placed in a cold furnace and then heated to about 290°. For crystals 1 mm in diameter the time of heating is 48 hours. The furnace is cooled to room temperature and the ampule is removed. The ampule is broken open and the melt is dissolved in 10% HCl solution, leaving a pool of gallium-silver metal and gallium(I) β-alumina crystals.

Properties

Gallium(I) β-alumina crystals are transparent and reddish-brown.[3] They become colorless when heated 16 hours at 540° in air. The lattice constants are $a = 5.600$ Å, $c = 22.718$ Å. The density is 3.51 g/cm³, and the compound is stable up to 750° in dry air. The material is not hygroscopic.

H. NITROSYL β-ALUMINA (ALUMINUM NITROSYL OXIDE)

$$NOCl + AlCl_3 \rightarrow [NO][AlCl_4]$$

$$Ag_2O \cdot 11Al_2O_3 + 2[NO][AlCl_4] \rightarrow [NO]_2O \cdot 11Al_2O_3 + 2Ag AlCl_4$$

Preparation

■**Caution.** *This preparation must be done in a hood because NOCl is very toxic.*

This material is prepared by allowing silver β-alumina to exchange with molten $[NO][AlCl_4]$.[4]

Using a long-stem funnel, 4 g of anhydrous $AlCl_3$ is placed in the bottom of a Vycor test tube 5 cm in diameter and 60 cm long that has previously been flushed with dry argon. The tube is then necked down about 20 cm from the bottom in preparation for sealing off. With argon flowing across the open mouth of the tube the $AlCl_3$ is sublimed from the tube bottom to above the necked-down section and the tube is sealed off at the neck. Nitrosyl chloride gas* is bled from the lecture bottle through Teflon and glass tubing into a small cold trap in Dry Ice where about 2.5 g (approximately 1.7 mL) of the NOCl is condensed. The Vycor tube containing the $AlCl_3$ and with argon flowing across the open mouth is immersed in Dry Ice and the NOCl is distilled through a long glass tube from the cold trap (bp −5°) and condensed in the bottom of the Vycor tube to contact the $AlCl_3$. The Vycor tube is placed in an ice-salt bath and allowed to stand 1 hour with occasional agitation to allow reaction, after which it is heated to 200° to melt the $[NO][AlCl_4]$.

The tube is cooled to −78° and 1.5 g of 0.1-mm-diameter dried (400°) silver

*Available from Matheson Gas Products, P.O. Box 85, East Rutherford, NJ 07073.

β-alumina crystals are added to the NOAlCl₄ through a long stem funnel. Additional NOCl, about 0.5 g (0.35 mL), is added to suppress the decomposition of [NO] [AlCl₄] , and the top of the Vycor tube is sealed off. The tube is heated to 200° for 24 hours. At the end of this time the tube is cooled and broken open in the hood and the melt is dissolved in distilled water containing ethylenediamine to dissolve the AgCl. (■**Caution.** *NO₂ fumes are given off.*) The crystals of nitrosyl β-alumina contain less than 0.5 wt % Ag.

Properties

Nitrosyl β-alumina crystals are colorless and nonhygroscopic. The lattice constants are a = 5.597 Å, c = 22.711 Å. The IR spectrum contains a strong absorption band at 2245 cm^{-1}, a frequency indicative of the N—O stretching motion of the nitrosonium ion. The material is thermally unstable. Particles 150-250 μ in size decompose above 400°. Particles smaller than 45 μ decompose above 150°. The density of the material is 3.22 g/cm^3.

References

1. J. T. Kummer, *Progr. Solid State Chem.*, **7**, 141 (1972).
2. Y. Y. Yao, J. T. Kummer, *J. Inorg. Nucl. Chem.*, **29**, 2453 (1967).
3. R. H. Radzilowski, *Inorg. Chem.*, **8**, 994 (1969).
4. R. H. Radzilowski, and J. T. Kummer, *Inorg. Chem.*, **8**, 2531 (1969).

Chapter Two

METAL ATOM SYNTHESES*

11. INTRODUCTION TO METAL ATOM SYNTHESES

Although the low-temperature codeposition of atoms with vapors of organic or inorganic substrates has been utilized in synthetic work for over 15 years,[1] new techniques of vaporization have led to direct syntheses of low valent complexes of the more refractory transition metals. In the present state of the art, metal atom chemistry is the quickest and cleanest method of preparing gram quantities of compounds such as B_2Cl_4 and $Cr(PF_3)_6$, as well as the only presently known method of preparing compounds such as $(\eta^6\text{-}C_6H_5Cl)_2Mo$. Several review articles on metal atom synthesis are available.[2-4]

Apparatus

Most metals vaporize as atoms, which are highly reactive as a result of the input of the heat of vaporization and the lack of steric restrictions. The basic strategy in metal atom synthesis is to codeposit the metal atoms with a large excess of reactant, thereby promoting reaction between the metal atom and the substrate and suppressing recombination to the bulk metal. As shown schematically in

*Based on detailed information supplied by K. Klabunde,** P. Timms,† P.S. Skell‡ and S. D. Ittel.§

**Department of Chemistry, University of North Dakota, Grand Forks, ND 58201.
†School of Chemistry, University of Bristol, Bristol BS8 1TS, England.
‡Department of Chemistry, Pennsylvania State University, University Park, PA 16802.
§Central Research and Development Dept., E. I. duPont de Nemours & Co., Wilmington, DE 19898.

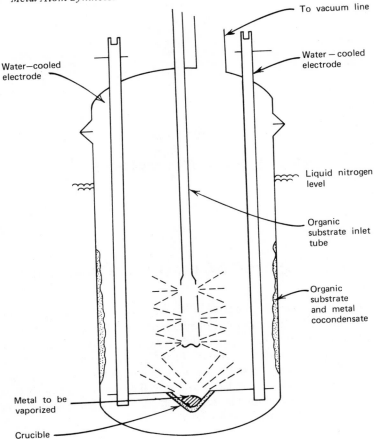

Fig. 1.   Schematic of a metal atom reactor.

Fig. 1, the metal atoms may be generated in an electrically heated crucible and co-condensed with the substrate on the cold walls of the reaction vessel. To minimize gas-phase reactions, a good vacuum must be maintained in the reactor during this codeposition. An alternative procedure is to condense the metal vapor into a well-stirred solution of the reactant in a suitable solvent cooled to a temperature at which the vapor pressure of the solution is $<10^{-3}$ torr. This method has special advantages for the preparation of unstable organometallic compounds and for reacting metal atoms with nonvolatile substrates.[2]

The initial cost of a metal atom system can total well over $5000; however, if high-vacuum equipment is already available the cost can be held under $1000. It

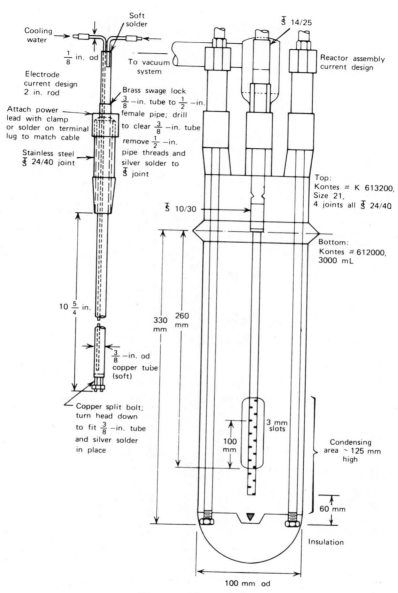

Fig. 2. Detail of a metal atom reactor.

61

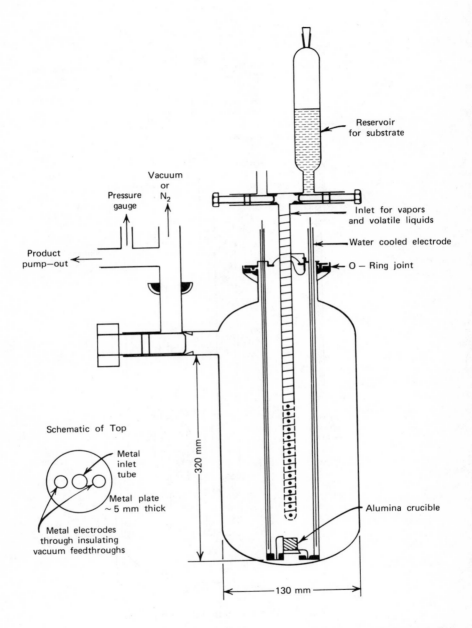

Fig. 3. Metal atom reactor. From Reference 1. Reproduced by permission.

also is possible to purchase complete systems, including reactor, vacuum system, and electrical supply.[*,†]

A detailed drawing of a simple metal atom reactor, largely build from commercially available parts, is given in Fig. 2. The main reaction chamber consists of a 3000-mL reaction flask and a four-necked top section. This reactor is suitable for all the experiments described in this chapter, with the possible exception of the molybdenum compounds. For syntheses of a practical scale with refractory metals (vaporization temperature greater than 2000°, e.g. Re, Mo, and W) a larger diameter reactor (140-178 mm) with a standard wall thickness (about 3.5 mm) is recommended to improve heat dissipation.

The construction of the alternate design for the metal atom reactor shown in Fig. 3 requires the services of a glassblower, but it has significant advantages in the work up of air-sensitive products. When a nonvolatile product is to be isolated, nitrogen gas is admitted, and the electrode and inlet assembly is replaced by an O-ring cap. At this stage the reactor can be taken into a dry box, or it can be handled like a large Schlenk flask, with the product being scraped down and transferred under a nitrogen flush.[5] Another alternative is to replace the inlet assembly by a large cold finger to receive sublimable products.

In both reactor designs, water-cooled electrodes extend down into the reactor chamber and a heating element is connected across the base of these electrodes. The metal sample may be placed in a crucible that fits into a wire-wound heater, both of which are commercially available.[‡,§] A more efficient source consists of an alumina crucible with an internal tungsten wire heater.[§] The heating efficiency of the source and cooling of the reactor are improved by wrapping the bottom of the crucible-heater assembly with refractory wool (Kaowool, $Al_2O_3 \cdot SiO_2$), which is replaced after each run.[ᵖ]

The power requirements for heating the crucible are adequately met by a 20-A variable transformer (120-V input, 0-120 V output) feeding a 1.5-kW step-down transformer designed for a 15-V output with a 120-V input. Cables to the reactor electrodes must be of adequate size to handle the large currents, which may range up to 300 A. A rough idea of the voltage and current requirements for specific metals may be obtained from Table I.

The reactor is evacuated with a two-or three-stage oil or mercury diffusion pump, backed by a rotary mechanical pump. A large removable liquid-nitrogen-

---

[*] Kontes/Martin, 1916 Greenleaf St., Evanston, IL 60204; resistive heating is employed in this unit.

[†] G. V. Planar, Sunbury, England; electron-beam or resistive heating is employed in this unit.

[‡] R. D. Mathis Co., 2840 Gundry Ave., Long Beach, CA 90806.

[§] Sylvania Emissive Products, Exeter, NH 03833.

[ᵖ] Morganite Ceramic Fibers, Neston, England; Zircar Products Inc., 110 N. Main Street, NY 10921, sells similar materials. Another source is Babcock and Wilcox, 5775-A Glenridge Dr., N.E., Atlanta, GA 30328.

TABLE I Conditions for the Evaporation of Metals

Metal	Temperature		Suitable Crucible	Approximate Conditions[a]		
	MP (°C)	Approx. Vap. (°C)		Source[b]	Amperage	Voltage
Cobalt	1495	1520	Alumina	0.5 mL	35-45	6-6.5
Copper	1084	1300	Alumina	0.5 mL	25-30	3.7-4
Chromium	1903	1500	W-boat	S-16 005	100-130	6.5-7
Iron	1538	1500	Chromia	0.5 mL	40-50	6-6.5
Magnesium	1244	480	Alumina	1.5 mL	35-45	1-1.5
Molybdenum	2610	2500	Mo wire	1.27 mm diam. 254 mm long	62	12-13
Nickel	1452	1500	Alumina	0.5 mL	35-45	6-6.5
Palladium	1552	1490	Alumina	0.5 mL	35-45	6-6.5
Platinum	1769	2100	Pt wire	1.0 mm diam. 150 mm long	73	13-14
Tungsten	3410	3200	W-wire			
Vanadium	1917	1850	W-boat	S-16.010	200	2.5
Zinc	419	320	Alumina	0.5 mL	25	2

[a]These conditions are given for uninsulated crucibles. Power requirements are reduced by about a factor of 2 when refractory wool insulation is used.

[b]The 0.5 or 1.5-mL crucibles are integrally heated units available from Sylvania Emissive Products. The S-16 tungsten boats are available from the R. D. Mathis Co. These specific products are mentioned only to provide guidelines.

cooled trap situated between the diffusion pump and the reactor contributes to the pumping efficiency and protects the pumps from volatiles emitted from the reactor. Pumping speeds at the reactor ranging from 0.5 to 200 L/sec have been used successfully. The lower pumping speeds are quite acceptable for some work, but higher speeds are desirable, and if permanent gases (CO, N_2, H_2, and CH_4) are evolved during the reaction the higher speeds are required. A vacuum system that is satisfactory for the preparations described in this chapter is outlined in Fig. 4.[5] The pumping system should be capable of achieving a vacuum of 10^{-4} torr, but pressures up to 10^{-2} torr usually can be tolerated during the reaction.* Typically the pressure during the reaction should be equal to or less than 5×10^{-3} torr as measured by a thermocouple gauge situated close to the reactor.

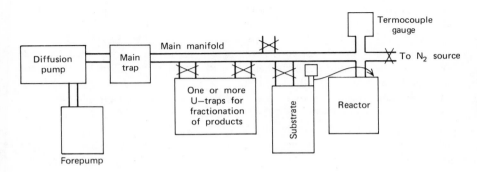

Fig. 4. Schematic vacuum system for metal atom reactions. X represents the stopcock or Teflon-in-glass valve. Satisfactory components (for a general discussion of vacuum line design see References 1 and 4): forepump, 25 L/min free air capacity; diffusion pump, 2 L/sec; main trap is removable and measures about 300 mm deep; main manifold has a diameter of about 25 mm, stopcock or valve in manifold should be at least 10 mm substrate container is removable container with 1-2 mm Teflon-in-glass needle valve connected to bottom of container. Connection between this needle valve and the reactor may be 1/8 in. od. Teflon tubing is used. Alternatively, the substrate may be added as shown in Fig. 3.

*The maximum pressure that should be tolerated in a metal atom reactor is a point of controversy among various workers in this field. High pressures favor reaction in the gas phase with respect to those in the matrix. Where different products are obtained from the gas and condensed phases, the former products begin to appear at pressures of 10^{-4} torr. The molybdenum atom syntheses described in this volume are best carried out under 10^{-4} torr and with apparatus described in synthesis number 16. Skell and co-workers consider this apparatus necessary and appropriate for all work.

General Procedure

1. *Setup.* A metal charge is placed in the evaporation source (crucible). Table I suggests some typical heated sources for use with different metals. Most conveniently, the metal is usually granules or small pieces. Powdered metals should be avoided because they are contaminated with oxide and tend to fly out of the crucible. The metal charge varies depending on the desired scale but normally is 0.5-1.0 g. The weighed crucible and charge are attached to the electrodes, and the crucible is then insulated with Kaowool. The reactor is assembled and the entire system is evacuated with the mechanical forepump. The main trap (Fig. 4) is cooled with liquid nitrogen to protect the pumps. Liquid ligand is placed in the proper container, attached to the manifold, and freeze-thaw degassed. A ten- to fifty-fold excess of ligand over metal is normally used. If a gaseous ligand is to be used it may be contained in a large bulb. If the material is not air free it is subjected to several cycles of freezing, pumping, and thawing. The diffusion pump is turned on and the reactor is pumped down to below 10^{-3} torr.

2. *Reaction.* The electrode cooling water is turned on followed by the power, and the crucible is slowly heated to a low dull red to allow the charge to degas. A dewar flask is placed around the reactor and filled nearly to the top with liquid nitrogen. A light layer of ligand is deposited on the walls and then the deposition is stopped. Over a period of several minutes the crucible is heated and the charge is occasionally viewed through a dark welder's glass. For metals that melt before vaporizing, it is best to melt them first and quickly back off on the power so that too rapid an evaporation does not occur. When the evaporation starts, some wall darkening occurs, and ligand deposition is then resumed. The introduction of *about* 50 mmole of ligand to *about* 3 mmole of metal in ½-1 hour is normal in exploratory work. Sometimes it is difficult to tell if metal evaporation has stopped. A common trick is to shut off the ligand supply for about 10 seconds and then introduce a "blast" of ligand. If the metal is still vaporizing, a "rainbow effect" is often seen on the inside walls of the reactor. The pressure in the reactor (indicated on the thermocouple gauge) should remain below 10^{-2} torr and preferably below 5×10^{-3} torr while ligand is being added. If the pressure rises during addition, ligand flow and metal evaporation rate should be decreased. After completion of the reaction, the reactor walls are usually dark; however, the appearance may vary considerably.

3. *Product recovery.* (■**Caution.** *Finely divided metals, such as those produced in the metal atom reactions, are often pyrophoric. Therefore, at the end of a reaction the reactor is always vented with an inert gas such as N_2. Residual pyrophoric metals may generally be rendered innocuous by the careful introduction of dilute HCl solution under a nitrogen purge. It is to be expected that the electrode assembly will be covered with pyrophoric metal which will glow or flame as the assembly is removed from the reactor. Care should be*

exercised to avoid ignition of organic vapors when the electrodes are first exposed to the atmosphere.) The following procedures are the most commonly used for working up the product: (1) the deposit is allowed to melt and is then removed along with excess substrate by means of a syringe or syphon tube *under a nitrogen blanket.* (2) The deposit is warmed to room temperature and excess ligand is pumped into the main trap or another trap on the vacuum line (Fig. 2). If desired, degassed solvent may then be introduced by trap-to-trap distillation or by syringe, under a nitrogen flush, and the solution may then be removed as outlined in procedure 1. Once the product is removed it is isolated and purified by normal techniques, involving Schlenk ware, vacuum line, or dry box.

Causes of Low Yields and Nonreproducibility

The success of chemical synthesis using metal atoms depends on a number of factors. Some of these factors are intrinsic to the chemistry being attempted, but other factors depend mainly on the experimental conditions. Some of the more common problems are discussed below.

1. *Substrate-to-metal ratio.* It is convenient to use as small a substrate/metal ratio as possible because less substrate is then required, separation of the product from excess substrate is facilitated, and problems of pressure increase due to rapid addition of substrate vapor are minimized. Low ratios give acceptable yields only when the reaction product is not catalytically decomposed by metal clusters. For example, 50% yields of $Cr(PF_3)_6$ can be obtained with a PF_3/Cr mole ratio of only 5:1. However, dismal yields of $Ti(C_6H_6)_2$ or $Fe(1,5-cod)_2$ would be obtained under the same conditions and substrate/metal ratios of at least 15:1 must be used.

2. *Mixing of atoms and molecules.* A well-designed metal atom apparatus will spray the ligand vapor onto the cold walls of the vacuum vessel such that the maximum concentration of ligand coincides with the position of the maximum concentration of metal vapor condensing on the walls. It is fairly easy to achieve this condition in the absence of much heat radiation from the metal evaporating source. Under good vacuum, the evaporating source emits both metal vapor and heat radiation in straight lines so that the area of the cold surface that receives a maximum flux of metal vapor also receives a large flux of heat radiation. Thus, one zone on the cold surface must cope with maxima of radiative heat flux from condensing ligand and metal. Not uncommonly, this zone ceases to condense the ligand as efficiently as other, colder parts of the vacuum vessel. The condensed ligand/metal ratio is lower in this zone and the yield of product falls. Solutions to the problem include: (*a*) insulating the metal evaporation source as well as possible so that only an unavoidable amount of heat radiation accompanies

metal evaporation and (*b*) using as large a diameter of reactor as available to reduce the flux of heat and condensing species. If only a smaller reactor is available, the scale of the reaction should be reduced. For cylindrical reactors, the scale of reactions that should be attempted varies with the square of the diameter.

3. *Rate of atom condensation.* The relationship between the yield of product and the ratio of ligand/metal atoms condensed on the cold surface is obvious. The more ligand, the more metal atom-metal atom collisions are avoided. However, in many reactions, particularly those in which the product is endothèrmic, for example, the preparation of dibenzenetitanium or bis(1,5-cyclooctadiene)iron, the absolute flux of metal atoms is also important. With high flux rates on the condensing surface, independent of the ligand/atom ratio, the product yield is low. With suitably low flux rates, the yields in such reactions can be very high. For the most critical reactions, the best flux of atoms corresponds to one monolayer of atoms per (cm^2 sec)—that is, 0.01 mmole/cm^2 hr. For less difficult reactions, flux rates 10 or even 100 times greater give satisfactory yields. For most reactions it seems that there are critical flux rates that, if exceeded at any time during a metal atom reaction or on any zone on the condensing surface, give low yields. The reason for this behavior of atoms is not properly understood, but it seems related to surface diffusion and surface accommodation factors. Undoubtedly, it is a major cause of nonreproducibility in metal atom reactions as metal atom flux rates are quite difficult to reproduce from one experiment and apparatus to another. The best solution is to reduce the atom flux rate, that is, to evaporate the same weight of metal over a long period, or to use a larger vacuum vessel to increase the surface area.

4. *Thermolysis on insulating wool.* Kaowool or other refractory wools are valuable for reduced radiative heat losses from a hot crucible. However, their use causes some increase in the extent of pyrolysis of substrate vapor by the crucible assembly and this, in rare instances, may spoil a metal atom synthesis. Only one example is known at present. The reaction of palladium atoms with benzyl chloride gives very low yields of η^3-benzylpalladium chloride when the palladium is evaporated from an alumina crucible insulated with Kaowool, but a 30-50% yield with an uninsulated crucible. It has been established that this is due to enhanced formation of product-destroying radicals on the hot Kaowool.

5. *Electron damage and electronic excitation effects.* Electron bombardment heating is suitable for the evaporation of all metals under vacuum, but its use in metal vapor synthesis can create problems because of electrons interacting with substrate molecules to form product-destroying ions or radicals. Electron

damage seems to be minimized by using an electrostatically focused electron gun of the Unvala design, with the metal target at high positive potential, the electron emitting filament at ground potential, and a pressure of less than 10^{-4} torr in the reaction vessel.[2,6] Evaporation of main group elements by either electron bombardment or arc evaporation methods may create atoms in excited electronic states with a chemistry different from that of the ground state atoms produced by simple thermal evaporation. This effect is rarely important for transition metal atoms, as the excited states of these atoms decay to the ground state very rapidly before the atoms condense on the surrounding cold surface.

6. *Temperature of the cold surface.* Some metal atom reactions, for example, those involving 1,5-cyclooctadiene, show a marked variability in yield depending on the true temperature of the condensing surface. Generally, the lower the surface temperature, the worse the yield of product. This may be due to a lower activation energy for the atom-atom reaction than the atom-ligand reaction, which at low temperatures would tend to favor the metal aggregation reaction. It may also be due to the form in which the ligand condenses at the different temperatures; both the rigidity of the condensate and the conformational distribution of the ligand could have an influence. These factors can be the cause of irreproducibility because there can be wide variations from one apparatus to another in the true temperature of the condensing surface even when the vacuum vessels are, in both cases, cooled in liquid nitrogen. The variations are caused by the different thermal conductivities of the reactor walls and the heat load on the surface from the furnace radiation and the heat of condensation of atoms and ligand. This problem is related to differences in the substrate-to-metal ratio discussed above, as both are concerned with the actual temperature of the condensing surface.

References

1. P. S. Skell and L. D. Westcott, *J. Amer. Chem. Soc.,* **85**, 1023 (1963).
2. P. L. Timms, in *Cryochemistry,* M. Moskovits and G. A. Ozin (eds.), Wiley-Interscience, New York, 1976, p. 61.
3. M. J. McGlinchey, and P. S. Skell, *ibid.,* p. 137.
4. K. J. Klabunde, *Acc. Chem. Res.,* 8, 393 (1975).
5. D. F. Shriver, *The Manipulation of Air-Sensitive Compounds,* McGraw-Hill Book Co., New York, 1969, Chapt. 1-4.
6. F. G. N. Cloke, M. L. H. Green, and G. E. Morris, *Chem. Commun.,* **1978**, 72.

12. BIS[1,3-BIS(TRIFLUOROMETHYL)BENZENE]CHROMIUM-(0), [1,3-$(CF_3)_2C_6H_4]_2Cr$

Submitted by K. J. KLABUNDE,* H. F. EFNER,* and T. O. MURDOCK*
Checked by J. A. GLADYSZ† and J. G. FULCHER†

The metal atom (vapor) technique has been extraordinarily useful for the preparation of a host of new bis(arene) metal sandwich compounds, including complexes of titanium, vanadium, chromium, molybdenum, and tungsten.[1-5] Wide variations in substituents can be tolerated (C_6H_5X, $C_6H_4X_2$ where X = H, F, Cl, CF_3, $CO(OCH_3)$, CH_3, C_2H_5, $CH(CH_3)_2$, $N(CH_3)_2$, $OCH_3)^5$, and large substituent effects have been noted.[4-6] A striking example is shown by the presence of two CF_3 groups on each arene ring, which allows the preparation of totally air- and heat-stable bis(arene)Cr(0) complexes.[4,5] This preparation is an especially well-suited "beginning" metal atom experiment since the product is air stable, soluble in alkanes, sublimable, and forms in high yield. Also m-$(CF_3)_2C_6H_4$ is fairly inexpensive and readily available.

This sandwich compound has not yet been obtained by more conventional synthetic methods.

Procedure

Before starting this synthesis the general procedure given in Section 11 should be reviewed. The reactor referred to in this synthetic procedure is shown in Fig. 2 of Section 11.

Chromium metal (about 1 g) in the form of granules or small pieces (not powder) is loaded into a tungsten boat evaporation source (Mathis Co., S-16-.010 W) and placed in position in a metal atom reactor such as that shown in Fig. 2 of Section 11. Copper adapter blocks are employed so that a good electrical connection to the electrodes is made. Alternately, the chromium metal can be placed in an integral tungsten-alumina crucible (Sylvania Emissive Products, CS-1002A, 0.5-mL capactiy) and secured in position with split bolts on the ends of the electrodes. The source is firmly secured to the electrodes (slight pressure with pliers) and the electrical current is turned on momentarily to make sure a good electrical connection has been made.

A Teflon-coated magnetic stirring bar is placed in the reactor flask, and the metal atom reactor is assembled and pumped down to less than 1×10^{-3} torr as described in Section 11. Thirty milliliters (20 mmole) of 1,3-bis(trifluoromethyl)-

*Department of Chemistry, University of North Dakota, Grand Forks, ND 58202.
†Department of Chemistry, University of California, Los Angeles, CA 90024.

benzene* is degassed (by several cycles of freezing at $-196°$, pumping, and thawing) and connected directly to the ligand inlet system. [There is no need for a bubbler or metering device; only a moderately sized valve, e.g., 10 mm Teflon in glass, is necessary to satisfactorily regulate the vaporization of the ligand. The ligand vaporizes from the liquid, recondenses slightly on passage into the inlet tube, but completely vaporizes again before reaching the bottom of the inlet tube; see Fig. 2 of Sec. 11.]

The evaporation source is heated to a dull red heat to drive off gases. The reactor flask is cooled to $-196°$ with liquid nitrogen, and about 3 mL of substrate is introduced to coat the walls. While substrate continues to distill into the reactor, the evaporation source is heated to a bright red-white heat (boat source: 150 A, 2.8 V; crucible source: 37 A, 6 V). (■**Caution.** *When the crucible is white hot it should be observed only through a dark welding glass.*) Chromium begins to vaporize and a yellowish-brown matrix forms. During the codeposition, which takes about 1 hour, the flow of substrate can be momentarily shut off and then turned back on; a rainbow coloration in the matrix is usually observed if Cr is still vaporizing. A total of 0.1-0.5 g Cr is vaporized (2×10^{-3} to 1×10^{-2} g-atom).

The reactor is allowed to warm to room temperature and then the excess substrate is removed under vacuum and recovered by condensation into a U-trap on the vacuum system. However, pumping should not be continued longer than necessary to remove the substrate because the product is relatively volatile and will sublime throughout the vacuum system. The reactor is filled with nitrogen gas and 50 mL of hexane is introduced by syringe. (■**Caution.** *Since pyrophoric metal particles may be present the reactor contents should always be handled under a blanket of nitrogen and the residues disposed of cautiously,* see p. 66.) The dark yellow-brown solution is stirred vigorously and the walls are scraped by means of an external hand magnet and the stirring bar in the reactor. The resulting solution is removed by means of a syringe and a long Teflon needle and placed in a sublimer.[†] The solvent is vaporized, the sublimer finger is cooled to $-78°$, and the product is sublimed at 40-60° at a pressure less than 1×10^{-2} torr. In a typical experiment, 0.10 g of chromium yields 0.43 g of product (44%). *Anal.* Calcd. for $C_{16}H_{18}F_{12}Cr$: C, 40.02; H, 1.68. Found: C, 40.05; H, 1.96.

Properties

This compound $[Cr[m-(CF_3)_2C_6H_4]_2]$ is yellow-green and air stable. It

---

*Available from PCR, Inc., P. O. Box 1466, Gainesville, FL 32602 [listed as 1,3-di(trifluoromethyl)benzene].

[†]Direct sublimation within the reactor is possible if an apparatus with a removable electrode assembly and insertable cold finger is available, such as that shown in Fig. 3 of Section 11.

melts at 91.5-92.5°, decomposes at 200° (sealed tube), and sublimes slowly at 25° (10 m torr). Mass spectrum at 70 eV: $C_6H_4(CF_3)_2Cr$, 22%; $C_6H_4C_2F_5Cr$, 20%; $C_6H_4(CF_3)_2$, 17%; $C_6H_4C_2F_5$, 100%; $C_6H_4C_2F_4$, 81%; $C_7H_4F_4$, 6%; $C_6H_4C_2F_3$, 10%; $C_6H_4CF_3$, 31%; $C_8H_3F_2$, 4%; C_6H_3Cr, 22%; CF_3, 9%; Cr, 96%. IR, in KBr pellet: 420 (w), 440 (m), 450 (w), 500 (s), 565 (m), 580 (m), 600 (w), 675 (s), 700 (m), 750 (w), 824 (s), 840 (s), 875 (w), 888 (m), 995 (m), 1050 (s), 1080 (s), 1120 (s), 1140 (s), 1168 (s), 1180 (s), 1280 (s), 1295 (s), 1335 (s), 1345 (s), 1400 (m), 1460 (w), 1495 (m), 1525 (w), 1545 (w), 1562 (w), 3120 (w) cm^{-1}. ^1H NMR in cyclohexane (tms standard) δ 5.38 ppm (singlet area 2), δ 5.05 ppm (doublet area 4), δ 4.83 ppm (triplet area 2). ^{19}F NMR in benzene is a singlet at 56.6 ppm upfield from $CFCl_3$.

References

1. P. L. Timms, *Chem. Commun.*, **1969**, 1033; R. Middleton, J. R. Hull, S. R. Simpson, C. H. Tomlinson, and P. L. Timms, *J. Chem. Soc., Dalton Trans.*, **1973**, 120.
2. P. S. Skell, D. L. Williams-Smith, and M. J. McGlinchey, *J. Am. Chem. Soc.*, **95**, 3337 (1973); M. P. Silvon, E. M. Van Dam, and P. S. Skell, *ibid.*, **96**, 1945 (1974).
3. M. T. Anthony, M. L. H. Green, and D. Young, *J. Chem. Soc., Dalton Trans.*, **1975**, 1419.
4. K. J. Klabunde and H. F. Efner, *Inorg. Chem.*, **14**, 789 (1975).
5. K. J. Klabunde, *Acc. Chem. Res.*, **8**, 393 (1975).
6. K. J. Klabunde, *Trans. N. Y. Acad. Sci.*, **295**, 83 (1977).

13. BIS(PENTAFLUOROPHENYL)(η^6-TOLUENE) NICKEL(II)

$$2Ni_{(g)} + 2C_6F_5Br + C_6H_5CH_3 \rightarrow (\eta^6\text{-}C_6H_5CH_3)Ni(C_6F_5)_2 + NiBr_2$$

Submitted by K. J. KLABUNDE,* B. B. ANDERSON,* and M. BADER*
Checked by R. J. CLARK†

The preparation of previously unknown, very unusual cobalt and nickel complexes[1] possessing both σ-bound R groups (C_6F_5) *and* a π-bound arene ligand is accomplished by the triple deposition of bromopentafluorobenzene,‡ toluene, and metal vapor. Because of the sensitivity of these compounds to various solvents and/or ligands in general, and to oxygen, the metal atom method is well

*Department of Chemistry, University of North Dakota, Grand Forks, ND 58202.
†Department of Chemistry, Florida State Univ., Tallahassee, FL 32306.
‡Available from Aldrich Chem. Co., 940 West Saint Paul Ave., Milwaukee, WI 53233 or PCR, Inc., P. O. Box 1466, Gainesville, FL 32602.

suited for their preparation. It is not yet known if they can be made by other means.

In the case of nickel, $(C_6F_5)NiBr$ is a reactive intermediate that can be trapped at $-80°$ by addition of Et_3P [to yield $C_6F_5NiBr(PEt_3)_2$]. However, on warming $(C_6F_5)NiBr$ in the presence of toluene (or other arenes) disproportionation takes place, giving the product. The product is soluble in toluene and so it is readily separable from the metal salt by-product. Bis-(pentafluorophenyl)($η^6$-toluene)-nickel is air sensitive and so must be handled in normal airless glassware.[2]

The metal-to-ligand and C_6H_5Br-to-$C_6H_5CH_3$ ratios are not very critical in this preparation.

Procedure

Bis(pentafluorophenyl)($η^6$-toluene)nickel(II) [Ni(II)(C_6F_5)_2($η^6$-C_6H_5CH_3$)] is prepared by the codeposition of nickel vapor with a bromopentafluorobenzene-toluene mixture. Approximately 50 mL of bromopentafluorobenzene (C_6F_5Br) is mixed with 15 mL of dry toluene, placed in a small round-bottomed flask with a stopcock attachment, and freeze-thaw degassed. A wire-wound aluminum oxide crucible (Sylvania Emissive Products CS-1002A, ½ mL capacity) is charged with about 2.5 g of nickel metal (pieces, not powder) and placed in position in a metal atom reactor such as that shown in Fig. 2 of Section 11. The system is evacuated and the nickel is degassed by gently warming the crucible to red heat for 10-15 minutes. (■**Caution.** *When the crucible is white hot, only view it through a dark welders glass.*) The reactor is cooled to $-196°$ and substrate is admitted as a vapor through a small stopcock until the walls are coated (about 3 g). The crucible is slowly warmed to a white heat and a good rate of vaporization of nickel is established (approximately 1 g/hr, about 40 A, 6 V). At the same time the substrate is introduced at a rate such that it is consumed in 1-1½ hours. During the reaction, the matrix should turn deep red. After completion of the reaction, the matrix is allowed to warm to room temperature. During this period the matrix should turn from red to light blue and finally to a black-red solution upon melting. The excess C_6F_5Br-toluene mixture is condensed into a liquid-nitrogen-cooled trap (the ligand mixture can be reused) on the vacuum line, and then the reactor is brought to 1 atm with nitrogen gas. (■**Caution.** *Always handle the reactor contents under nitrogen.*) Toluene is freshly distilled from benzophenone ketyl under nitrogen. While maintaining a nitrogen flush on the reactor, four 20-mL portions of the dry deoxygenated toluene are introduced by syringe to wash down the black residue. This process is aided by a magnetic stirring bar that is manipulated by an external hand magnet to stir and scrape down the product, which is removed by means of a syringe equipped with a long Teflon needle. The individual (or combined) washes are immediately filtered under nitrogen using typical Schlenk-type airless glassware.[2] A fine-porosity filter frit is used. The washes should be dark red. Crystals are obtained by vacuum evaporation of the solvent to approximately 15 mL and cooling to about $-20°$ overnight. Further product can be obtained from the mother liquor by a second solvent reduction and crystallization. The yield is approximately 3.8 g (8 mmole) or 30% based on 1.5 g of nickel vaporized (26 g-atm). *Anal.* Calcd. for $(C_6F_5)_2(C_6H_5CH_3)Ni$: C, 47.06; H, 1.66; F, 39.17. Found: C, 47.10; H, 1.70; F, 39.20.

Properties

Bis(pentafluorophenyl)(η^6-toluene)nickel(II) forms dark red-brown crystals that are sensitive to air and moisture. The crystals darken at $125°$ and melt at 137-$140°$. IR (KBr pellet) 3120 (w), 2940 (w), 1640 (w), 1615 (w), 1535 (m, sh), 1505 (vs), 1470 (vs), 1440 (vs, sh), 1390 (m), 1360 (m), 1280 (w, sh), 1260 (w), 1215 (w), 1180 (w), 1120 (w), 1060 (s), 1040 (m, sh), 1005 (w), 985 (w), 960 (vs), 875 (w), 800 (s), 790 (s), 730 (w) cm^{-1}. NMR (CDCl$_3$, TMS standard) singlet δ 2.23 ppm, 3H; two broad peaks δ 6.67, 7.00 ppm, 5H. The compound is soluble and stable in dry deoxygenated toluene but decomposes in pentane (even if dry and deoxygenated). Hydrolysis with water and small amounts of glacial acetic acid yields pentafluorobenzene, toluene, and a nickel(II) salt. A toluene solution of the compound reacts with triethylphosphine to yield the light-yellow compound bis(pentafluorophenyl)bis(triethylphosphine)nickel(II) as a powder that melts at $210°$.

References

1. B. B. Anderson, C. L. Behrens, L. Radonovich, and K. J. Klabunde, *J. Am. Chem. Soc.,* 98, 5390 (1976).
2. D. F. Shriver, *The Manipulation of Air-Sensitive Compounds,* McGraw Hill Book Co., New York, 1969.

14. TETRACHLORODIBORANE(4) (DIBORON TETRACHLORIDE)

$$2Cu(atoms) + 2BCl_3 \rightarrow 2CuCl + B_2Cl_4$$

Submitted by PETER L. TIMMS*
Checked by STEVEN D. ITTEL[†] and MARK KUCK[‡]

*School of Chemistry, University of Bristol, Bristol, BS8 ITS, United Kingdom.
[†]Central Research and Development Laboratory, E. I. du Pont de Nemours & Co., Wilmington, DE 19898.
[‡]Eastern Research Center, Stauffer Chemical Co., Dobbs Ferry, NY 10522.

Tetrachlorodiborane(4) was first reported in 1925 by Stock, who prepared the compound by striking an arc between zinc electrodes in liquid boron trichloride.[1] Better yields have been obtained by later workers using boron trichloride in the vapor phase and an electrical discharge between mercury[2] or copper electrodes.[3] However, the rate of formation of tetrachlorodiborane(4) is low, at best about 1.0 g/hr.

The preparation of tetrachlorodiborane(4) using copper atoms can be carried out on a scale that gives several grams an hour using a simple form of metal-vapor synthesis apparatus.[4,5] Copper vapor and boron trichloride are condensed simultaneously on a surface cooled by liquid nitrogen. The yield of B_2Cl_4 increases with the ratio of BCl_3/Cu condensed, but a 7:1 ratio gives a satisfactory yield without making it too difficult to separate the product from a huge excess of BCl_3. The yield is not very dependent on the rate of condensation of copper atoms and rates of up to 10^{-7} mole/cm^2 sec can be used.

Procedure

Figure 1 is a sketch of a complete apparatus for the preparation of tetrachlorodiborane(4). Boron chlorides are very corrosive and attack many elastomers and vacuum greases, so only Teflon, Viton, and halocarbon greases* should be used for seals and lubrication in the vacuum system.

The scale of preparation of B_2Cl_4 that can be attempted depends on the size of the metal atom reactor available. For a 100-mm-diameter reactor (as in Fig. 2 of Section II), evaporation of 10-12 g of copper and condensation with about 140 g of BCl_3 in a 1-hour run is the useful limit; this yields about 4 g of B_2Cl_4. The maximum scale for a preparation is roughly proportional to the square of the reactor diameter; thus, as repeatedly demonstrated at the University of Bristol, a 160-mm-diameter reactor can be used to evaporate 20-30 g of copper and make 10 g of B_2Cl_4 in an hour run. The following detailed description of the procedure is based on using a reactor with a diameter of 100 mm. If a larger reactor is available the quantities can be scaled-up accordingly.

Copper is best evaporated from an alumina crucible of 2-5 mL capacity, resistance heated either by an external molybdenum or tungsten wire spiral or, preferably, by imbedded molybdenum or tungsten wire (See Sec. 11). If the crucible is insulated with a 10-mm layer of a refractory wool, for example, Kaowool (see Sec. 11), a power input of 180-250 W will heat the crucible to 1400° and evaporate the copper. The power required to give the desired rate of evaporation has to be determined for a new crucible by a trial evaporation in the absence of BCl_3. The volt and amp settings then found should be valid for many runs with the same crucible. An ordinary grade of soft copper rod is satisfactory

*Obtainable from Halocarbon Products Corp., 82 Burlews Court, Hackensack, NJ, 07601.

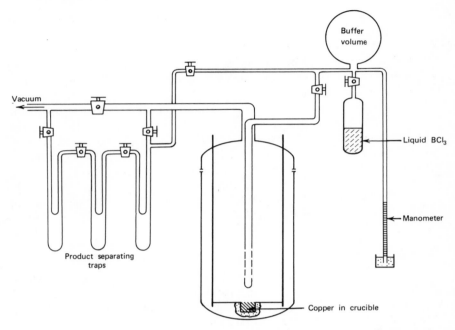

Fig. 1. Schematic of apparatus for making B_2Cl_4 from BCl_3 + Cu vapor.

source of metal for the experiment.

■**Caution**. *Boron trichloride is highly toxic and corrosive. Excess BCl_3 must be vented in a good hood.*

Boron trichloride is most conveniently added using a thick-walled glass reservoir or lecture bottle containing liquid BCl_3 at room temperature (vapor pressure 2 atm at 33°) and attached to a gas reservoir of known volume as shown in Fig. 1. The gas reservoir is initially filled with BCl_3 and is then shut off from the liquid reservoir by closing the intervening Teflon-and-glass or metal needle valve. The gas is let into the atom reactor through a Teflon-and-glass or other controllable valve, which is adjusted until the rate of fall of pressure in the reservoir corresponds to the required rate of addition of BCl_3. The valve between the liquid and gas reservoirs is then opened so that the rate of loss of BCl_3 to the reactor is balanced by evaporation from the liquid. Reagent quality BCl_3 should be used (>99.5% pure*), further purified only by vacuum pumping at its melting point to remove excessive amounts of HCl.

To carry out a preparation of B_2Cl_4, the crucible is charged with 12-15 g of

*Obtainable from Matheson Gas Products, P.O. Box 85, East Rutherford, NJ 07073.

copper (conveniently in the form of a piece of rod, rough-filed to fit the contours of the crucible) and mounted between the electrodes. The reaction vessel is put in place and the assembly is evacuated. The BCl_3 reservoirs should contain 130-150 g (1.1-1.3 moles) of liquid and gas.

With a vacuum of $<10^{-3}$ torr, the crucible is heated to about 800° (dull orange color) by increasing the applied voltage over 15 minutes. The reaction vessel is cooled in liquid nitrogen (-196°), and BCl_3 is slowly admitted to the reactor (about 2 mmole/min) to give an initial layer of the compound on the cold walls before evaporation of copper commences. The crucible is heated further until the copper melts (1084°); slight gas evolution is usually observed during melting. The flow of BCl_3 is increased to 18-22 mmole/min, and power to the crucible is increased to the level that gives the desired evaporation rate of 2.5-3.5 mmole/min; the surface of the molten metal should be in vigorous agitation. (■**Caution.** *The molten metal should be observed through a welders glass until copper begins to deposit on the inside of the reactor reducing the glare.*) The heat of condensation of the copper and BCl_3 cause vigorous boiling of liquid nitrogen around the reactor. The liquid is topped up every few minutes; typically, 3-5 L are boiled-off in a 1-hour run. The condensate on the reactor walls appears a patchy white, grey, and copper color. The BCl_3 condenses initially as a liquid and this may form rivulets which run down the reactor walls before it solidifies.

After an hour, the BCl_3 flow is stopped and the voltage applied to the crucible is reduced to zero over 10-15 minutes. The liquid nitrogen is removed from around the reactor, which is allowed to warm slowly to room temperature. The liberated vapors are passed through two traps cooled by Dry Ice-acetone slush (about -85°) and a trap at -196° backed by the vacuum pumps. A mixture of BCl_3 and B_2Cl_4 condenses in the first two traps and BCl_3 in the last, which must be large enough to hold 150 g of BCl_3 without blocking. As described by Wartik,[3] B_2Cl_4 is slightly volatile at -85° and the liquid condensed in the first two traps should not be pumped on too long. Final purification of B_2Cl_4 from BCl_3 requires that the mixture be passed backwards and forwards several times through a trap at -80 to -85°. A more complete separation can be achieved using a low-temperature distillation column.[6] As described in Reference 3, the purity is considered adequate when the vapor pressure of the liquid is 43 torr at 0°. The yield is about 4 g, 30% of theoretical based on the copper evaporated.

The boron trichloride can be reused in the experiment with only brief pumping at its melting point to remove excessive amounts of HCl. Traces of B_2Cl_4 left in the BCl_3 have no adverse effects in subsequent runs. The reactor should be filled with nitrogen after all volatiles have been pumped out, opened to the air, and quickly put into a fume hood. The residues containing Cu, CuCl, and traces of boron chlorides, may fume slightly in air but they are not pyrophoric.

References

1. A. Stock, A. Brandt, and H. Fischer, *Chem. Ber.,* **58**, 635 (1925).
2. A. G. Massey, D. S. Urch and A. K. Holliday, *J. Inorg. Nucl. Chem.,* **28**, 365 (1966).
3. T. Wartik, R. Rosenburg and W. B. Fox, *Inorg. Synth.,* **10**, 118 (1967).
4. P. L. Timms, *Chem. Commun.,* **1968**, 1525 and *J. Chem. Soc. Dalton,* **1972**, 830.
5. P. L. Timms, in *Cryochemistry,* M. Moskovits and G. A. Ozin (eds), John Wiley & Sons, Inc., New York, 1976, Chap. 3.
6. Described in D. F. Shriver, *The Manipulation of Air-Sensitive Compounds,* McGraw-Hill Book Co., New York, 1969, p. 91.

15.   AN ACTIVE CADMIUM-TOLUENE SLURRY AND ITS USE IN THE PREPARATION OF ETHYLIODOCADMIUM

$$Cd \text{ vapor} + \langle C_6H_4 \rangle - CH_3 \xrightarrow{-196°} (Cd)-(\langle C_6H_4 \rangle - CH_3)_n \tag{1}$$

$$(Cd)_x-(\langle C_6H_4 \rangle - CH_3)_y \xleftarrow{warm}$$

$$(Cd)_x-(\langle C_6H_4 \rangle - CH_3)_y + C_2H_5I \rightarrow C_2H_5CdI \tag{2}$$

Submitted by THOMAS O. MURDOCK* and KENNETH J. KLABUNDE*
Checked by C. KING† and K. IRGOLIC†

When metal vapors are cocondensed with weakly coordinating solvents, weakly complexed "solvated metal atoms" are formed at the low temperatures employed (usually −196°). Upon warming, these highly colored, weakly complexed materials decompose relatively cleanly to metal particles and solvent.[1] However, solvation of these small particles and the presence of chemically bound solvent fragments impede growth of the particles. Thus very small particles of high surface area result, which are very reactive.[2] Highly active metals can be produced in bis(2-methoxyethyl)ether (diglyme), 1,4-dioxane, tetrahydrofuran (THF), toluene, mesitylene, hexane, pentane, and other solvents. A host of metals have been studied, including zinc, cadmium, aluminum, indium, germanium, tin, lead, bismuth, copper, and nickel.

*Department of Chemistry, University of North Dakota, Grand Forks, ND 58202.
†Department of Chemistry, Texas A & M University, College Station, TX 77843.

Generally, smaller particles are obtained with polar, more highly solvating solvents. However, these solvents do not necessarily yield the most active metal slurries. The reactivities vary, and the metal slurries can be "fine tuned" somewhat for use in specific types of reactions. For example, nickel particles from pentane are very active as hydrogenation catalysts, whereas nickel particles from THF are not active as hydrogenation catalysts but are very active in alkyl halide reactions.

In the present context, it has been found that active cadmium slurries can be employed to prepare RCdX species.[3] The novelty of the present method is that: (1) R_2Cd is formed only in very low yield compared with RCdX, (2) the preparations can be carried out under mild conditions; and (3) any type of solvent, polar or nonpolar can be used; thus the RCdX species can be prepared in the nonsolvated state.

Procedure

A metal atom reactor such as that shown in Fig. 2 (Sec. 11) is assembled. An aluminum oxide crucible heated by an externally wrapped tungsten coil (Mathis Co. B10 heater and C5 crucible) is charged with approximately 20 g of cadmium pieces, weighed, and placed in proper position. The system is evacuated to 10^{-3} to 10^{-4} torr. Dry toluene (50 to 75 mL) is degassed by alternate cycles of freezing, pumping, and thawing in a sample tube, which is then connected to the reactor system. After cooling the reactor to $-196°$ with liquid nitrogen, about 5 mL of toluene is introduced to coat the walls. The toluene is admitted as a vapor through a medium-sized stopcock (a needle valve or bubbler is not necessary). The crucible is slowly warmed until the cadmium melts (tungsten wire, dull red). Toluene is admitted and cadmium is vaporized simultaneously for about 1 hour. During this time the toluene charged is used up and 5-10 g of cadmium is deposited. The black deposit is allowed to melt, and the resultant slurry is stirred magnetically. (The cadmium-toluene slurry is not easily taken up by a syringe but it can be removed by simply disassembling the reactor with a good nitrogen purge and pouring the slurry with N_2 purge into any desired container. In the procedure given here, the slurry is not removed but is allowed to react with RX in the metal atom reactor.) During warm-up the initial very finely divided metal sinters somewhat to form larger particles.

The reactor is again cooled to $-196°$ and 0.05 mole of C_2H_5I (7.8 g, 3.99 mL) is distilled directly into the reactor containing the slurry. (■**Caution.** *Alkyl iodides are potential carcinogens and should be handled in a good hood.*) The upper portion of the reactor is wrapped with three to four turns of Tygon tubing with water flowing through it, which serves as a means of condensing the toluene-alkyl halide vapor. Heat is applied and the slurry is allowed to reflux overnight. In about 15-20 hours, the cadmium metal is consumed, yielding a colorless solution with a white precipitate. Removal of toluene and other volatiles under vacuum leaves yellow-white RCdX; yield: 55%.

The product may be analyzed by hydrolysis with excess 10% HCl on a vacuum system. The ethane produced is passed through a trap held at $-78°$ (Dry Ice), is condensed in another trap at $-196°$ (liquid nitrogen), and finally is measured as a gas by pressure, volume, and temperature determination.

Properties

The finely divided cadmium may be stored for several months either as a slurry or in a dry state. (A dry 7 month-old sample was just as reactive as a fresh sample.) Strict anhydrous, airless conditions must be maintained, however. The cadmium crystallite sizes are variable but are generally in the 100-1000 Å range. Particle sizes range up to several microns, with some particles even much larger. (Non-transition metals yield more nonuniform particle distributions than transition metals, which form stronger complexes with the solvent).

Ethyliodocadmium is a *white* nonvolatile solid that is insoluble in pentane, toluene, and chloroform but very soluble in acetone and THF. It is very sensitive to air. In a sealed ampule it darkens slightly at 145° and decomposes at 160-170°. NMR [$(CD_3)_2CO$, TMS standard]: δ 1.00 ppm (triplet, area 3), δ 1.00 ppm (triplet, area 3), δ 0.50 ppm (quartet, area 2), $J_{CH_3\text{-}CH_2} = 6.9$ Hz.

References

1. K. J. Klabunde, H. F. Efner, L. Satek, and W. Donnelly, *J. Organomet. Chem.*, **71**, 309 (1974).
2. K. J. Klabunde, H. F. Efner, T. O. Murdock, and R. Ropple, *J. Am. Chem. Soc.*, **98**, 1021 (1976).
3. T. O. Murdock and K. J. Klabunde, *J. Org. Chem.*, **41**, 1076 (1976).

16. BIS(η^6-CHLOROBENZENE)MOLYBDENUM, BIS(η^6-N,N-DIMETHYLANILINE)MOLYBDENUM, AND (1-3:6-7:10-12-η-2,6,10-DODECATRIENE-1,12-DIYL)NICKEL

Submitted by J. E. DOBSON,* R. J. REMICK,* B. E. WILBURN,* and P. S. SKELL*
Checked by E. P. KUNDIG† and P. L. TIMMS†

*Department of Chemistry, Pennsylvania State University, University Park, PA 16802.
†School of Chemistry, University of Bristol, Bristol BS8 1TS, England.
‡Editorial cutting was required. The authors may be contacted for further details.

The direct vaporization of refractory metals from wire loops can be accomplished with large, thin-walled reactor vessels. The high cooling efficiency of thin walls and increased distance between vaporization source and cold surface permit efficient heat dissipation, which is necessary with refractory metals.‡

Bis(η^6-*N,N*-dimethylaniline)molybdenum has been prepared in good yield by cocondensation of molybdenum atoms with a fifty-fold excess of *N,N*-dimethylaniline vapor on a liquid nitrogen cooled surface. This method has been extended to the synthesis of other molybdenum arene complexes and is at present the only synthetic route to such compounds.

The increased cooling efficiency of thin-walled reactors also has permitted the use of more volatile substrates in near molar quantities. (1-3:6-7:10-12-η 2,6,10-Dodecatriene-1,12-diyl)nickel has been prepared in multiple gram quantities by cocondensation of nickel vapor and 1,3-butadiene. This method has provided a clean one-step route to this complex, which was first isolated and identified by Wilke et al.[1] as an intermediate in the cyclotrimerization of 1,3-butadiene by nickel catalysts.

A pumping system capable of maintaining a vacuum of 10^{-4} torr is necessary for the molybdenum reactions, to minimize gas phase collisions of the metal atoms and pyrolysis of the substrate. To maintain this low pressure a diffusion pump with a capacity greater than 50 L/sec is required. Between the pump and reactor is a high-throughput trap and the assembly is connected with short lengths of 25 mm Pyrex tubing. A power supply which is suitable for molybdenum and other refractory metals consists of 70 amp 10 to 1 stepdown transformer controlled by a 10 amp 220 volt Variac. Details on the reactor design are given in Figure 1.

A. BIS(η^6-CHLOROBENZENE)MOLYBDENUM

$$2C_6H_5Cl + Mo \rightarrow Mo(C_6H_5Cl)_2$$

Procedure

■**Caution.** *Molybdenum, vaporizing at 2400°, produces an intense white light and should not be observed directly.*

A 178-mm-diameter 400-mm-long reactor vessel with walls 5 mm thick or less is fitted with a shielded, heated inlet system as described in Fig. 1. A 230-mm length of 1.27-mm-diameter (50-mil) molybdenum wire* is shaped into a "V," weighed, and connected between two water-cooled electrodes. A 30-mL sample of chlorobenzene is placed in a 100-mL round-bottomed flask and attached to the vacuum system. A Teflon-in-glass needle valve assembly is connected between the flask and a vacuum manifold and the chlorobenzene is degassed by

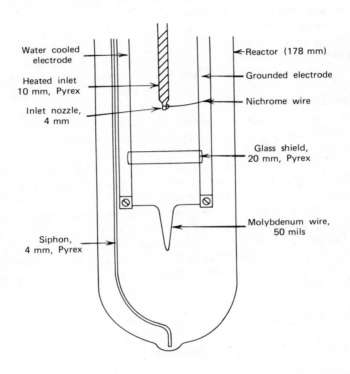

Fig. 1. Detail of reaction zone of the metal-atom reactor. Suitable reactor dimensions are 15-18 cm diameter, 5 mm wall thickness and 36-46 cm depth. The water-cooled electrodes are 7.5 cm apart. The central substrate inlet tube, a 6 mm od Pyrex slightly constricted at the end, extends 5 cm below the liquid nitrogen level. A 14 mm od Pyrex tube which serves as a substrate deflector is positioned 5 cm below the inlet nozzle and is suspended horizontally between the electrodes. A built-in Pyrex syphon tube extends to the bottom of the reactor for the removal of air sensitive products under an inert atmosphere.

*Molybdenum wire available from H. Cross Co. 363 Park Ave., Weekhawken, NJ 07087.

the freeze-pump-thaw method. The Teflon needle valve is closed and the flask assembly is connected to the reactor inlet system. The reactor is evacuated to a pressure of 10^{-6} torr, as measured on an ionization gauge, and outgassed for several hours. The flask containing the substrate is heated to 50° with an oil bath or heating mantle. The reactor is then cooled with liquid nitrogen, the nichrome ribbon surrounding the inlet system is energized, and the inlet is warmed sufficiently to prevent condensation of the substrate vapors. The needle valve is slowly opened and a thin film of substrate is deposited on the inner surfaces of the reactor. The molybdenum wire is then gradually energized over a 3-minute period to an ultimate current of 63 A. As the vaporization continues, the current is observed to decrease on its own, on occasion reaching the low fifties before burnout. Chlorobenzene (30-mL) is cocondensed with 130 mg (1.35 mmole) of molybdenum over a 30 minute period. (Quantities in excess of 500 mg of molybdenum have been evaporated in carefully controlled reactions.) At no time should the internal pressure of the reactor be allowed to exceed 10^{-4} torr. At pressures above 10^{-4} torr the mean free path of substrate molecules is sufficiently short to result in pyrolysis of substrate on the molybdenum wire. Pyrolysis of substrate causes a weight increase of the wire and premature burnout due to molybdenum carbide formation. When the reaction is over, the light-green matrix is melted and the excess chlorobenzene is pumped off. The reactor is then recooled to $-196°$ and 150 mL of dry degassed diethyl ether is added through the inlet system. The liquid nitrogen is removed and a pan of warm water is placed around the bottom 5 cm of the reactor, which causes refluxing of the ether condensate on the cool walls of the reactor, washing down the product. The light-green solution is siphoned into a Schlenk tube and the ether is removed by vacuum pumping. Alternatively, the product may be recovered by direct crystallization from diethyl ether at $-78°$. The compound is extracted several times with dry degassed benzene under an inert atmosphere. The benzene extract is filtered through a frit and frozen. Benzene is removed by sublimation, leaving 240 mg (55%) of a light-green powder.

Properties

Bis(η^6-chlorobenzene)molybdenum is an extremely air-sensitive compound and is pyrophoric. High-resolution mass spectrometery shows a measured mass of 321.9230 for $^{98}Mo(C_6H_5Cl)_2$ (calculated value 321.9214) and a mass of 319.9208 for $^{96}Mo(C_6H_5Cl)_2$, (calculated value 319.9205). IR: 1470 (m), 1440 (sh), 1410 (s), 1395 (sh), 1250 (m), 1070 (sh), 1030 (s), 960 (m), 850 (sh), 840 (w), 790 (m), 760 (m), 720 (m), 670 (m), 640 (sh) cm^{-1}. NMR: τ 4.92 (doublet); τ 5.63 (multiplet). The compound decomposes at 92° under nitrogen in a sealed capillary tube.

B. Bis(η^6-*N,N*-DIMETHYLANILINE)MOLYBDENUM

$$2C_6H_5N(CH_3)_2 \; + \; Mo \; \rightarrow \; Mo[C_6H_5N(CH_3)_2]_2$$

Procedure

■**Caution.** *N,N-Dimethylaniline is a hazardous substance and should be handled only in a fume hood. Molybdenum vaporizing at 2400° produces an intense white light and should not be observed directly.*

A large-diameter reactor vessel is prepared in the same manner as described in the previous synthesis. Dry *N,N*-dimethylaniline is degassed in a 100-mL, round-bottomed flask fitted with a Teflon-in-glass needle valve, and the flask is connected to the inlet system. The reactor is evacuated and outgassed for several hours at 10^{-6} torr and is then cooled with liquid nitrogen. The 100-mL flask containing the *N,N*-dimethylaniline is heated to 70° with an oil bath or heating mantle and the inlet heating system is warmed sufficiently to prevent condensation of the substrate in the inlet. A thin film of substrate is deposited on the inner walls of the reactor and the molybdenum wire is slowly energized to a current flow of 63 A. Over a 50-minute period, 24 mL of *N,N*-dimethylaniline is codeposited with 400 mg (4.16 mmole) of molybdenum on the liquid nitrogen cooled surfaces. At no time should the internal pressure be allowed to exceed 10^{-4} torr. Internal pressure can best be controlled by adjusting the rate of substrate influx. The product is removed from the reactor by the same technique used in the bis(η^6-chlorobenzene)molybdenum synthesis. The Schlenk tube containing the final product is extracted with dry heptane under an inert atmosphere. The heptane extract is filtered through a frit and slowly cooled to −50°. A 585-mg quantity (42% yield) of bis(η^6-*N,N*-dimethylaniline)molybdenum is crystallized out of the heptane.

Properties

Bis(η^6-*N,N*-dimethylaniline)molybdenum is an air-sensitive, dark red-brown solid with needlelike crystals. The compound melts at 90.5-91.0° under nitrogen.

High-resolution mass spectrometery shows a measured mass of 340.0821 for $^{98}MoC_{16}H_{22}N_2$ (calculated mass 340.0837). IR: 3020 (s), 2950 (s), 2840 (sh), 2810 (s), 2770 (s), 1660 (w), 1600 (m), 1480 (s), 1442 (s), 1420 (s), 1290 (s), 1190 (m), 1150 (m), 1120 (s), 1090 (sh), 1042-1038 (m), 980 (m), 925 (m), 840 (w), 762 (m), 720 (m) cm^{-1}. NMR: τ 5.42 (multiplet); τ 7.76 (singlet).

C. (1-3:6-7:10-12-η-2,6,10-DODECATRIENE-1,12-DIYL)NICKEL

$$3C_4H_6 + Ni \rightarrow Ni(C_{12}H_{18})$$

Procedure

A large-diameter, thin-walled reactor similar to that described in the previous two syntheses is fitted with an alumina-coated tungsten wire crucible charged with 100-mesh nickel powder.* Alternatively, nickel may be smoothly evaporated from a triple stranded tungsten coil that has been electroplated with nickel. Matheson C.P. grade 1,3-butadiene is dried by vacuum condensation from stock held in a trap cooled by Dry Ice into a trap cooled by liquid nitrogen. In a similar manner the 1,3-butadiene is charged into a flask fitted with a Teflon needle valve capable of withstanding several atmospheres of pressure. The pressure flask containing the 1,3-butadiene is fitted to the inlet system of the reactor and the reactor is evacuated and outgassed for several hours at 10^{-6} torr. The nickel-containing crucible is then heated to a dull red to outgas the metal, and the reactor is cooled with liquid nitrogen. A thin film of 1,3-butadiene is coated on the walls and the crucible current is then increased to that necessary to vaporize nickel. In a typical run 1.1 g (19.5 mmole) of nickel is cocondensed with approximately 20 g of 1,3-butadiene over a period of 50 minutes. The internal reactor pressure may fluctuate in the 10^{-4} torr region, but under no conditions should the pressure be allowed to exceed 10^{-3} torr. The condensate that forms is initially light pink and gradually darkens to red, finally becoming nearly opaque to the crucible radiation. After the reaction is complete, the liquid nitrogen is removed and an additional 20 g of 1,3-butadiene is rapidly condensed into the cold reactor flask immediately after the matrix melts. This both aids in washing all of the organometallic product to the bottom of the flask and provides a larger liquid pool of excess butadiene. This solution is aged for about 1 hour at $-22°$ (CCl$_4$ slush bath) to ensure maximum conversion of any intermediate species to the desired products. Excess butadiene is subsequently removed by vacuum transfer and pentane is introduced through the inlet system to act as a solvent for the product. The product is siphoned from the reactor through a fritted-glass filter of medium porosity and into a Schlenk-type receiver. The residue after stripping pentane away from the filtrate is a deep-red oil, which is recrystallized twice from butane at $-78°$ to yield 2.0 g (39%) of the red-orange product.

*99.9+% pure metal can be obtained from Alfa Products, Ventron Corp., P.O. Box 299, Danvers, MA 01923.

Properties

(1-3:6-7:10-12-η-2,6,10-dodecatriene-1,12-diyl)nickel is an air-sensitive compound with a melting point of 15-16°; the literature value is 1°. Although the melting point of this product found by the present authors is considerably higher than reported in Reference 2 it was reproducible from run to run. However, the mass and NMR[1] spectra accorded well with the literature and it seems possible that a variation of isomer distribution, as discussed in Reference 2, accounts for the difference.

The NMR spectra are complex and definitive assignments have not been made. For quick identification the NMR spectra reproduced in Reference 1 should be consulted.

References

1. B. Bogdanovic, P. Heimbach, M. Kroner, and G. Wilke, *Justus Liebig's Ann. Chem.,* 727, 143 (1969).
2. J. J. Havel, Ph.D. Thesis, The Pennsylvania State University, University Park, PA, 1972.

Chapter Three

TRANSITION METAL COMPOUNDS AND COMPLEXES

17. HYDRIDO(TRIPHENYLPHOSPHINE)COPPER(I)

$$3(C_6H_5)_3P + CuCl \rightarrow CuCl[P(C_6H_5)_3]_3$$

$$CuCl[P(C_6H_5)_3]_3 + Li[AlH_4] \rightarrow$$

$$2[CuHP(C_6H_5)_3] \ LiAlH_3Cl + 8(C_6H_5)_3P$$

Submitted by R. D. STEPHENS*
Checked by Y-C LIN,† B. A. MATRANA,† H. D. KAESZ,†
W. L. GALDFELTER‡ and G. L. GEOFFROY‡

Hydridic copper complexes have been discussed in the literature for a very long time. Recently they have been shown to have a variety of interesting chemical and structural properties. Unfortunately, a good workable synthesis leading to stable isolated compounds does not currently exist in the literature.

The procedures described below have consistently given stable, crystalline products for a number of tertiary phosphine copper hydride complexes. These complexes can be prepared by the careful reaction of lithium tetrahydrido-aluminate(1-) with tertiary phosphine copper halide complexes. The reactions are run in etheral solvents under an inert atmosphere. The resultant products are

*State of California, Department of Health, 2151 Berkeley Way, Berkeley, CA 94704.
†Department of Chemistry, University of California, Los Angeles, CA 90024.
‡Department of Chemistry, The Pennsylvania State University, University Park PA 16802.

deep-red solids that can be crystallized in fine needles. The crystalline product is stable indefinitely under an inert atmosphere at room temperature. The procedure has been successful using tertiary aryl phosphines, but not tertiary alkyl phosphines. The tertiary alkyl phosphines give rise to viscous deep-red oils that resist crystallization and decompose readily.

Although a crystal structure has been reported for the hexamer,[1] at least two distinctly different crystalline forms exist and molecular weight studies indicate that monomers, dimers, and trimers exist in equilibrium in solution. For these reasons, the complexes are referred to herein as in an indefinite state of aggregation, that is $[CuH(PR_3)]$.

Procedure

The procedures described below require rigorous exclusion of water and oxygen. All solvents should be distilled and stored under an inert atmosphere. Tetrahydrofuran (THF) and diethyl ether are initially dried with calcium hydride or 4A molecular sieves, bubbled with nitrogen, and then heated to reflux with sodium wire plus 0.1 g of benzophenone per 500 mL of ether. After a deep blue color develops the solvent is distilled from the sodium benzophenone ketyl.

Triphenylphosphine, recrystallized from ethanol, 3.72 g (14.3 mmole) is added under nitrogen flush to 35 mL of dry THF in a 250-mL, two necked, round-bottomed flask fitted with a nitrogen inlet tube, dropping funnel, and magnetic stirrer. Dry purified $CuCl$,[2]* 0.473 g (4.78 mmole) is added under nitrogen flush to this solution. The solution is first homogeneous and colorless, but after 1 minute of stirring, a thick white precipitate forms.

A diethyl ether solution of $Li[AlH_4]$ is prepared by extracting $Li[AlH_4]$ into 100 mL of dry diethyl ether by means of a Soxhlet apparatus topped with a T-tube going to a bubbler and nitrogen source and fitted with a 250-mL, one-necked, round-bottomed flask with a side arm. The concentration of $Li[AlH_4]$ in the diethyl ether is determined by weight loss of $Li[AlH_4]$ in the thimble, or alternatively, by hydrolysis.[3] The concentration of $Li[AlH_4]$ is of critical importance in this experiment. An excess of hydride leads to a highly unstable product. If less than 1 mole of $Li[AlH_4]$ is used per mole of CuCl, a stable product with respectable yield is obtained. For example, if the extracted $Li[AlH_4]$ is 0.21 M, 17 mL of the $Li[AlH_4]$ solution (3.57 mmole; $Li[AlH_4]$/CuCl ratio 0.75:1) is transferred by syringe to the dropping funnel. The $Li[AlH_4]$ solution is slowly added over a 10-minute period to the previously prepared THF suspension of $CuCl[P(C_6H_5)_3]_3$. The solution, which immediately turns deep red, is stirred for 10-15 minutes, and then a nitrogen flush is main-

*Available from Alfa Products, Ventron Corp., P.O. Box 299, Danvers, MA 01923.

tained on the flask while 50 mL of dry diethyl ether is introduced by syringe. The mixture is cooled in an ice bath for 15-20 minutes, causing the precipitation of a reddish-orange crystalline product that is separated from the supernatant liquid by filtration under nitrogen. If the cooling period is less than 15 minutes, only a gray-brown precipitate is formed. The product is purified by first washing with dry diethyl ether and then dissolving it in 60 mL of dry benzene. After filtration, the benzene is stripped off under reduced pressure. The diethyl ether wash and crystallization from benzene are repeated, giving 0.82 g of [CuH P(C$_6$H$_5$)$_3$], 52% yield. *Anal.* Calcd. for [CuH P(C$_6$H$_5$)$_3$]: C, 66.14; H, 4.93; P, 9.48; Cu, 19.45. Found: C, 66.2; H, 5.1; P. 9.7; Cu, 19.9; Cl < 0.1.

The *p*-tolylphosphine complex is prepared in a similar manner, except that its greater solubility in THF and diethyl ether precludes precipitation with diethyl ether. The product is isolated by removing the THF/diethyl ether reaction solvent under vacuum and extracting the residue with toluene (dried by distillation from sodium). The toluene extracts are filtered and diluted with an equal volume of hexane (dried by distillation from sodium). Upon slow cooling, fine red needles are deposited.

Properties

Neither of these two products gives definite melting points. [CuHP(C$_6$H$_5$)$_3$] darkens around 115° and melts with decomposition between 125 and 140°. [CuHP(*p*-CH$_3$C$_6$H$_4$)$_3$] darkens with partial melting at 140-145°.

The complexes are soluble in aromatic solvents and in THF. The tri-*p*-tolyl-phosphine complex has limited solubility in ether, and the tri-phenyl phosphine complex is insoluble in diethyl ether. Both complexes are insoluble in hexane and related solvents and decompose in chlorinated solvents. Limited solubility is achieved in *N,N*-dimethylformamide, but moderate decomposition occurs.

The two complexes have absorption maxima at 524.0 nm, with concentration-dependent extinction coefficients. This concentration dependence is thought to result from a concentration-dependent distribution of molecular aggregates. Infrared spectra show only the bands associated with the ligand, the most prominent of which are 3300-3460, 1458, 1412, 1075, 730, 682, and 480 cm^{-1} (KBr disk). Raman bands for the triphenylphosphine complex appear at 180, 124, and 90 cm^{-1}.

References

1. M. R. Churchill, S. A. Bezman, J. A. Osborne, and J. Wormald, *Inorg. Chem.,* **11**, 1818 (1972).
2. R. N. Keller and D. H. Wycoff, *Inorg. Synth.,* **2**, 1 (1940).
3. H. C. Brown, *Organic Syntheses via Boranes,* John Wiley & Sons, Inc., New York, 1975, p. 241.

18.   TETRAKIS(ACETONITRILE)COPPER(I) HEXAFLUOROPHOSPHATE*

$$Cu_2O + 2HPF_6 \xrightarrow{CH_3CN} 2[Cu(CH_3CN)_4][PF_6] + H_2O$$

Submitted by G. J. KUBAS[†]
Checked by B. MONZYK[‡] and A. L. CRUMBLISS[‡]

The $[Cu(CH_3CN)_4]^+$ cation was originally isolated as a nitrate salt in 1923 by the reduction of silver nitrate with copper powder in acetonitrile.[1] Preparations of the perchlorate and tetrafluoroborate salts have since appeard in the literature,[2-5] but the existence of the complex and its potential usefulness as a synthetic reagent are generally not appreciated. The cation is stabilized by large anions, and although the preparation of the $[PF_6]^-$ salt is described here, virtually any large-anion salt can be similarly prepared. Many synthetic options exist also (e.g., $Cu + Ag[PF_6]$), but the following method, based on References 4 and 5, is simple, economical, and avoids Ag(I) contamination. The preparation can be carried out in glassware open to the atmosphere until final drying of the product.

Procedure

■**Caution.** *The following procedures should be carried out in a well-ventilated hood because of the toxicity of acetonitrile and the HF fumes evolved from HPF_6.*

To a magnetically stirred suspension of 4.0 g (28 mmole) of copper(I) oxide in 80 mL of acetonitrile in a 125-mL Erlenmeyer flask is added 10 mL of 60-65% HPF_6 (about 113 mmole of HPF_6) in 2-mL portions. The reaction is very exothermic and may cause the solution to boil. However, the reaction temperature is not critical, and the warming is beneficial in that the product remains dissolved. After addition of the final portion of HPF_6, the solution is stirred for about 3 minutes and is then filtered hot through a medium-porosity frit to remove small amounts of undissolved black solid (some white $[Cu(CH_3CN)_4]$-$[PF_6]$ may begin to crystallize before filtration; if so, it is washed through the frit with a minimum amount of CH_3CN). The pale-blue solution is cooled in a

*This work performed under the auspices of the U.S. Energy Research and Development Administration.

[†]University of California, Los Alamos Scientific Laboratory, Los Alamos, NM 87545.

[‡]Department of Chemistry, Duke University, Durham, NC 27706.

freezer to about $-20°$ (addition of an equal volume of diethyl ether and cooling to $0°$ yields equivalent results) for several hours, whereupon a blue-tinged white microcrystalline precipitate of $[Cu(CH_3CN)_4] [PF_6]$ forms. The solid is collected by filtration, washed with diethyl ether, and immediately redissolved in 100 mL of CH_3CN. A small amount of blue material, presumably a Cu^{2+} species, remains undissolved and is removed by filtration. To the filtrate (which may still retain a slight blue coloration) is added 100 mL of diethyl ether, and the mixture is allowed to stand for several hours at $-20°$. The precipitated complex may still retain a bluish cast, in which case a second recrystallization may be necessary if high purity is desired. This second recrystallization is carried out using 80 mL each of CH_3CN and diethyl ether. The product is pure white and is dried *in vacuo* for about 30 minutes immediately after being washed with diethyl ether. The yield is 12.5 g (60%) and is dependent on recrystallization losses. *Anal.* Calcd. for $C_8H_{12}N_4PF_6Cu$: C, 25.8; H, 3.3; N, 15.0; P, 8.3; Cu, 17.0. Found: C, 25.9; H, 3.3; N, 15.1; P, 8.1; Cu, 16.7.

Properties

Tetrakis(acetonitrile)copper(I) hexafluorophosphate is a free-flowing, white, microcrystalline powder that does not darken upon long-term storage in an inert atmosphere. Exposure to air for longer than about 1 hour results in minor surface oxidation, due to the slightly hygroscopic nature of the complex. The acetonitrile ligands are firmly bound and cannot be removed at a significant rate by pumping *in vacuo* at ambient temperature. Approximate CH_3CN dissociation pressures are 5 torr at $80°$ and 25 torr at $110°$. The infrared spectrum in Nujol shows absorptions at 2277 (m) and 2305 (m) cm^{-1} due to CH_3CN, and at 850 (vs) and 557 (s) cm^{-1} due to $[PF_6]^-$.

The complex is moderately soluble in polar solvents and is remarkably stable to air oxidation in CH_3CN solution. It does not react with halide ions to give CuX (X = Cl, Br, I) in CH_3CN solution, but the coordinated acetonitriles can be displaced in other solvents or even in CH_3CN in certain cases. For example, copper(I) phenoxide, $Cu[OC_6H_5]$ can be precipitated upon addition of ethanolic $Na[OC_6H_5]$ to an acetonitrile soluton of $[Cu(CH_3CN)_4] [PF_6^-]$.[6] Thus the complex is especially suited for nonaqueous-media syntheses of Cu(I) compounds. An interesting synthetic application has been described by Maspero et al.,[7] wherein $[Cu(CH_3CN)_4]^+$ was allowed to react with $[Rh(C_2H_4)_2Cl]_2$ and C_2H_4 in dichloromethane to give $[Rh(C_2H_4)_3(CH_3CN)_2]^+$ and CuCl. A recent X-ray structure[8] of $[Cu(CH_3CN)_4] [ClO_4]$ revealed nearly ideal tetrahedral coordination of copper by the almost linear CH_3CN molecules.

References

1. H. H. Morgan, *J. Chem. Soc.,* **1923**, 2901.

2. B. J. Hathaway, D. G. Holah, and J. D. Postlethwaite, *ibid.,* **1961**, 3215.
3. G. D. Davis and E. C. Makin, Jr., *Sep. Purif. Meth.,* 1, 199 (1972).
4. P. Hemmerich and C. Sigwart, *Experentia,* 19, 488 (1963).
5. E. Heckel, German Patent 1,230,025 (1966); *Chem. Abstr.,* 66, 46487e (1967).
6. P. G. Eller and G. J. Kubas, *J. Am. Chem. Soc.,* 99, 4346 (1977).
7. F. Maspero, E. Perrotti, and F. Simonetti, *J. Organomet. Chem.,* 38, C43 (1972).
8. I. Csoregh, P. Kierkegaard, and R. Norrestam, *Acta Cryst.,* B31, 314 (1975).

19. COORDINATION COMPLEXES OF COPPER(I) NITRATE

Submitted by HENRY J. GYSLING*
Checked by GREGORY J. KUBAS†

Copper(I) coordination complexes with various ligands containing heavy main group V or VI donor atoms have generally been prepared by reaction of a stoichiometric amount of the ligand with a copper(I) halide in a nonaqueous solvent such as chloroform.[1-3] Procedures for the preparation of copper(I) halides (CuCl,[4] CuBr,[4] CuI[5]), a tributylphosphine complex of CuI, $[CuI[P(C_4H_9)_3]_4]$,[6] and a 1,4-butadiene complex of CuCl, $[ClCu(C_4H_6)CuCl]$,[7] have been described in previous volumes of this series. Copper(I) chloride and bromide are quite sensitive to aerial oxidation, but they can be prepared conveniently in good yields in air and stored under an inert atmosphere indefinitely.[8]

Such coordination complexes with copper(I) chloride, and to a lesser extent copper(I) bromide, undergo metathetical reactions with a variety of other anionic ligands (e.g., SCN^-, $SeCN^-$, N_3^-, BH_3CN^-, BH_4^-, $B_3H_8^-$, $B_5H_9^-$).[1-3,9,10]

An alternative, and generally more convenient, synthetic route to Cu(I) coordination complexes utilizes ligand reduction reactions in which an excess of the ligand stabilizes the copper(I) state. Analogous reactions have been reported in previous volumes for the synthesis of zero-valent phosphite complexes of nickel, palladium, and platinum[11] and $CuNH_4MoS_4$.[12]

Ligand reduction reactions of copper(II) chloride by a varieyt of organophosphines have been reported to give a wide range of $(CuCl)_aL_b$ complexes.[1-3] Analogous reduction reactions with copper(II) nitrate give products that contain weakly coordinated nitrate in the solid state (e.g., $[Cu[P(C_6H_5)_3]_2(NO_3)]$[13] and $[Cu[P(C_6H_{11})_3]_2(NO_3)]$,[14] bidentate nitrate; $[Cu[P(C_6H_5)_2CH_3]_3(NO_3)]$,[15] unidentate nitrate).

Copper(I) nitrate complexes with organophosphines, arsines, and stibines can be readily prepared in good yield in alcohol solution by such ligand reduction reactions.[9] In the cases of the arsines and stibines, addition of copper powder

*Research Laboratories, Eastman Kodak Company, Rochester, NY 14650.
†Los Alamos Scientific Laboratory, Los Alamos, NM 87545.

as a supplementary reducing agent is useful. These complexes are convenient reagents for the introduction of other anionic ligands by metathetical reactions.

A. NITRATOBIS(TRIPHENYLPHOSPHINE)COPPER(I)

$$2Cu(NO_3)_2 \cdot 3H_2O + 5P(C_6H_5)_3 \rightarrow 2[Cu[P(C_6H_5)_3]_2(NO_3)] +$$

$$OP(C_6H_5)_3 + 2HNO_3 + 5H_2O$$

Procedure

To a solution of triphenylphosphine (21 g, 0.08 mole) in 200 mL of hot methanol is added $Cu(NO_3)_2 \cdot 3H_2O$ (4.9 g, 0.02 mole) as a solid in portions. No special precautions are required for the exclusion of air throughout this synthesis. The Cu(II) salt dissolves immediately with the formation of a clear colorless solution, but near the end of the addition a heavy white precipitate forms. After addition of all the $Cu(NO_3)_2 \cdot 3H_2O$ the reaction suspension is heated at reflux for 5 minutes and then cooled to ambient temperature. The white precipitate is filtered, washed well with ethanol and diethyl ether, and air dried to give 11.2 g (86%) of $[Cu[P(C_6H_5)_3]_2(NO_3)]$. This crude powder is analytically pure, but it can be recrystallized from hot methanol (10 g complex/L) to give white needles when the filtered solution is cooled overnight in a freezer compartment (mp 248°, dec.). *Anal.* Calcd. for $C_{36}H_{30}CuNO_3P_2$: (MW = 650.14): C. 66.45; H, 4.65; N, 2.15; Cu, 9.8. Found: C, 66.6; H, 4.4; N, 2.1; Cu, 10.0.

Properties

The $[Cu[P(C_6H_5)_3]_2(NO_3)]$ complex obtained by this procedure is readily soluble in chloroform, dichloromethane, acetonitrile, and N,N-dimethylformamide at room temperature and is moderately soluble in hot methanol, ethanol, benzene, and tetrahydrofuran. The complex exhibits a broad absorption maximum at 300 nm (powder reflectance spectrum). Unlike a number of organophosphine complexes of copper(I) halides,[16] $[Cu[P(C_6H_5)_3]_2(NO_3)]$ exhibits no fluorescence. The conductivity (10^{-3} M solution) in dichloromethane (Λ_M = 1.2 ohm^{-1} cm^2 mole^{-1}) is typical of a nonelectrolyte, whereas in N,N-dimethylformamide a value (Λ_M = 63.5 ohm^{-1} cm^2 mole^{-1}) corresponding to a 1:1 electrolyte[17] is found. Weak coordination of the nitrate anion is indicated by the solid-state infrared spectrum of the complex[18,19] and has been confirmed by a single-crystal X-ray diffraction study.[13] It has been reported that the tris complex, $[Cu[P(C_6H_5)_3]_3(NO_3)]$, is obtained when the above reaction is carried out in ethanol, but upon recrystallization from methanol the bisphosphine complex is reportedly obtained.[19]

The facile dissociation of the nitrate ligand in this complex and other analogous copper(I) nitrate complexes (Table I) makes them useful reagents for the introduction of other anionic ligands through metathetical reactions. In these reactions the copper(I) nitrate complex is generally dissolved in chloroform, and an alcohol solution of the sodium salt of the appropriate anion is added. After stirring at room temperature for about 1 hour the solution is filtered through a fine-porosity glass frit and concentrated to give the desired product, which can then be recrystallized from an appropriate solvent.

TABLE I   Organophosphine Complexes of Copper(I) Nitrate

Complex	Calcd. (Found)				mp (°C)	Recryst. Solvent
	C	H	N	Cu		
$Cu[P(p\text{-}CH_3C_6H_4)_3]_2NO_3$	68.64	5.77	1.91	8.65	219	Methanol
MW = 734.30	(68.2)	(5.9)	(1.8)	(8.4)		
$Cu[P(p\text{-}CH_3OC_6H_4)_3]_2NO_3$	60.70	5.10	1.69	7.66	178	Ethanol
MW = 830.3	(60.6)	(5.3)	(2.2)	(7.6)		
$Cu[P(C_6H_5)_2(p\text{-}CH_3C_6H_4)]_2NO_3$	67.3	5.05	2.07	9.37	195	Ethanol/ methanol (1:2)
MW = 678.16	(67.5)	(4.9)	(2.2)	(8.8)		
$Cu[P(m\text{-}CH_3C_6H_4)_3]_2NO_3$	68.64	5.77	1.91	8.65	150	Ethanol
MW = 734.30	(68.4)	(6.1)	(1.8)	(8.3)		
$Cu[CH_3C[CH_2P(C_6H_5)_2]_3]NO_3$	65.64	5.24	1.87	8.47	295	Dichloromethane
MW = 750.19	(65.4)	(5.11)	(2.1)	(8.1)		

B. NITRATOTRIS(TRIPHENYLSTIBINE)COPPER(I)

$$Cu(NO_3)_2 \cdot 3H_2O + Cu + 6Sb(C_6H_5)_3 \rightarrow 2[Cu[Sb(C_6H_5)_3]_3(NO_3)] + 3H_2O$$

Procedure

To a solution of $Cu(NO_3)_2 \cdot 3H_2O$ (6.0 g, 0.025 mole) in 700 mL of methanol in a 2-L, round-bottomed flask are added $Sb(C_6H_5)_3$ (53.0 g, 0.15 mole) and copper powder (7 g, 0.11 g-atom). The resulting suspension is refluxed for 15 minutes and filtered hot. The residue on the filter is extracted with 500 mL of chloroform to give a pale-green solution, which is filtered and concentrated to dryness on a rotary evaporator. The resulting pale-green residue is washed well with diethyl ether and air dried to give 17.8 g of yellow powder. Another 5 g of crude product is obtained by cooling the filtrate from the reaction suspension in a freezer compartment overnight. The combined solids are recrystallized by dissolving in 325 mL of hot toluene and filtering the resulting solution through a medium-porosity glass frit. Upon cooling the dark-yellow filtrate to room temperature, a crop of yellow crystals is deposited. The crystals are filtered, washed with three 150-mL portions of diethyl ether, and vacuum dried over P_4O_{10} to give 16.2 g (27.3%) of $[Cu[Sb(C_6H_5)_3]_3(NO_3)]$. The complex decomposes above 150° with darkening, a brown gummy melt being formed by 175°. *Anal.* Calcd. for $C_{54}H_{45}CuNO_3Sb_3$: C, 54.70; H, 3.83; N, 1.18; Cu, 5.36; Sb, 30.83. Found: C, 55.1; H, 3.9; N, 1.1; Cu, 5.2; Sb, 30.6.

Properties

The $[Cu[Sb(C_6H_5)_3]_3(NO_3)]$ complex is soluble in chloroform, acetonitrile, benzene, and *N,N*-dimethylformamide. In the last two solvents, oxidation of the copper(I) complex is evidenced by the formation of green solutions on standing. Thermal analyses (DSC and TGA in air) show a weight loss above 141° associated with an exothermic reaction.

The infrared spectrum of this compound, when compared with that of $Cu[Sb(C_6H_5)_3]_3Cl$, shows new absorption at 1569 (w, sp), 1430 (the sharp, strong band in the chloro analogue is considerably broadened), 1290 (vs, sp), 1025 cm^{-1} (s, sp), 810 (m, sp) and 233 (w) cm^{-1}. The destruction of the double degeneracy of the E' (NO$_2$ asym str; ν_3 = 1390 cm^{-1}) of ionic nitrate and the separation of 140 cm^{-1} between the two components (i.e., 1430 and 1290 cm^{-1}) support a monodentate bonding mode of the nitrate in the complex.[18,19]

The stoichiometry and IR spectrum of $[Cu[Sb(C_6H_5)_3]_3(NO_3)]$ are therefore consistent with a four-coordinate derivative containing unidentate nitrate. This complex has been shown to be useful for the preparation of complexes of the type $[Cu(N,N)_2]NO_3$ (*N,N*-bidentate nitrogen ligand such as 1,10-phenanthroline).[19]

C. NITRATOTRIS(TRIPHENYLARSINE)COPPER(I)

$$Cu(NO_3)_2 \cdot 3H_2O + Cu + 6As(C_6H_5)_3 \rightarrow 2[Cu[As(C_6H_5)_3]_3(NO_3)] + 3H_2O$$

Procedure

To a solution of triphenylarsine (29.4 g, 0.096 mole) in 350 mL of methanol under a nitrogen atmosphere are added $Cu(NO_3)_2 \cdot 3H_2O$ (3.9 g, 0.016 mole) and 6.3 g of copper powder. [It is essential to use copper powder rather than any other form of the metal. The copper powder obtained from Eastman Organic Chemicals (Catalog No. P1804) gives satisfactory results.] The resulting suspension is heated at reflux for 2 hours and filtered under nitrogen. The residue in the Schlenk filter is vacuum dried to give 39.0 g of crude complex. When dry, the complex can be handled in air. The crude product is recrystalliz-ed from 1.5 L of boiling methanol under nitrogen to give a crop of white crystals upon cooling the filtered solution in a freezer compartment overnight. The solid is filtered and vacuum dried at 110°C for 4 hours to give 21.1 g of white crystals (63.2%).

The $[Cu[As(C_6H_5)_3]_3(NO_3)]$ complex, which is air stable in the solid state, does not exhibit a sharp melting point but decomposes to a gum above 169°. *Anal.* Calcd. for $C_{54}H_{45}As_3CuNO_3$: C, 62.05; H, 4.2; N, 1.34; Cu, 6.09. Found: C, 62.1; H, 4.2; N, 1.2; Cu, 6.0.

The complex $[Cu[As(C_6H_5)_3]_3(NO_3)]$ is soluble in chloroform and moderately soluble in hot methanol, toluene, and acetone. It dissolves in *N,N*-dimethylform-amide with decomposition. This complex has been previously prepared by a similar reaction using the less-convenient reducing agent tributylphosphine[19] rather than copper metal. The infrared spectrum of this complex ($\nu_1 = 1294\,cm^{-1}$; $\nu_2 = 1035\,cm^{-1}$; $\nu_4 = 1423\,cm^{-1}$; $\nu_6 = 812\,cm^{-1}$) supports a unidentate coordina-tion mode of the nitrate ion in the solid state.[18,19] Conductivity measurements indicate that the complex is a 1:1 electrolyte in *N,N*-dimethylformamide[17] (Λ_M = 68 ohm^{-1} cm^2 mole^{-1}) but undissociated in dichloromethane (Λ_M = 0.3 ohm^{-1} cm^2 mole^{-1}). The triphenylarsine ligands in this complex can be displaced readily by bidentate ligands such as 1,10-phenanthroline and 2,2'-bipyridine to give complexes of the type $Cu[As(C_6H_5)_3]_a(N,N)_bNO_3$ ($a = b = 1; a = 0, b = 2$).[19]

D. TETRAHYDROBORATOBIS (TRIPHENYLPHOSPHINE)COPPER(I)

$$[Cu[P(C_6H_5)_3]_2(NO_3)] + NaBH_4 \rightarrow [Cu[P(C_6H_5)_3]_2(BH_4)] + NaNO_3$$

Procedure

To a solution of $[Cu[P(C_6H_5)_3]_2(NO_3)]$ (6.5 g, 0.01 mole) in 100 mL of dichloromethane is added a solution of NaBH$_4$ (0.42 g, 0.011 mole) dissolved in 50 mL of ethanol. No special precautions for the exclusion of air are necessary during the synthesis. The reaction solution is stirred for 30 minutes at room temperature and filtered through a fine-porosity glass frit, and the clear colorless filtrate is concentrated to dryness on a rotary evaporator. The residue is recrystallized from 200 mL of 3:1 chloroform-cyclohexane to give 4.8 g (80% yield) of $[Cu[P(C_6H_5)_3]_2(BH_4)]$, mp 165° (dec.). Heptane can be used in the recrystallization instead of cyclohexane with equivalent results. *Anal.* Calcd. for $C_{36}H_{34}BCuP_2$: C, 71.6; H, 5.63; Cu, 10.54, B, 1.13. Found: C, 71.5; H, 5.7; Cu, 10.3; B, 1.2.

Properties

The $[Cu[P(C_6H_5)_3]_2(BH_4)]$ complex prepared by this metathetical reaction is readily soluble in chloroform, dichloromethane, benzene, and tetrahydrofuran and moderately soluble in acetonitrile and N,N-dimethylformamide. It decomposes in warm acetone and N,N-dimethylformamide. This metalloborane has also been prepared by a metathetical reaction between $[Cu[P(C_6H_5)_3]_3CL]$ and $NaBH_4$.[20] The X-ray crystallographic analysis of this compound has shown the coordination geometry of the copper atom to be quasitetrahedral with two hydrogen atoms of the borohydride group bridging the copper atom and boron atom.[21] The solid-state thermal decomposition of this complex has been shown to produce $Cu(O)$, H_2, $P(C_6H_5)_3$, and $P(C_6H_5)_3 \cdot BH_3$.[22] The catalytic nature of this thermal decomposition reaction has found utility in various imaging processes.[23,24]

References

1. F. H. Jardin, *Adv. Inorg. Radiochem.*, **17**, 116 (1973).
2. *Transition Metal Complexes of Phosphorus, Arsenic and Antimony Ligands*, C. A. McAuliffe (ed.), John Wiley & Sons, Inc., New York, 1973.
3. J. T. Gill, J. J. Mayerle, P. S. Welcker, D. F. Lewis, D. A. Ucko, D. J. Barton, D. Stowens, and S. J. Lippard, *Inorg. Chem.*, **15**, 1155 (1976).
4. R. N. Keller and H. D. Wycoff, *Inorg. Synth.*, **2**, 1 (1946).
5. G. B. Kauffman and R. P. Pinnell, *Inorg. Synth.*, **6**, 3 (1960); **11**, 215 (1968).
6. G. B. Kauffman and L. A. Teter, *Inorg. Synth.*, **7**, 9 (1963).
7. J. R. Doyle, P. E. Slade, and H. B. Jonassen, *Inorg. Synth.*, **6**, 217 (1960).
8. D. F. Shriver, *The Manipulation of Air-Sensitive Compounds*, McGraw-Hill Book Co., New York, 1969.
9. H. J. Gysling, U.S. Patents 3,859,092 (1975); 3,860,500 (1975).
10. M. Kabesova, M. Dunaj-Jurco, M. Serator, J. Gazo, and J. Garaj, *Inorg. Chim Acta*, **17**, 161 (1976).
11. M. Meier and F. Basolo, *Inorg. Synth.*, **13**, 112 (1972); D. Titus, A. A. Orio, and H. B. Gray, *Inorg. Synth.*, **13**, 117 (1972).
12. M. J. Redman, *Inorg. Synth.*, **14**, 95 (1973).
13. G. G. Messmer and G. J. Palenik, *Inorg. Chem.*, **8**, 2750 (1969).
14. W. A. Anderson, A. J. Carty, G. J. Palenik, and G. Schreiber, *Can. J. Chem.*, **49**, 761 (1971).
15. M. Matthew, G. J. Palenik, and A. J. Carty, *Can. J. Chem.*, **49**, 4119 (1971).
16. R. F. Ziolo, S. Lipton, and Z. Dori, *Chem. Commun.*, **1970**, 1124; R. F. Ziolo, Ph.D. Thesis, Temple University, 1971.
17. W. J. Geary, *Coord. Chem. Rev.*, **7**, 81 (1971).
18. C. C. Addison and B. M. Gatehouse, *J. Chem. Soc.*, **1960**, 613.
19. F. H. Jardine, A. G. Vohra, and F. J. Young, *J. Inorg. Nucl. Chem.*, **33**, 2941 (1971).
20. S. J. Lippard and D. A. Ucko, *Inorg. Chem.*, **7**, 1051 (1968).
21. S. J. Lippard and K. M. Melmed, *Inorg. Chem.*, **6**, 2223 (1967).
22. J. M. Davidson, *Chem. Ind. (Lond.)*, **1964**, 2021.
23. D. R. Schultz, U.S. Patent 3,505,093 (1970).
24. H. J. Gysling and R. S. Vinal, U.S. Patent 3,859,092 (1975).

20. (DIMETHYL PHOSPHITO) COMPLEXES OF PLATINUM(II)

Submitted by ROGER P. SPERLINE* and D. MAX ROUNDHILL*
Checked by A. J. CARTY† and D. K. JOHNSON†

Substituted phosphinous acid (R_2POH) and phosphorous acid [$(RO)_2POH$] exist in solution primarily as the tetracoordinate tautomers $R_2P(H)O$ and $(RO)_2P(H)O$. Deprotonation yields a potentially ambidentate anionic ligand, which, with platinum(II), coordinates through the phosphorus atom. The parent acid also can coordinate through the phosphorus atom by way of the electron pair of the tricoordinate phosphorus atom. Frequently, a complex contains the ligand in each formal valence type, and in this case the proton is symmetrically bonded between the oxygen atoms on the two ligands.[1-3] This proton between the coordinated dialkyl phosphito and phosphorous acids is acidic and can be replaced by BF_2[4] or divalent first row transition metal ions.[5,6] Replacement of the coordinated phosphorous acid by a tertiary phosphine or arsine ligand leads to platinum(II) complexes having the phosphite ligand coordinated solely as a tetracoordinate anionic ligand. These complexes form trimetallic cationic complexes with first row transition metal ions.[5]

A. CHLORO(DIMETHYL HYDROGEN PHOSPHITE-*P*) (DIMETHYL PHOSPHITO-*P*) (TRIPHENYLPHOSPHINE) PLATINUM(II)

$$K_2PtCl_4 + 4P(OCH_3)_3 + 4H_2O \rightarrow Pt[OP(OCH_3)_2]_2[HOP(OCH_3)_2]_2 +$$

$$2KCl + 4CH_3OH + 2HCl$$

$$Pt[OP(OCH_3)_2]_2[HOP(OCH_3)_2]_2 + cis\text{-}PtCl_2[P(C_6H_5)_3]_2 \rightarrow$$

$$2PtCl[OP(OCH_3)_2][HOP(OCH_3)_2][P(C_6H_5)_3]$$

Procedure

This procedure is a modification of the published method[7] for the preparation of $Pt[OP(OCH_3)_2]_2[HOP(OCH_3)_2]_2$, which subsequently is converted to $PtCl[OP(OCH_3)_2][HOP(OCH_3)_2][P(C_6H_5)_3]$ by a method similar to the one used for the triethylphosphine analogue.[8] We have found that it is necessary

*Department of Chemistry, Washington State University, Pullman, WA 99164.
†Guelph-Waterloo Centre, Waterloo Campus, Department of Chemistry, University of Waterloo, Waterloo, Ontario N2L 3G1, Canada.

to purify commercial samples of potassium tetrachloroplatinate(II) to avoid reduction to platinum metal with a simultaneous decrease in yield of desired product. Satisfactory samples of potassium tetrachloroplatinate(II) can be obtained by two recrystallizations of commercial samples from 0.1 M potassium chloride solution.

Purified potassium tetrachloroplatinate(II) (1 g, 2.2 mmole) is dissolved in 3 mL of deionized water acidified by 1 drop of 6 M H_2SO_4. Trimethyl phosphite is added to this solution (use a well-ventilated hood) in 0.1-mL portions, with continuous vigorous stirring, to raise the temperature of the reaction mixture to about 60°, and then it is added at a rate that maintains this temperature. A total of 3 mL is added and all the potassium tetrachloroplatinate(II) is dissolved. This operation is carried out in a 25 mL beaker with a 7-mm stirring bar and the temperature is monitored with a small thermometer. When precipitation of white solid begins, deionized water (5 mL) is added and the mixture is stirred for 1 hour. If precipitation begins before 3 mL of trimethyl phosphite has been added, an additional 5 mL of deionized water should be added. The crude product is filtered from the cool solution by means of a medium-porosity sintered-glass filter and washed first with cold methanol (three 5-mL portions) and then with diethyl ether (three 5-mL portions). This product is extracted through the filter with dichloromethane (about 10 mL) into a 100-mL, round-bottomed flask, leaving unreacted potassium tetrachloroplatinate(II) on the filter. To the dichloromethane solution is added methanol (25 mL) and the volume of the solution is reduced to 10 mL under reduced pressure without heating, using a rotary evaporator. On standing, the compound precipitates as colorless needles, which are filtered off and washed with two 5-mL portions of diethyl ether. Yield 0.389 g (25%); mp 176-178°. The filtrate and diethyl ether wash are combined to precipitate on standing a second crop of compound. Yield 0.583 g (38%); mp 175-178°. Combined yield 0.972 g (63%). If necessary, the compound can be recrystallized by dissolving it in the minimum volume of dichloromethane and then precipitating it by slow addition of diethyl ether. This intermediate compound is bis(dimethyl hydrogen phosphite-*P*)bis(dimethyl phosphito-*P*)platinum(II). Phosphorus-31 NMR spectrum: δ (85% H_3PO_4) = 90.65; $^1J_{PPt}$ = 3461 Hz.

Bis(dimethyl hydrogen phosphite-*P*)bis(dimethyl phosphito-*P*)platinum(II) (0.420 g, 0.66 mmole) and *cis*-[dichlorobis(triphenylphosphine)platinum(II)] (0.524 g, 0.66 mmole) are refluxed in toluene (25 mL) for 30 hours. The colorless reaction mixture is filtered while hot using vacuum to remove any unreacted *cis*-[dichlorobis(triphenylphosphine)platinum(II)]. The filtrate is transferred to a 100-mL, round-bottomed flask and evaporated to dryness on a rotary evaporator, using gentle heating if necessary. The cream-colored residue is dissolved in a mixture of dichloromethane (15 mL) and methanol (15 mL) and the solution is allowed to concentrate slowly under water aspirator vacuum on a rotary evaporator without heating until the volume of the solution is 10 mL. The flask is removed from the evaporator, stoppered, and allowed to remain at room temperature for 1 hour. The colorless crystals of product are collected by vacuum filtration and washed with diethyl ether (50 mL). The complex is dried at 80°/1 torr for 30 minutes. Yield 0.501 g (53%), mp 167-168°. A second crop of product can be obtained by evaporation of the methanol solution to 5 mL on the rotary evaporator. Total yield 0.592 g (63%). *Anal.* Calcd. for $C_{22}H_{28}ClO_6P_3Pt$: C, 37.1; H, 3.97; Cl, 4.98; P, 13.1. Found: C, 37.3; H, 4.07; Cl, 5.05; P, 13.1%.

Properties

Chloro(dimethyl hydrogen phosphite-*P*)(dimethyl phosphito-*P*)(triphenylphosphine)platinum(II) is very soluble in chloroform and dichloromethane, soluble in toluene and methanol, and insoluble in diethyl ether, ligroin, and water. The ^{1}H NMR spectrum shows resonances for the methoxy groups at τ 6.83 and 6.25 with $^3J_{POCH}$ = 12 Hz. The 36-line ^{31}P NMR spectrum has chemical shifts from 85% H_3PO_4 at δ 29.1 ($^1J_{PPt}$ = 2179 Hz), δ 89.4 ($^1J_{PPt}$ = 4110 Hz), and 60.1 ($^1J_{PPt}$ = 5209 Hz). The complex gives a trimetallic compound $[PtCl[OP(OCH_3)_2]_2P(C_6H_5)_3]_2Co$ on heating in toluene with bis[(2,4-pentanedionato)]cobalt(II) and a pentametallic compound $\{PtCl[OP(OCH_3)_2]_2-P(C_6H_5)_3\}_4Th$ using a similar procedure with tetrakis[(2,4-pentanedionato)]-thorium(IV) in mesitylene solution.

B. BIS(DIMETHYL PHOSPHITO-*P*) [*o*-PHENYLENEBIS(DIMETHYLARSINE)] PLATINUM(II)

$$Pt[OP(OCH_3)_2]_2[HOP(OCH_3)_2]_2 + o\text{-}C_6H_4[As(CH_3)_2]_2 \rightarrow$$

$$2HOP(OCH_3)_2 + Pt[OP(OCH_3)_2]_2[o\text{-}C_6H_4[As(CH_3)_2]_2]$$

Procedure

This procedure is similar to that used for the synthesis of bis(dimethyl phosphito-*P*)[ethylenebis(diphenylphosphine)] platinum(II).[8] To a solution of bis(dimethyl hydrogen phosphite-*P*)bis(dimethyl phosphito-*P*)platinum(II) (0.976 g, 1.5 mmole) in benzene (50 mL) is added *o*-phenylenebis(dimethylarsine)*[9] (0.441 g, 1.5 mmole). The solution is heated at reflux with stirring for 1 hour and then allowed to cool to room temperature. The colorless compound precipitates and is collected by suction filtration on a medium-porosity fritted glass. The complex is washed with benzene (two 5-mL portions) and then

*The compound can be purchased from Strem Chemicals Inc., Danvers, MA 01923.

diethyl ether (two 5-mL portions) and is dried at 80° under vacuum for 15 minutes. Yield 0.758 g (69%), mp 240° (there is some decomposition between 235 and 240°). If necessary, the complex can be recrystallized by dissolving it in the minimum volume of dichloromethane and precipitating by the slow addition of benzene. *Anal.* Calcd. for $C_{14}H_{28}As_2O_6P_2Pt$: C, 24.1; H, 4.18; P, 8.86. Found: C, 23.9; H, 4.04; P, 9.00%.

Properties

The compound is very soluble in chloroform and dichloromethane and insoluble in benzene, diethyl ether, and ligroin. The ^1H NMR spectrum shows a resonance for the methoxy groups at τ 6.905 ($^3J_{POCH}$ = 12 Hz) with additional resonances from *o*-phenylenebis(dimethylarsine) centered at τ 8.70 and 2.90. The complex forms the trimetallic compounds $[\{Pt[OP(OCH_3)_2]_2[o\text{-}C_6H_4[As(CH_3)_2]_2\}_2M]$ $(ClO_4)_2$ [M = Cu, Co] on treatment with the appropriate metal perchlorate salt.

References

1. K. R. Dixon and A. D. Rattray, *Can. J. Chem.*, **49**, 3997 (1971).
2. P. C. Kong and D. M. Roundhill, *J. Chem. Soc., Dalton Trans.*, **1974**, 187.
3. A. J. Carty, S. E. Jacobson, R. T. Simpson, and N. J. Taylor, *J. Am. Chem. Soc.*, **97** , 7254 (1975).
4. W. B. Beaulieu, T. B. Rauchfuss, and D. M. Roundhill, *Inorg. Chem.*, **14**, 1732 (1975).
5. R. P. Sperline and D. M. Roundhill, *Inorg. Chem.*, **16**, 2612 (1977).
6. K. R. Dixon and A. D. Rattray, *Inorg. Chem.*, **16**, 209 (1977).
7. T. N. Itskovich and A. D. Troitskaya, *Tr. Kazan, Khim. Teknol. Inst.*, **18**, 59 (1953); *Chem. Abstr.*, **51**, 11148g (1957).
8. A. Pidcock and C. R. Waterhouse, *J. Chem. Soc., A*, **1970**, 2080.
9. R. D. Feltham and W. Silverthorn, *Inorg. Synth.*, **10**, 159 (1967).

21. TWO-COORDINATE PHOSPHINE COMPLEXES OF PALLADIUM(0) AND PLATINUM(0)

Submitted by T. YOSHIDA* and S. OTSUKA*
Checked by D. G. JONES,[†] J. L. SPENCER,[†] P. BINGER,[‡]
A. BRINKMANN,[‡] and P. WEDEMANN[‡]

Two-coordinate complexes still remain a rarity in transition metal chemistry.

*Department of Chemistry, Faculty of Engineering Science, Osaka Universtiy, Toyonaka, Osaka, Japan 560.

[†]Department of Inorganic Chemistry, University of Bristol, Bristol BS8 1TS, England; checked Sections A-C.

[‡]Max-Plank-Institute für Kohlenforschung, D-4330 Mülheim/Ruhr, Germany; checked Sections D and E.

Recently a few bicoordinate complexes of palladium(0) and platinum(0), ML_2[1-7] [M = Pd, Pt; L = PPh(t-Bu)$_2$ or P($cyclo$-C$_6$H$_{11}$)$_3$] have been reported, and X-ray studies reveal that they have almost linear structures.[1,2,6,8] The PdL$_2$ complexes were prepared by a reaction of the phosphine with (η^5-C$_5$H$_5$)(η^3-C$_3$H$_5$)Pd[1,2] or [η^3-(2-methylallyl)PdCl]$_2$.[3] The former reaction may involve an incipient formation of Pd(η^1-C$_3$H$_5$)(η^1-C$_5$H$_5$)L$_2$ (L = phosphines) followed by reductive elimination of organic moieties as C$_8$H$_{10}$.[9] The compound Pd[P($cyclo$-C$_6$H$_{11}$)$_3$]$_2$ was also prepared by removing the ethylene molecule from Pd(C$_2$H$_4$)[P($cyclo$-C$_6$H$_{11}$)$_3$]$_2$[7] or by reduction of Pd(acac)$_2$ (acac = acetylacetonato) with AlEt$_3$ in the presence of the phosphine.[4] The preparative procedure described here is a slight modification of the first method.[1,2]

The platinum analogues PtL$_2$ [L = PPh(t-Bu)$_2$ or P($cyclo$-C$_6$H$_{11}$)$_3$] may be prepared by reduction of the corresponding dichloro compounds with sodium amalgam or sodium naphthalene.[2] The compound Pt[P($cyclo$-C$_6$H$_{11}$)$_3$]$_2$ is also accessible from Pt(cod)$_2$ (cod = 1,5-cyclooctadiene) and P($cyclo$-C$_6$H$_{11}$)$_3$[5] or from Pt(η^3-allyl)[P($cyclo$-C$_6$H$_{11}$)$_3$]$_2$[+] and t-BuONa.[6]

■**Caution.** *All the phosphines, (η^5-C$_5$H$_5$)(η^3-C$_3$H$_5$)Pd, and bicoordinate complexes are air sensitive. The phosphines and (η^5-C$_5$H$_5$)(η^3-C$_3$H$_5$)Pd are malodorous materials with unknown physiological effects. Therefore all the manipulations described here should be carried out under a dry nitrogen atmosphere using Schlenk-tube techniques[10,11] and in a well ventilated hood. A hypodermic syringe is employed for weighing and transferring PPh(t-Bu)$_2$ and P(t-Bu)$_3$. All the solvents are dried with sodium metal (except methanol) and distilled under nitrogen.*

A. BIS(DI-*tert*-BUTYLPHENYLPHOSPHINE)PALLADIUM(0)

$$(\eta^5\text{-C}_5\text{H}_5)(\eta^3\text{-C}_3\text{H}_5)\text{Pd} + 2\text{PPh}(t\text{-Bu})_2 \rightarrow \text{Pd}[\text{PPh}(t\text{-Bu})_2]_2 + \text{C}_8\text{H}_{10}$$

Procedure

A 50-mL Schlenk flask containing a magnetic stirring bar is charged with (η^3-allyl)(η^5-cyclopentadienyl)palladium[12] (0.40 g, 1.89 mmole), toluene (10 mL), and di-*tert*-butylphenylphosphine[13] (0.89 g, 4.00 mmole). The deep-red solution is heated at 70-75° for 1 hour. The resulting pale-brown solution is concentrated *in vacuo* to dryness and the pale-brown crystals that separate are washed with methanol (five 3-mL portions). The crude crystals are dissolved in hot hexane (25 mL). The solution is filtered and the filtrate is concentrated *in vacuo* to a quarter of the original volume to give pale-yellow crystals. For complete crystallization the solution is cooled overnight at −35°. The mother liquor is removed with a syringe, and the crystals are washed at −35° with cold hexane (two 5-mL portions) and dried *in vacuo*. Yield 0.91 g (86%), mp 160-162°

(under N_2 in a sealed capillary tube). *Anal.* Calcd. for $C_{28}H_{46}P_2Pd$: C, 61.06; H, 8.41. Found: C, 60.81; H, 8.46.

Properties

The properties of bis(di-*tert*-butylphenylphosphine)palladium(0) are described with those of the other bicoordinates complexes (see below).

B. BIS(TRICYCLOHEXYLPHOSPHINE)PALLADIUM(0)

$$(\eta^5\text{-}C_5H_5)(\eta^3\text{-}C_3H_5)Pd \; + \; 2P(cyclo\text{-}C_6H_{11})_3 \; \rightarrow \; Pd[P(cyclo\text{-}C_6H_{11})_3]_2 \; + \; C_8H_{10}$$

Procedure

In a 50-mL Schlenk flask containing a magnetic stirring bar are placed (η^3-allyl)(η^5-cyclopentadienyl)palladium[12] (0.34 g, 1.60 mmole) and a toluene solution (15 mL) of tricyclohexylphosphine*[14] (0.99 g, 3.54 mmole). The dark-red mixture is stirred with heating at 75-80° for 3 hours. The brown solution is concentrated *in vacuo* to dryness. The brown crystalline solid is washed with MeOH (two 10-mL portions) to remove a slight excess of the phosphine. The solid is dissolved in hot toluene (5 mL), and methanol (5 mL) is added to give crystals. After standing in a freezer (−35°) overnight, the crystals are isolated by removing the mother liquor with a syringe, washed with MeOH (five 2-mL portion), and dried *in vacuo*. The off-white crystals thus obtained are pure enough to prepare organopalladium complexes. Yield: 0.84 g (79%). Analytically pure, colorless crystals can be obtained by recrystallization from a toluene (5 mL)-methanol (5 mL) mixture. Mp 185-189° (under N_2 in a sealed capillary tube). *Anal.* Calcd. for $C_{36}H_{66}P_2Pd$: C, 64.79; H, 9.99. Found: C, 64.76; H, 9.97.

Properties

The properties of bis(tricyclohexylphosphine)palladium(0) are described with those of the other bicoordinate complexes (see below).

C. BIS(TRI-*tert*-BUTYLPHOSPHINE)PALLADIUM(0)

$$(\eta^5\text{-}C_5H_5)(\eta^3\text{-}C_3H_5)Pd \; + \; 2P(t\text{-}Bu)_3 \; \rightarrow \; Pd[P(t\text{-}Bu)_3]_2 \; + \; C_8H_{10}$$

Procedure

This compound is prepared by a procedure similar to the one described above

---

*Tricyclohexylphosphine is available from Strem Chemicals Inc., Box 212, Danvers, MA 01923.

for the di-*tert*-butylphenylphosphine compound, employing $(\eta^3$-allyl)(η^5-cyclo-pentadienyl)palladium[12] (0.21 g, 1.0 mmole) and tri-*tert*-butylphosphine[15] (0.46 g, 2.3 mmole).* The product is obtained in 60% yields as colorless crystals. Mp 150-153° (dec. in air). *Anal.* Calcd. for $C_{24}H_{54}P_2Pd$: C, 56.40; H, 10.67. Found: C, 56.62; H, 10.73.

Properties

The properties of bis(tri-*tert*-butylphosphine)palladium(0) are described with those of the other bicoordinate complexes (see below).

D. BIS(DI-*tert*-BUTYLPHENYLPHOSPHINE)PLATINUM(0)

$$K_2PtCl_4 + 2PPh(t\text{-Bu})_2 \rightarrow trans\text{-}PtCl_2[PPh(t\text{-Bu})_2]_2 + 2KCl$$

$$trans\text{-}PtCl_2[PPh(t\text{-Bu})_2]_2 + 2Na(Hg) \rightarrow Pt[PPh(t\text{-Bu})_2]_2 + 2NaCl$$

Procedure

A 100-mL Schlenk flask containing a magnetic stirring bar is charged successively with K_2PtCl_4 (1.0 g, 2.3 mmole), deoxygenated water (5 mL), EtOH (10 mL), and di-*tert*-butylphenylphosphine (1.06 g, 4.8 mmole). The mixture is stirred at room temperature for 40 hours, and the colorless solid is filtered in air, washed successively with H_2O (10 mL) and EtOH (20 mL), and dried *in vacuo*. A yield of 1.6 g (98%) is obtained. The crude *trans*-[dichlorobis(di-*tert*-butyl-phenylphosphine)platinum(II)] [16] (0.95 g, 1.34 mmoles) thus obtained, 1% sodium amalgam (50 g), and tetrahydrofuran (15 mL) are placed successively in a 100-mL Schlenk flask containing a magnetic stirring bar. The mixture is stirred vigorously at room temperature for 22 hours. The gray suspension is transferred with a syringe to a filtration funnel[17] fitted to a 100-mL Schlenk flask and filtered through a filter paper. The sodium amalgam is washed with hexane (15 mL). The combined filtrate and washings are concentreated *in vacuo* to dryness. The solid residue is dissolved in hexane (30 mL) and the solution is filtered. The pale-yellow filtrate is concentrated to one-third of the original volume. After standing in a freezer ($-35°$), the colorless crystals are isolated by removing the solution with a syringe, washed at $-35°$ with cold hexane (5 mL), and dried *in vacuo*. A yield of 0.79 g (92%) is obtained, mp 171-174° (under N_2

---

*Since tri-*tert*-butylphosphine is low melting (mp 30°), it is recommended to weigh the phosphine liquidfied at 50° employing a syringe preheated in a oven (60°). If the phosphine solidifies in the syring, it can be melted by heating with a heat gun.

in a sealed capillary tube). *Anal.* Calcd. for $C_{28}H_{46}P_2Pt$: C, 52.57; H, 7.25. Found C, 52.68; H, 7.05.

Properties

The properties of bis(di-*tert*-butylphenylphosphine)platinum(0) are described with those of the other bicoordinate complexes (see below).

E. BIS(TRICYCLOHEXYLPHOSPHINE)PLATINUM(0)

$$K_2PtCl_4 + 2P(cyclo\text{-}C_6H_{11})_3 \rightarrow trans\text{-}PtCl_2[P(cyclo\text{-}C_6H_{11})_3]_2 + 2KCl$$

$$trans\text{-}PtCl_2[P(cyclo\text{-}C_6H_{11})_3]_2 + 2[C_{10}H_8^-]Na^+ \rightarrow$$

$$Pt[P(cyclo\text{-}C_6H_{11})_3]_2 + 2NaCl + 2C_{10}H_8$$

Procedure*

A 100-mL Schlenk flask containing a magnetic stirring bar is charged with K_2PtCl_4 (1.0 g, 2.3 mmole), deoxygenated water (5 mL), and an ethanol solution (40 mL) of tricyclohexylphosphine[14] (1.5 g, 5.4 mmole). The mixture is stirred at room temperature for 15 hours, and the colorless solid is filtered, washed successively with H_2O (10 mL) and EtOH (20 mL) in air, and dried *in vacuo*. A yield of 1.8 g (95%) is obtained. The crude *trans*-PtCl$_2$[P(*cyclo*-C$_6$H$_{11}$)$_3$]$_2$[2] (1.0 g, 1.2 mmole) thus obtained is placed in a 50-mL Schlenk flask containing a magnetic stirring bar, and a 0.33 *M* tetrahydrofuran solution (15 mL) of sodium naphthalene prepared from sodium (0.5 g) and naphthalene (2.2 g) in THF (50 mL) is added. The mixture is stirred at room temperature for 5 hours and the resulting brownish-green solution is concentrated *in vacuo*. The dark-brown solid residue is extracted with hot hexane (twenty 2-mL portions) (50-55°) and the extract is transferred into a filtration funnel[17] fitted with a sublimation apparatus and filtered through a filter paper. (■**Caution.** *The residue, which is insoluble in hexane, is pyrophoric and should be treated with EtOH under a nitrogen atmosphere before it is discarded.*) The filtrate is concentrated *in vacuo* and the resulting solid is heated at 50-70° *in vacuo* (10^{-3} torr) for 10 hours to remove the naphthalene by sublimation. It is then dissolved in hot hexane (30 mL) and the solution is filtered as above. Concentration of the filtrate *in vacuo* to 5 mL gives pale-yellow crystals. After standing in a freezer

---

*Sodium amalgam can also be employed as a reducing agent. However, in this case a prolonged heating (50 hr at 55-60°) is required because of insolubility of *trans*-PtCl$_2$[P(*cyclo*-C$_6$H$_{11}$)$_3$]$_2$ in tetrahydrofuran.

($-35°$) overnight, the crystals (0.40 g) are isolated by removing the mother liquor with a syringe, washed with hexane (three 2-mL portions), and dried *in vacuo*. Additional crystals (0.10 g) are obtained on concentration of the combined mother liquor and washings to 1 mL. Total yield 55%.* Colorless crystals are obtained by recrystallization from hexane. Mp 204-208° (under N_2 in a sealed capillary tube). *Anal.* Calcd. for $C_{36}H_{66}P_2Pt$: C, 57.19; H, 8.80. Found: C, 57.11; H, 8.98.

Properties

The bicoordinate complexes described here are soluble in benzene and hexane. The $Pd[P(t\text{-}Bu)_3]_2$ complex is stable in air in the solid state, whereas the other complexes are unstable and give the dioxygen complexes MO_2L_2[18] [M = Pd, Pt; L = $PPh(t\text{-}Bu)_2$, $P(cyclo\text{-}C_6H_{11})_3$]. In the case of palladium complexes, the formation of dioxygen complexes is readily detectable by the developement of a pale-green color. All of the two-coordinate complexes can be stored under dry nitrogen for more than a year. The linear structure of the $P(t\text{-}Bu)_3$ and $PPh(t\text{-}Bu)_2$ complexes is readily deducible from a 1:2:1 triplet of the *tert*-butyl proton signal (Table I). Mass spectra of PtL_2 [L - $PPh(t\text{-}Bu)_2$ or $P(cyclo\text{-}C_6H_{11})_3$] show the corresponding parent and fragment ions. $[M\text{-}(R\text{-}1)]^+$, $[M\text{-}2(R\text{-}1)]^+$, $[M\text{-}3(R\text{-}1)]^+$, and MP_2^+, where R is the alkyl substituent of the phosphine.[2] As expected from the high degree of coordinative

Table I ^{1}H NMR Spectra of Two-Coordinate Complexes

	Chemical Shift (ppm, Me$_4$Si)^a	$^3J_{H\text{-}P}$ $+^5J_{H\text{-}P}$	Area	Assignment
$Pd[P(t\text{-}Bu)_3]_2$	1.51 (t)	12.0		t-Bu
$Pd[PPh(t\text{-}Bu)_2]_2{}^b$	1.48 (t)	12.7	9	t-Bu
	8.40 (m)^c		1	o-H
	6.94-7.30 (m)		d	m- and p-H
$Pd[P(cyclo\text{-}C_6H_{11})_3]_2$	0.70-2.60 (m)			$cyclo\text{-}C_6H_{11}$
$Pt[PPh(t\text{-}Bu)_2]_2{}^b$	1.56 (t)	13.5	9	t-Bu
	8.46 (m)^c		1	o-H
	6.90-7.30 (m)		d	m- and p-H
$Pt[P(cyclo\text{-}C_6H_{11})_3]_2$	0.70-2.60 (m)			$cyclo\text{-}C_6H_{11}$

^{a}Measured in benzene-d_6 at 22.5°.
^{b}Measured in toluene-d_7 at 22.5°.
^{c}At $_{\mp}71°$ the ortho proton signals of $Pd[PPh(t\text{-}Bu)_2]_2$ and $Pt[PPh(t\text{-}Bu)_2]_2$ are observed at δ 7.55 (m), 9.33 (m) and 7.40 (m), 9.38 ppm (m), respectively.

*The checkers found that for unknown reasons only two of five experiments gave this compound. In the other experiments they got $Pt[P(cyclo\text{-}C_6H_{11})_3]_3$ with metallic palladium.

unsaturation, the bicoordinate complexes, particularly the platinum complexes, show an enhanced reactivity toward small molecules and weak protonic acids, for example, alcohol, and π-acids like maleic anhydride.[18]

References

1. M. Matsumoto, H. Yoshioka, K. Nakatsu, T. Yoshida, and S. Otsuka, *J. Am. Chem. Soc.,* **96**, 3322 (1974).
2. S. Otsuka, T. Yoshida, M. Matsumoto, and K. Nakatsu, *ibid.,* **98**, 5850 (1976).
3. A. Musco, W. Kuran, A. Silvani, and M. W. Anker, *Chem. Commun.,* **1973**, 938.
4. K. Kudo, M. Hidai, and Y. Uchida, *J. Organomet. Chem.,* **56**, 413 (1973).
5. M. Green, J. A. Howard, J. L. Spencer, and F. G. A. Stone, *Chem. Commun.,* **1975**, 3.
6. A. Immirzi, A. Musco, and P. Zambelli, *Inorg. Chim. Acta.,* **13**, L13 (1975).
7. R. van der Linde and R. O. der Jongh, *J. Chem. Soc., D,* **1971**, 563.
8. A. Immirzi and A. Musco, *Chem. Commun.,* **1974**, 400.
9. H. Werner and A. Kühn, *Angew. Chem. Int. Ed.,* **16**, 412 (1977).
10. D. F. Shriver, *The Manipulation of Air-Sensitive Compounds,* McGraw-Hill Book Co., New York, 1969.
11. R. B. King, in *Organometallic Syntheses,* Vol. 1, J. J. Eisch and R. B. King (eds.), Academic Press Inc., New York, 1965.
12. Y. Tatsuno, T. Yoshida, and S. Otsuka, *Inorg. Synth.,* **19**, 220 (1979).
13. B. E. Mann, B. L. Shaw, and R. M. Slade, *J. Chem. Soc., A,* 2976 (1971).
14. K. Issleib and A. Brack, *Z. Anorg. Allg. Chem.,* **277**, 258 (1954).
15. H. Hofmann and P. Schellenbeck, *Chem. Ber.,* **100**, 692 (1967).
16. A. J. Cheney, B. E. Mann, B. L. Shaw, and R. M. Slade, *J. Chem. Soc., A,* **1971**, 3833.
17. A filtration funnel employed for preparation of $(\eta^5\text{-}C_5H_5)(\eta^3C_3H_5)Pd^{12}$ is satisfactory, see p. 222 of this volume.
18. S. Otsuka and T. Yoshida, *J. Am. Chem. Soc.,* **99**, 2134 (1977).

22. THREE-COORDINATE PHOSPHINE COMPLEXES OF PLATINUM(0)

Submitted by T. YOSHIDA*, T. MATSUDA*, and S. OTSUKA*
Checked by G. W. PARSHALL † and W. G. PEET†,
P. BINGER‡, A. BRINKMANN‡, and P. WEDEMANN‡

Tris(triethylphosphine)platinum was originally prepared by vacuum thermolysis

*Department of Chemistry, Faculty of Engineering Science, Osaka University, Toyonaka, Osaka, Japan 560.

†Central Research Department, E. I. duPont de Memours & Co., Wilmington, DE 19898; checked Section A.

‡Max-Planck-Institute für Kohlenforschung, D-4330 Mülheim/Ruhr, Germany; checked Section B.

of Pt(PEt$_3$)$_4$, but the latter was synthesized from Pt(B$_3$H$_7$)(PEt$_3$)$_2$, which is an inconvenient starting material.[1,2] A direct synthesis from readily available starting materials is the reaction of K$_2$PtCl$_4$ with KOH and PEt$_3$ in alcohol,[3] but the product is difficult to isolate from the reaction mixture in pure form. The procedure described here is based on the vacuum thermolysis of Pt(PEt$_3$)$_4$, as obtained by a procedure outlined in this volume.[4]

The triisopropylphosphine analogue, Pt[P(i-Pr)$_3$]$_3$, is obtained by reducing *trans*-PtCl$_2$[P(i-Pr)$_3$]$_2$ with sodium amalgam in the presence of P(i-Pr)$_3$[5] as described here.

■**Caution.** *All zero-valent platinum compounds and trialkylphosphines employed here are extremely air sensitive and should be handled in a dry nitrogen or argon atmosphere. The trialkylphosphines are malodorous and toxic, and should be handled with care, in a well-ventilated hood. All solvents should be dried (except ethanol) and distilled under nitrogen.*

A. TRIS(TRIETHYLPHOSPHINE)PLATINUM(0)

$$Pt(PEt_3)_4 \xrightarrow{\text{in vacuo}} Pt(PEt_3)_3 + PEt_3$$

Procedure

A 15-mL Schlenk flask is evacuated and refilled with nitrogen three times. Tetrakis(triethylphosphine)platinum[4] (0.66 g, 1 mmole) is charged by the Schlenk-tube techniques.[6] Under a nitrogen flow the flask is connected to a vacuum line through a liquid nitrogen U-trap. The flask is heated at 50-60° at reduced pressure (5 torr) for 6 hours to give an orange-red viscous oil. Yield 0.49 g (90%). *Anal.* Calcd. for C$_{18}$H$_{45}$P$_3$Pt: C, 39.3; H, 8.3. Found: C, 38.9; H, 8.2.

Properties

The properties of tris(triethylphosphine)platinum(0) are described with those of tris(triisopropylphosphine)platinum(0).

B. TRIS(TRIISOPROPYLPHOSPHINE)PLATINUM(0)

$$K_2PtCl_4 + 2P(i\text{-}Pr)_3 \rightarrow trans\text{-}PtCl_2[P(i\text{-}Pr)_3]_2 + 2KCl$$

$$trans\text{-}PtCl_2[P(i\text{-}Pr)_3]_2 + P(i\text{-}Pr)_3 + 2Na(Hg) \rightarrow Pt[P(i\text{-}Pr)_3]_3 + 2NaCl$$

Procedure

To a 50-mL nitrogen-flushed Schlenk flask containing a magnetic stirring bar is added K_2PtCl_4 (0.50 g, 1.2 mmole), deoxygenated water (3 mL), $P(i-Pr)_3$[7] (0.42 g, 2.6 mmole), and ethanol (3 mL). The mixture is stirred at room temperature for 2 hours and the resulting colorless solid to *trans*-$PtCl_2[P(i-Pr)_3]_2$ is filtered, washed with ethanol, and dried *in vacuo*. (■**Caution.** *The amalgamation of sodium is highly exothermic. Small pieces of sodium must be added to mercury behind a shield.*[8]). The crude *trans*-$PtCl_2[P(i-Pr)_3]_2$[4] (0.6 g, 1.0 mmole) is placed in a 50 mL Schlenk flask containing a stirring bar. A 20-g sample of 1% sodium amalgam and 10 mL of dried tetrahydrofuran[9] containing 0.24 g (1.5 mmole) of $P(i-Pr)_3$ are added successively. The mixture is stirred at room temperature for 10 hours. The red solution is transferred with a syringe into a filtration funnel (see Fig. 1, Reference 10) and filtered through a filter paper. The sodium amalgam is washed with dried and degassed pentane (two 10-mL portions). The combined filtrate and washings are concentrated under reduced pressure (7 torr) to dryness. The solid residue is extracted with pentane (two 10-mL portions) and the extract is filtered as above. The filtrate is concentrated under reduced pressure to a quarter of the original volume. After standing at $-35°$ overnight, the pale-yellow crystals are isolated by removing the solution with a syringe, washing with pentane (three 2-mL portions) at $-78°$, and drying at $-35°$ under reduced pressure (7 torr). Yield 0.33-0.41 g (48-60%), mp 60-62° (under nitrogen in a sealed tube). *Anal.* Calcd. for $C_{27}H_{63}P_3Pt$: C, 47.97; H, 9.39. Found: C, 48.09; H, 9.51.

Properties

The three-coordinate complexes PtL_3 [L = PEt_3, $P(i-Pr)_3$] are extremely unstable toward air and should be kept under dry nitrogen in a freezer. They are readily soluble even in saturated aliphatic hydrocarbons. The ^1H NMR spectrum of $Pt(PEt_3)_3$ measured in benzene-d_6 shows two broad signals at δ 1.76 (CH_2) and 1.16 ppm (CH_3), while that of $Pt[P(i-Pr)_3]_3$ shows signals at δ 1.86 (CH) and 1.24 ppm (CH_3). In contrast to $Pt(PEt_3)_3$, which does not dissociate the coordinate phosphine, $Pt[P(i-Pr)_3]_3$ readily liberates $P(i-Pr)_3$ even in the solid state (K_d = 4.0 × 10^{-2} M in heptane). They are strong nucleophiles and readily react with hydrogen and weak protonic acids, for example, C_2H_5OH and H_2O. With hydrogen $Pt(PEt_3)_3$ gives $PtH_2(PEt_3)_3$,[1] while $Pt[P(i-Pr)_3]_3$ affords *trans*-$PtH_2[P(i-Pr)_3]_2$.[11] Oxidative addition of alcohol to $Pt(PEt_3)_3$ is reversible to give $[PtH(PEt_3)_3]OC_2H_5$,[1] but with $Pt[P(i-Pr)_3]_3$ it is irreversible and gives *trans*-$PtH_2[P(i-Pr)_3]_2$.[11] They add H_2O reversibly to give the strong hydroxo bases $[PtH(PEt_3)_3]OH$[1] and $\{PtH(S)[P(i-Pr)_3]_2\}OH$ (S = solvent).

References

1. D. H. Gerlach, A. R. Kane, G. W. Parshall, J. P. Jesson, and E. L. Muetterties, *J. Am. Chem. Soc.,* **93**, 3543 (1971).
2. L. J. Guggenberger, A. R. Kane, and E. L. Muetterties, *ibid.,* **94**, 5665 (1972).
3. R. G. Pearson, W. Louw, and J. Rajaram, *Inorg. Chim. Acta.,* **9**, 251 (1974).
4. T. Yoshida, T. Matsuda, and S. Otsuka, *Inorg. Synth.,* **19**, 110 (1979).
5. S. Otsuka, T. Yoshida, M. Matsumoto, and K. Nakatsu, *J. Am. Chem. Soc.,* **98**, 5850 (1976).
6. D. F. Shriver, *Manipulation of Air-Sensitive Compounds,* McGraw-Hill Book Co., New York, 1969.
7. A. H. Cowley and M. W. Taylor, *J. Am. Chem. Soc.,* **91**, 2915 (1969).
8. W. B. Renfrow, Jr., and C. R. Hauser, *Org. Synth.,* Coll. Vol. 2, 609 (1943).
9. *Inorg. Synth.,* **12**, 317 (1970).
10. S. Otsuka, T. Yoshida, and Y. Tatsumo, *Inorg. Synth.,* **19**, 222 (1979).
11. S. Otsuka and T. Yoshida, *J. Am. Chem. Soc.,* **99**, 2134 (1977).

23. TETRAKIS(TRIETHYLPHOSPHINE)PLATINUM(0)

$$K_2PtCl_4 + 4PEt_3 + 2KOH + C_2H_5OH \rightarrow$$

$$Pt(PEt_3)_4 + 4KCl + CH_3CHO + 2H_2O$$

Submitted by T. YOSHIDA*, T. MATSUDA*, and S. OTSUKA*
Checked by G. W. PARSHALL[†] and W. G. PEET[†]

Tetrakis(triethylphosphine)platinum(0) has been prepared by two routes: (1) treatment of $Pt(B_3H_7)(PEt_3)_2$ with PEt_3[1,2] and (2) reduction of *cis*-$PtCl_2(PEt_3)_2$ with potassium[3] or sodium amalgam in the presence of PEt_3. The procedure described here is a direct synthesis from K_2PtCl_4, PEt_3, and potassium hydroxide in alcohol that was originally developed by Pearson et al.[4] for the preparation of $Pt(PEt_3)_3$.

Procedure

■**Caution.** *Triethylphosphine and tetrakis(triethylphosphine)platinum(0) are extremely air-sensitive. The phosphine is malodorous and toxic. Therefore all*

*Department of Chemistry, Faculty of Engineering Science, Osaka University, Toyonaka, Osaka, Japan 560.
[†]Central Research Department, E. I. du Pont de Nemours and Co., Wilmington, DE, 19898.

manipulations should be carried out in a nitrogen atmosphere and in a well-ventilated hood. All solvents should be degassed with an inert gas.

A 50-mL Schlenk flask containing a magnetic stirring bar is charged with a solution of KOH (0.7 g. 12.5 mmole) dissolved in a 30:1 EtOH-H_2O mixture (31 mL) and PEt$_3$ (3.0 mL, 20 mmole). To the mixture a solution of K_2PtCl_4 (1.5 g, 3.6 mmole) in H_2O (10 mL) is added dropwise by syringe over a period of 5 minutes. The mixture is stirred at room temperature for 1 hour and then at 60° for 3 hours. The colorless solution is concentrated to dryness *in vacuo* (5 torr) at room temperature under stirring. The reddish, oily-solid residue is extracted with hexane (two 15-mL portions) and the extract is filtered by the Schlenk-flask filtration method.[5] The orange filtrate is concentrated to a quarter of the original volume under reduced pressure (5 torr). The concentrate is treated with PEt$_3$ (0.5 mL, 3.4 mmole) and cooled at −78° (Dry Ice-acetone) for 4 hours. The colorless crystals that separate are isolated by removing the solution with a syringe, washing with hexane (two 3 mL portions) at −78°, and drying *in vacuo* (5 torr) at −40°. Yield: 2.0 g (85%); mp 47-48° (under nitrogen in a sealed capillary tube). *Anal.* Calcd. for $C_{24}H_{60}P_4Pt$: C, 43.2; H, 9.1. Found: C, 42.6; H, 8.9.

Properties

Tetrakis(triethylphosphine)platinum(0) is extremely air sensitive and readily soluble in saturated aliphatic hydrocarbons. The complex can be stored under dry nitrogen in a freezer (−35°) for several months. The complex readily loses one of the coordinated phosphine molecules to give Pt(PEt$_3$)$_3$[6] (dissociation constant (K_d) in heptane is 3.0×10^{-1}). The ^{1}H NMR spectrum measured in benzene-d_6 shows two multiplets at δ 1.56 (CH$_2$) and 1.07 ppm (CH$_3$). Tetrakis-(triethylphosphine)platinum(0) is a strong nucleophile and reacts readily with chlorobenzene and benzonitrile to give σ-phenyl complexes PtX(Ph)(PEt$_3$)$_2$ (X = Cl, CN).[7] Oxidative addition of EtOH affords [PtH(PEt$_3$)$_3$]$^+$.

References

1. D. H. Gerlach, A. R. Kane, G. W. Parshall, J. P. Jesson, and E. L. Muetterties, *J. Am. Chem. Soc.,* 93, 3543 (1971).
2. L. J. Guggenberger, A. R. Kane, and E. L. Muetterties, *ibid.,* 94, 5665 (1972).
3. R. A. Schunn, *Inorg. Chem.,* 15, 208 (1976).
4. R. G. Pearson, W. Louw, and J. Rajaram, *Inorg. Chim. Acta.,* 9, 251, (1974).
5. D. F. Shriver, *Manipulation of Air-Sensitive Compounds,* McGraw-Hill Book Co., New York, 1969.
6. T. Yoshida, T. Matsuda, and S. Otsuka, *Inorg. Synth.,* 19, 108 (1979).
7. G. W. Parshall, *J. Am. Chem. Soc.,* 96, 2360 (1974).

24. BARIUM TETRACYANOPLATINATE(II) TETRAHYDRATE*

$$2Ba(CN)_2 + Pt + 6H_2O \xrightarrow{\text{alternating current}}$$

$$Ba[Pt(CN)_4] \cdot 4H_2O + Ba(OH)_2 + H_2 \uparrow$$

Submitted by ROBERT L. MAFFLY[†] and JACK M. WILLIAMS[‡]
Checked by R. N. RHODA [§]

The yellow-green compound $Ba[Pt(CN)_4] \cdot 4H_2O$ is highly useful in the syntheses of a series of partially oxidized tetracyanoplatinates. The appropriate metal sulfate is added to precipitate $BaSO_4$ and obtain aqueous solutions of the metal platinum cyanide.[1-2] The procedure described here, first reported by A. Brochet and J. Petit,[3] is suitable for the rapid direct synthesis of very pure $Ba[Pt(CN)_4] \cdot 4H_2O$ with a considerable savings of time and effort. More important, however, is the fact that this electrolytic procedure at no time involves a solution containing potassium ion, which, because of the stability of its partially oxidized platinum compounds $K_{1.75}[Pt(CN)_4] \cdot 1.5H_2O$ and $K_2[Pt(CN)_4]Br_{0.33} \cdot 3.0H_2O$, is always an impurity of major concern.

Procedure

The $Ba(CN)_2$ used** contains metal cation impurities limited to sodium, aluminum, strontium, and 0.05% potassium, plus traces of iron, magnesium and lithium.[#] Each platinum electrode is a 90-100 cm^2 heavy sheet with purity $>$ 99%. During the electrolysis, half of each electrode is submerged in the solution. All water used is distilled, and all filters are medium-pore fritted-glass filters. The electrolysis apparatus consists of a variable ac voltage supply with an ac ammeter included in the circuit.

*Research performed under the auspices of the Division of the Basic Energy Sciences of the U.S. Department of Energy.
†Research participant sponsored by the Argonne Center for Educational Affairs from Whitman College, Walla Walla, WA.
‡Correspondent: Chemistry Division, Argonne National Laboratory, Argonne, IL 60439.
§ Paul D. Merica Research Laboratory, International Nickel Co., Inc., Sterling Forest-Suffern, NY 10901.
**Available from ICN-K&K Laboratories, Inc., 121 Express St., Plainview, NY 11803.
#We wish to thank J. P. Faris for the emission spectrographic work performed at Argonne National Laboratory, Argonne IL.

■**Caution.** *Because of the extremely poisonous nature of cyanide, these steps should be carried out in a well-ventilated fume hood using protective gloves, clothing, and face shield. Although the magnetic stirrer and voltage regulator should be adequately grounded, care should be taken not to come in contact with the electrodes or connecting wire while the reaction is in progress.*

Initially, 200 g (1.06 moles) $Ba(CN)_2$ is placed in a 1000-mL beaker, 600 mL of H_2O is added, and stirring is initiated by means of a magnetic stirrer. Two platinum electrodes of known weight are introduced into the mixture, typically separated by 7 cm, and an alternating current is applied such that the current remains at 5 A and the applied voltage somewhere between 15 and 20 v. The volume is kept at 600 mL by the addition of water and the current is kept at 5-6 A through this part of the procedure for approximately 45-50 hours. The mixture ranges from a white to a yellow-green color as the platinum is oxidiz ˑd.

The electrolysis is discontinued when the mixture visibly begins to change to a grayish color as a result of the formation of platinum black. After the platinum electrodes are disconnected, they are washed in concentrated HCl until clean, and weighed (Pt loss = 30-35 g).

The murky solution is stirred, heated to boiling, and immediately filtered. The filtrate is transferred to a clean 1000-mL beaker and allowed to cool overnight, resulting in yellow-green $Ba[Pt(CN)_4] \cdot 4H_2O$ crystals. The beaker is placed in an ice bath for 1 hour to cause remaining $Ba[Pt(CN)_4] \cdot 4H_2O$ to crystallize. The crystals are collected by filtration, and the filtrate is saved. The saved filtrate is evaporated to 300 mL on a steam bath and immediately filtered hot. This solution is allowed to cool in an ice bath, resulting in another crystallization of $Ba[Pt(CN)_4] \cdot 4H_2O$. These crystals are collected on a filter and added to those from the first isolation. The filtrate is disposed of in a waste platinum receptacle. The material is purified by adding 4 mL of H_2O per 10 g of solid and heated with stirring for 15 minutes in a boiling water bath. The beaker is then placed in an ice bath for 30 minutes and the crystals are filtered out. This procedure is repeated four times. The highly pure product* represents 90-100% yield based on the platinum lost from the electrodes.

Properties

The compound $Ba[Pt(CN)_4] \cdot 4H_2O$ is a colorless crystalline material with a measured density of 3.09 g/mL (calculated density 3.13 g/mL). The first 12

---

*Typical impurity levels as measured by emission spectrographic analysis are: Na, 0.03-0.1%; Sr, 0.01-0.1%; Al, 0.01%; K not detected (<0.001%).

d-spacings from X-ray powder diffraction are: 8.70 (mw), 6.90 (m), 5.80 (s), 4.60 (mw), 4.40 (mw), 4.30 (m), 3.73 (w), 3.21 (ms), 2.96 (ms), 2.91 (ms), and 2.77 (mw) Å.

References

1. T. F. Cornish and J. M. Williams, *Inorg. Synth.,* **19**, 10 (1979).
2. R. L. Maffly, J. A. Abys, and J. M. Williams, *Inorg. Synth.,* **19**, 6 (1979).
3. A. Brochet and J. Petit, *C. R.,* **138**, 1095 (1904).

25. TRANS PHOSPHINE COMPLEXES OF PLATINUM(II) CHLORIDE

Submitted by CHAO-YANG HSU,* BRIAN T. LESHNER,* and MILTON ORCHIN*
Checked by MICHEL LAURENT†

Heretofore, the most common method for the preparation of the useful platinum(II) complexes of the type [PtCl$_2$L$_2$], where L is a tertiary phosphine, consisted of the reaction between potassium tetrachloroplatinate(II) and tertiary phosphines.[1] When trialkylphosphines are used, the reaction usually leads to a mixture of cis- and trans isomers,[2,3] and when triarylphosphines are employed[4] only the cis isomers are obtained. The preparation of pure trans complexes by a simple, convenient procedure is highly desirable.

trans-[Dichlorobis(triphenylphosphine)platinum(II)] has been prepared by the reaction of *trans*-[chlorohydridobis(triphenylphosphine)platinum(II)] with mercury(II) chloride,[5] by photochemical isomerization of the cis isomer,[6] and by the oxidative elimination of chlorine from tetrakis(triphenylphosphine)-platinum(0) under carefully controlled reaction conditions.[7] *trans*-[Dichlorobis-(tributylphosphine)platinum(II)] has been prepared by the thermal isomerization of the corresponding cis isomer.[3]

The following procedures describe the direct preparations of *trans*-[PtCl$_2$\{P(C$_6$H$_5$)$_3$\}$_2$] and *trans*-[PtCl$_2$[P(n-C$_4$H$_9$)$_3$]$_2$] by the reaction between the appropriate tertiary phosphine and potassium trichloro(ethylene)-platinate(II)**

*Department of Chemistry, University of Cincinnati, Cincinnati, OH 45221.
†Department of Chemistry, Northwestern University, Evanston, IL 60201.
**We thank Engelhardt Industries, Inc. for a generous supply of platinum.

A. *trans*-[DICHLOROBIS(TRIPHENYLPHOSPHINE)PLATINUM(II)]

$$K[PtCl_3(C_2H_4)] + 2P(C_6H_5)_3 \rightarrow$$

$$trans\text{-}[PtCl_2\{P(C_6H_5)_3\}_2] + KCl + C_2H_4$$

Procedure

In a 50-mL, round-bottomed flask containing a Teflon-coated magnetic stirring bar are placed 0.77 g (2.0 mmole) of anhydrous potassium trichloro(ethylene)-platinate(II) and 15 mL of acetone. This anhydrous complex may be prepared according to the procedure reported by Hartley.[8] An alternate procedure may also be used.[9] On stirring, a clear yellow solution is formed, whereupon a solution of 1.0 g (3.8 mmole) of triphenylphosphine in 5 mL of acetone is added dropwise (conveniently by syringe) over about 5 minutes. (Because trans isomers are readily isomerized to cis isomers by excess phosphine or by elevated temperature, the phosphines are chosen as the limiting reagents and the reaction is conducted at room temperature.) Light-yellow crystals separate immediately upon the addition of phosphine. After stirring for about 10 minutes, the crystals are filtered by suction and washed successively with 10-mL portions of water, ethanol, and diethyl ether. The resulting pale-yellow crystals are dried *in vacuo* overnight. Yield: 1.4 g (91%, based on triphenylphosphine); mp: 312-314° (dec.). *Anal.* Calcd. for $PtCl_2[P(C_6H_5)_3]_2$: C, 54.69; H, 3.83; Cl, 8.97. Found: C, 54.49; H, 3.91; Cl, 8.84.

Properties

$trans\text{-}[PtCl_2\{P(C_6H_5)_3\}_2]$ is a pale-yellow crystalline solid, stable in air. It is soluble in benzene, chloroform, and dichloromethane and insoluble in alcohols, acetone, and water. It can be recrystallized from benzene/methanol. The infrared spectrum of the trans isomer shows only one Pt-Cl stretching band, $v = 344$ cm^{-1}, whereas the spectrum of the cis isomer shows two corresponding absorptions at 319 and 295 cm^{-1} (Nujol mull).[10]

B. *trans*-[DICHLOROBIS(TRIBUTYLPHOSPHINE)PLATINUM(II)]

$$K[PtCl_3(C_2H_4)] + 2P(n\text{-}C_4H_9)_3 \rightarrow$$

$$trans\text{-}[PtCl_2\{P(n\text{-}C_4H_9)_3\}_2] + KCl + C_2H_4$$

Procedure

■**Caution.** *$P(n\text{-}C_4H_9)_3$ is toxic and flammable.*

To a stirring solution of 0.77 g (2.0 mmole) of potassium trichloro(ethylene)-platinate(II) in 10 mL of methanol is added dropwise (with syringe) 0.80 g (0.98 mL or 3.9 mmole) of tributylphosphine over a period of about 3 minutes, and the reaction solution is stirred for about 10 minutes. The precipitated potassium chloride is filtered and washed with two 5-mL portions of methanol. The filtrate and washings are transferred to a 50-mL, round-bottomed flask, and the volume is reduced to 5 mL under reduced pressure (rotatory evaporator). The remaining solution, which may contain some crystals, is cooled in an ice-water bath for about 30 minutes. The yellow crystals are then filtered and air-dried. The crystals are further purified by grinding them under water to remove occluded potassium chloride. The crystals are again filtered and then dried *in vacuo* overnight. Yield 0.66 g (50%, based on tributylphosphine); mp: 64.5-65.5°. The conversion is about 85% based on the volume of ethylene evolved, but because of the high solubility of the complex the yield of isolated material is considerably lower. *Anal.* Calcd. for $PtCl_2[P(C_4H_9)_3]_2$: C, 42.98; H, 8.12; Cl, 10.57. Found: C, 42.91; H, 8.24; Cl, 10.48.

Properties

trans-$[PtCl_2\{P(n\text{-}C_4H_9)_3\}_2]$ is a yellow crystalline solid. The melting point is appreciably lowered by traces of impurities.[3] This complex is highly soluble in benzene, acetone, ethanol, and most other organic solvents. The infrared spectrum of the cis isomer shows two absorptions $\nu_{(Pt-Cl)}$ at 308 and 285 cm^{-1}, whereas that of the trans isomer has only one absorption $\nu_{(Pt-Cl)}$ = 338 cm^{-1} (Nujol mull).

References

1. K. A. Jensen, *Z. Anorg. Allgem. Chem.*, **229**, 242 (1936).
2. G. W. Parshall, *Inorg. Synth.*, **12**, 27 (1970).
3. G. B. Kauffman and L. A. Teter, *Inorg. Synth.*, **7**, 245 (1963).
4. J. C. Bailar, Jr. and H. Itatani, *Inorg. Chem.*, **4**, 1618 (1965).
5. A. D. Allen and M. C. Baird, *Chem. Ind. (Lond.)*, **1965**, 139.
6. S. H. Mastin and P. Haake, *Chem. Commun.*, **1970**, 202.
7. T. W. Lee and R. C. Stoufer, *J. Am. Chem. Soc.*, **97**, 195 (1975).
8. F. R. Hartley, *Organomet. Chem. Rev. A*, **6**, 119 (1970).
9. P. B. Chock, J. Halpern, and F. E. Paulik, *Inorg. Synth.*, **14**, 90 (1973).
10. D. M. Adams and P. J. Chandler, *J. Chem. Soc., A*, **1969**, 588.

26. TRIS(ETHYLENEDIAMINE)RUTHENIUM(II) AND TRIS(ETHYLENEDIAMINE)RUTHENIUM(III) COMPLEXES

Submitted by P. J. SMOLENAERS* and J. K. BEATTIE*
Checked by J. N. ARMOR†

Although resolution of the tris(ethylenediamine)ruthenium(III) ion is mentioned in an obituary of Werner,[1] no synthesis of any tris(ethylenediamine)ruthenium species appears to have been published until that of $[Ru(en)_3][ZnCl_4]$ was described by Lever and Bradford in 1964.[2] Their method of reducing hydrated ruthenium(III) chloride with zinc dust in an aqueous solution of ethylenediamine remains the general method of preparation of $[Ru(en)_3]^{2+}$ salts.[3,4] The use of zinc as the reducing agent generally leads to the isolation of the relatively insoluble tetrachloro- or tetrabromozincate salts; contamination with $Zn(en)_3^{2+}$ salts can occur. To obtain a more soluble salt, Beattie and Elsbernd[5] prepared $[Ru(en)_3][ZnBr_4]$, complexed the Zn^{2+} with ethylenediaminetetraacetic acid, and isolated $[Ru(en)_3]Br_2$. This procedure has undesirable features, however, and is unsuitable for preparation of the even more soluble chloride salt. In the procedure described here, zinc is removed by precipitation of zinc dianthranilate [zinc bis(2-aminobenzoate)] and $[Ru(en)_3]Cl_2$ is isolated by the addition of acetone in which any excess lithium anthranilate reagent is soluble. From the very soluble chloride other less soluble salts can readily be obtained by metathesis in aqueous solution.

The tris(ethylenediamine)ruthenium(III) species is obtained by oxidation of $[Ru(en)_3]^{2+}$ with, for example, iodine[4] or bromine.[6] The oxidizing agent and conditions employed must be chosen carefully to avoid further oxidation of the ethylenediamine ligand to coordinated diimine.[7] In the present procedure solid silver anthranilate is used to oxidize $[Ru(en)_3][ZnCl_4]$, and $[Ru(en)_3]Cl_3$ is isolated. In this heterogeneous procedure the desired $[Ru(en)_3]Cl_3$ is the only soluble product and can easily be separated from the insoluble silver, silver chloride, and zinc dianthranilate. Other less soluble $[Ru(en)_3]^{3+}$ compounds can be obtained easily from the soluble chloride.

*School of Chemistry, University of Sydney, N.S.W. 2006, Australia.
†Allied Chemical Corp., Chemical Research Center, P.O. Box 1021 R, Morristown, NJ 07960.

A.   TRIS(ETHYLENEDIAMINE)RUTHENIUM(II) TETRACHLOROZINCATE

$$RuCl_3 \cdot nH_2O \; + \; en \xrightarrow{\;Zn\;} [Ru(en)_3](ZnCl_4)$$

■**Caution.** *Ethylenediamine has an irritating vapor, is harmful by skin absorption, and is flammable. Breathing of vapor and contact with skin and eyes should be avoided.*

Procedure

A 100-mL flask containing 2.0 g (approximately 7.6 mmole) of hydrated ruthenium trichloride and 30 mL (about 125 mmole) of 25% aqueous ethylenediamine solution is purged with argon and the mixture is then gently refluxed for about 50 minutes. A total of about 1 g (15 mmole) of powdered zinc is added in small amounts during this time. After the mixture becomes bright orange, it is heated at reflux for 30 minutes more with further additions of zinc dust. The hot mixture is filtered in an argon atmosphere using Schlenk techniques, the filtrate is cooled in ice, and ice-cold, deoxygenated, concentrated HCl is added slowly by syringe until pH 2 is obtained (approximately 18 mL of 10 M acid is used). The resulting yellow-brown precipitate is collected on a filter under an inert atmosphere and washed with deoxygenated ethanol and deoxygenated, peroxide-free diethyl ether, yielding 3.0 g (80%). This crude product can be recrystallized under an argon atmosphere by dissolving in the minimum amount (about 50 mL) of warm deoxygenated 0.01 M trifluoroacetic acid, cooling the solution, and adding acetone to precipitate the pale-yellow complex, which is washed with ethanol and diethyl ether and dried on a vacuum line for 8 hours. *Anal.* Calcd. for $[Ru(C_2H_8N_2)_3][ZnCl_4]$: C, 14.75; H, 4.95; N, 17.20; Cl, 29.03; Ru, 20.68. Found: C, 14.96; H, 4.98; N, 17.10; Cl, 28.82; Ru, 20.42.

B.   TRIS(ETHYLENEDIAMINE)RUTHENIUM(II) CHLORIDE

$$[Ru(en)_3][ZnCl_4] \; + \; 2Li(an) \longrightarrow$$

$$[Ru(en)_3]Cl_2 \; + \; Zn(an)_2 \; + \; 2LiCl$$

(an = anthranilate = 2-aminobenzoate)

Procedure

A solution of lithium anthranilate (0.82 M) is prepared by adding a solution of 0.206 g (4.92 mmole) of LiOH·H$_2$O in 6.0 mL of H$_2$O to an equivalent amount of solid anthranilic acid (0.674 g, 4.92 mmole). The resulting solution is deoxygenated and kept in the dark to avoid decomposition. Excess LiOH must be avoided; the solution should be neutral or slightly acidic.

The crude [Ru(en)₃] [ZnCl₄] (1.0 g) (2.0 mmole) is placed in a deoxygenated flask and a stoichiometric amount (5.0 mL) of 0.82 M deoxygenated lithium anthranilate solution is added. The resulting white precipitate of zinc dianthranilate is removed by filtration on a fine frit under an inert atmosphere and washed with two 1.5-mL portions of deoxygenated water. Addition of 20 mL of acetone to the combined washings and light-brown filtrate results in formation of a yellow precipitate. An additional 20 mL of acetone is added and the precipitate is collected by filtration under reduced argon pressure. The precipitate is washed with two 1.5-mL portions of deoxygenated acetone and two 5-mL portions of diethyl ether and is then dried under a stream of argon and, if it is not to be recrystallized immediately, on a vacuum line overnight. This yellow product (0.58 g, 80%) can be reprecipitated under an inert atmosphere by dissolving in a minimum amount (approximately 3 mL) of warm deoxygenated 0.01 M trifluoroacetic acid, cooling, and adding acetone to give canary-yellow hexagonal plate crystals. *Anal.* Calcd. for [Ru(C₂H₈N₂)₃] Cl₂: C, 20.46; H, 6.87; N, 23.85; Cl, 20.13; Ru, 28.69. Found: C, 20.19; H, 6.99; N, 23.62; Cl, 19.86; Ru, 28.50.

C. TRIS(ETHYLENEDIAMINE)RUTHENIUM(III) CHLORIDE

$$2[Ru(en)_3][ZnCl_4] + 4Ag(an) \longrightarrow$$

$$2[Ru(en)_3]Cl_3 + 2Ag + 2AgCl + 2Zn(an)_2$$

■**Caution.** *Skin contact with silver salts and solutions should be avoided.*

Procedure

Solid silver anthranilate is prepared as follows. A solution of 2.0 g (50 mmole) of NaOH in 10 mL of water is stirred for 10 minutes with a slight excess of anthranilic acid (7.0 g, 51 mmole). The excess undissolved anthranilic acid is removed by filtration and a solution of 8.5 g (50 mmole) of AgNO₃ in 6 mL of water is added to the filtrate. The slowly precipitated silver anthranilate is collected by filtration, washed with water, ethanol, and acetone, and dried first in the air and then on a vacuum line for 8 hours. The off-white solid is light and heat sensitive.

The silver anthranilate (1.30 g, 5.33 mole) is slowly added with shaking to an ice-cold, deoxygenated slurry of 1.30 g (2.66 mmole) of [Ru(en)₃] [ZnCl₄] in 15 mL of deoxygenated 10^{-4} M trifluoroacetic acid. The resultant heterogeneous mixture is agitated for several minutes, preferably in an ultrasonics bath. The solids are removed by filtration under an inert atmosphere and washed with two 2.5-mL portions of deoxygenated 10^{-4} M trifluoroacetic acid. The combined light-brown filtrate and washings are cooled in an ice bath and 30 mL of deoxygenated acetone is added, causing some precipitation. An additional 30 mL of acetone is added and the resulting white solid is collected, washed with two 2.5-mL portions of deoxygenated ethanol, two 5-mL portions of acetone, and two 5-mL portions of diethyl ether, and dried in a vacuum for 8 hours. Yield 0.98 g (95%). Care must be taken not to prolong the above preparation or to use excess silver anthranilate.

The compound can be purified by reprecipitation from the minimum amount of 0.1 M HCl by the addition of acetone or by recrystallization from dilute hydrochloric acid. *Anal.* Calcd. for $[Ru(C_2H_8N_2)_3]Cl_3$: C, 18.59; H, 6.24; N, 21.67; Cl, 27.43; Ru, 26.07. Found: C, 18.65; H, 6.40; N, 21.43; Cl, 27.53; Ru, 25.85.

Properties

The canary-yellow crystals of $[Ru(en)_3]Cl_2$ can be transferred in the air. Solutions and the incompletely dried solid are readily air oxidized. The pale-yellow $[Ru(en)_3]Cl_3$ is an air-stable solid, but neutral solutions are air sensitive. Its electronic absorption spectrum[4] can be used to characterize $[Ru(en)_3]^{2+}$, with two absorption maxima in 0.01 M trifluoroacetic acid solution: $\epsilon_{370} = 120$ M^{-1} cm^{-1} and $\epsilon_{304} = 1020$ M^{-1} cm^{-1}. It is difficult to detect small amounts of $[Ru(en)_3]^{3+}$ in the solution, however, since its absorption spectrum in this region consists of a single weaker band,[4] $\epsilon_{310} = 355$ M^{-1} cm^{-1}, and a shoulder with $\epsilon_{240} = 458$ M^{-1} cm^{-1}. A common impurity is $[Ru(en)_2(diimine)]^{2+}$ in which the coordinated ethylenediamine ligand has undergone a four-electron oxidation.[7] This is readily detected by its intense absorption band with $\epsilon_{448} \sim 7000$ M^{-1} cm^{-1}. Pale-yellow solutions of $[Ru(en)_3]^{2+}$ can be prepared in deoxygenated 0.01 M trifluoroacetic acid in concentrations up to 0.9 M. To reduce any $[Ru(en)_3]^{3+}$ impurity the solution is stirred over zinc amalgam for 20-30 minutes. Any $[Ru(en)_2(diimine)]^{2+}$ impurity is not reduced by this procedure, however, and produces a yellow-brown solution.

Studies of these complexes include reports of NMR,[5,8] EPR,[9] ORD and CD,[10] and ESCA[7] spectra, and chemical studies of the reactions of $[Ru(en)_3]^{2+}$ with oxygen[11] and Fe(III)[4] and of electron exchange between the two oxidation states.[4]

References

1. P. Karrer, *Helv. Chim. Acta,* **3**, 196 (1920).

2. F. M. Lever and C. W. Bradford, *Platinum Met. Rev.,* **8**, 106 (1964).
3. A. D. Allen and C. V. Senoff, *Can. J. Chem.,* **43**, 888 (1965).
4. T. J. Meyer and H. Taube, *Inorg. Chem.,* **7**, 2369 (1968); T. J. Meyer, Ph.D. Thesis, Stanford University, 1967.
5. J. K. Beattie and H. Elsbernd, *J. Am. Chem. Soc.,* **92**, 1946 (1970).
6. H. J. Peresie and J. A. Stanko, *Chem. Commun.,* **1970**, 1674.
7. B. C. Lane, J. E. Lester and F. Basolo, *Chem. Commun.,* **1971**, 1618; H. Elsbernd and J. K. Beattie, *J. Chem. Soc., A,* **1970**, 2598.
8. H. Elsbernd and J. K. Beattie, *J. Am. Chem. Soc.,* **91**, 4573 (1969); H. Elsbernd, Ph. D. Thesis, University of Illinois, 1969.
9. J. A. Stanko, H. J. Peresie, R. A. Bernheim, R. Wang, and P. S. Wang, *Inorg. Chem.,* **12**, 634 (1973).
10. H. Elsbernd and J. K. Beattie, *Inorg. Chem.,* **8**, 893 (1969).
11. J. R. Pladziewicz, T. J. Meyer, J. A. Broomhead, and H. Taube, *Inorg. Chem.,* **12**, 639 (1973).

27. MANGANESE DIPHOSPHATE (MANGANESE PYROPHOSPHATE)

$$Na_4P_2O_7 + 2(MnCl_2 \cdot 2.03KCl) \longrightarrow$$

$$Mn_2P_2O_7 + 4NaCl + 4.06KCl$$

Submitted by B. DURAND* and J. M. PARIS*
Checked by E. KOSTINER[†] and M. H. RAPPOSCH[†]

Owing to its stability, manganese diphosphate is very convenient as a calibrant for magnetic susceptibility measurements. Its magnetic susceptibility follows the Curie-Weiss law between −80 and +485°.

It is generally prepared by thermal decomposition in air of ammonium manganese phosphate monohydrate $Mn(NH_4)(PO)_4 \cdot H_2O$. Yet, as was shown by Etienne and Boulle,[1] structureless phases of complex composition appear before the crystallization of the diphosphate (presence of mono, di-, tri-, and polyphosphate anions). This is a quite general phenomenon, observed not only during the thermal decomposition of ammonium metallic phosphates $M_2NH_4PO_4$, but also during the decomposition of hydrogen metallic phosphates M_2HPO_4 and hydrated diphosphates $M_2P_2O_7 \cdot nH_2O$. A ratio of MO/-P_2O_5 = 2 is characteristic of all these compounds. The unavoidable formation of these intermediate phases and their often incomplete transformation into diphosphate represent a major difficulty in the preparation of pure divalent

*University of Lyon 1, 43 Boulevard du 11 November 1918, 69621 Villerbanne, France.
[†]Institute of Materials Science, The University of Connecticut, Storrs, CT 06268.

metal diphosphates. The synthesis described here involves the double decomposition reaction between the sodium diphosphate and a molten divalent manganese salt. This method avoids the formation of intermediate phases.

Preparation of Starting Materials

Sodium diphosphate is prepared by the following procedure. Disodium hydrogen phosphate dihydrate, $Na_2HPO_4 \cdot 2H_2O$, (7.12 g, 40 mmole) is placed in a 60-mL platinum crucible and heated in air at 400° for 12 hours (Product Merck for analysis K 6580). According to Porthault,[2] this method leads to a tetrasodium diphosphate, $Na_4P_2O_7$, the purity of which is better than 99.5%. No impurities can be found by chromatographic or potentiometric analysis.

The potassium and manganese double chloride is prepared by the following procedure. The eutectic, $MnCl_2 \cdot 2.03KCl$,[3] is prepared by mixing potassium chloride reagent grade (12.11 g, 0.1624 mole) with manganese(II) chloride $MnCl_2 \cdot 4 H_2O$ reagent grade (15.83 g, 0.800 mole). The mixture is heated, in a 250-mL Pyrex beaker, in air at 200° for 3 hours.

Procedure

The eutectic $MnCl_2 \cdot 2.03KCl$ and the phosphate $Na_4P_2O_7$ are quickly mixed by milling in air in an agate mortar so as to avoid as much as possible the hydratation of the salt. To get a rapid and complete reaction an excess of eutectic is used: 5.32 g $Na_4P_2O_7$ (2 × 10^{-2} mole) for 22.17 g of eutectic (8 × 10^{-2} mole).

The reaction mixture is placed in a 60-mL platinum crucible, heated in a nonoxidizing atmosphere at 500° in a regulated furnace (temperature rise: 150°/hr), and held there for 24 hours. The nonoxidizing atmosphere is produced by a stream of pure nitrogen that has been dried by passage over P_2O_5.

After it is cooled, the solidified mixture is washed with distilled water to remove excess manganese salt and the alkaline salt formed. After filtration, the manganese diphosphate is dried by heating in air at 200°.

Properties

The manganese diphosphate produced by this method is a pale-pink powder composed of crystallites between 0.2 and 0.8 m in diameter. It is stable in air and slightly soluble in water.

The compound crystallizes in the monoclinic system with unit-cell dimensions of $a = 6.64$ Å; $b = 8.58$ Å; $c = 4.54$ Å; $\beta = 102.8°$. These data agree well with those reported by Lukaszewicz.[4] Manganese diphosphate can be identified according the following powder diffraction lines:

$d(Å)$	I/I_0	d	I/I_0	d	I/I_0	d	I/I_0	d	I/I_0
5.16	50	3.08	100	2.58	88	2.16	25	2.06	68
4.43	39	3.03	25	2.37	30	2.14	38	2.04	36
4.30	32	2.93	95	2.21	12	2.08	35	1.97	29
3.11	100	2.61	83	2.17	91	2.07	55	1.93	58

Anal.[5,6] Calcd.: Mn, 38.72; P, 21.82. Found: Mn, 38.65; P, 21.90.

The magnetic susceptibility, measured with a vibrating sample magnetometer (Foner), calibrated with nickel (purity 99.99% — specific susceptibility at $293°K$ 55 cgs/g) at $293 ± 0.2°K$ with a magnetic field of 17,700 G, is 101.39 $± 0.20 \ 10^{-6}$ cgs/g. This value may be compared with $101.65 ± 0.20 \ 10^{-6}$ cgs/g for manganese diphosphate prepared by thermal decomposition of the ammonium manganese phosphate. The value calculated from Reference 7 is $101.96 ± 0.20$ cgs/g.

References

1. J. Etienne and A. Boulle, *Bull. Soc. Chim. Fr.,* **1968**, 1805.
2. M. Porthault, Thesis Lyon, Order No. 285, p. 49 (1962).
3. H. J. Seifert and F. W. Koknat, *Z. Anorg. Allg. Chem.,* **341**, 269 (1965).
4. K. Lukaszewicz and R. Smaskiewicz, *Rocz. Chem.,* **35**, 741 (1961).
5. J. J. Lingane and R. Karplus, *Ind. Eng. Chem. Anal. Ed.,* **18**, No. 3, 191 (1946).
6. G. Charlot, *Les méthodes de la Chimie Analytique,* Analyse Minerale Quantitative, 4th ed., Masson & Cie, Paris, 1961, **18**, 846-47.
7. G. Foex, C. J. Gorter, and L. J. Smith, *Tables de constantes sélectionnées. Diamagnétisme et paramagnétisme. Relaxation paramagnétique,* Masson et Cie, Paris, 1957.

28. ELECTROCHEMICAL SYNTHESIS OF CHROMIUM(III) BROMIDE; A FACILE ROUTE TO CHROMIUM(III) COMPLEXES

Submitted by JACOB J. HABEEB* and DENNIS G. TUCK*
Checked by WILLIAM E. GEIGER, Jr.,† and WILLIAM BARBER†

Chromium(III) bromide can be prepared directly by treating the metal with bromine in a sealed tube at temperatures variously reported[1,2] as being between 750 and 1000°. The direct electrochemical synthesis described below leads to

*Department of Chemistry, University of Windsor, Windsor, Ontario, N9B 3P4, Canada.
†Department of Chemistry, University of Vermont, Burlington, VT 05401.

the rapid production of the anhydrous compound in gram quantities at room temperature. This product reacts readily with a variety of mono- and bidentate ligands. The preparation of cationic and anionic coordination complexes of chromium(III) is described; the examples selected are but a few of the many such compounds that can presumably be obtained in this way.

Some other synthetic applications of the oxidation of positively charged metals in electrochemical cells have been described previously.[3-6]

A. ANHYDROUS CHROMIUM(III) BROMIDE

$$Cr_{(anode)} + 3/2\ Br_{2\,(sol)} \longrightarrow CrBr_{3(s)}$$

Procedure

■**Caution.** *This synthesis involves liquid bromine, which should be handled with care in a well-vented hood, and it is best that the whole experiment be performed in such a hood.*

The electrochemical cell is set up in a 200-mL tall-form beaker, with a tightly fitting rubber stopper to support the electrodes. The cathode is a platinum wire connected to a 2 × 2 cm Pt sheet; the exact form of the cathode is not critically important, provided that sufficient flow of current can occur. The anode consists of a selected flat piece of chromium metal (99.999%)*, weighing approximately 2 g (in the original work, this metal had one shiny and one carbuncular surface.) The chromium is supported in a cage made by winding platinum wire around it, or it can be placed on a flat coil of platinum wire; the platinum should not be welded to chromium, nor should adhesives be used. The anode and cathode are placed 1-2 cm apart. The solution phase consists of 100 mL of benzene-methanol (3:1, v/v) containing 2 g of bromine.†

The dc power supply should be capable of delivering up to 100 V and 500 mA. (In the original work, a Coutant LQ 50/50 unit was used.) (■**Caution.** *Care is needed in the operation of cells at such voltages, and warning notices should be posted.*) In a typical experiment, an applied voltage of 42 V gives a current of 200 mA. The current does not change significantly throughout an experiment lasting about 3 hours. As the experiment proceeds, a dark-green oil settles in the bottom of the cell, and after 2-3 hours the whole liquid phase turns green. The reaction time apparently is not critical. Some chromium metal (0.37 g in the

---

*Available as lumps from Alfa Products, Ventron Corp. P.O. Box 299, Danvers, MA 01923.

†The checkers report that magnetic stirring is recommended, and this has been found helpful in other electrochemical preparations studied by the original authors.

original work) remains at the end of the experiment, and the smell of bromine can still be detected, although the color of the liquid phase masks the color of any bromine present.

The cell is then dismantled and flushed with a slow stream of dry nitrogen for 15-30 minutes to remove any unreacted bromine. The resultant liquid is transferred to a suitable vessel (e.g., 100-mL, round-bottomed flask) in a dry nitrogen atmosphere (glove bag or dry box) and pumped on a vacuum line for 2 hours at room temperature. The resultant thick oil is washed with dry diethyl ether (two 10-mL portions) and sucked dry on a sinter disc by means of a trapped aspirator vacuum (all under nitrogen) to give a dark-green powder of chromium(III) bromide. Both the green oil and the resultant solid can be successfully used in preparing chromium(III) complexes; examples of each procedure are given below.

Yield 8.5 g (93%, based on chromium loss at the anode). *Anal.* Calcd. for $CrBr_3$: Cr, 17.8; Br, 82.2. Found: Cr (atomic absorption), 17.8; Br ($AgNO_3$/KCNS titration), 81.8%.

B. TRIS(ETHYLENEDIAMINE)CHROMIUM(III) BROMIDE, $[Cr(en)_3]Br_3$

$$CrBr_3 + 3en \longrightarrow [Cr(en)_3]Br_3$$

(en = ethylenediamine)

The preparation of salts containing the $[Cr(en)_3]^{3+}$ cation from anhydrous chromium sulfate has been described previously in *Inorganic Syntheses*,[7,8] and the merits of this, and other, methods have been reviewed.[9] A more rapid route to this cation involves refluxing $CrCl_3 \cdot 6H_2O$ in methanol with ethylenediamine and zinc metal, which allows the substitution to proceed by way of the kinetically labile chromium(II) species.[10] All of these preparations yield hydrated salts; the procedure described below leads to anhydrous $[Cr(en)_3]Br_3$.

Procedure

Approximately 0.5 g (1.7 mmole) of $CrBr_3$, as the dark-green oil derived in Section A, is added dropwise to 15 mL of colorless ethylenediamine (reagent grade). An exothermic reaction occurs, following which the mixture is allowed to cool to room temperature, resulting in the formation of yellow crystals. These are collected, washed with diethyl ether (2 × 25 mL), and dried *in vacuo*. Yield 0.74 g (1.55 mmole, 90%). *Anal.* Calcd. for $C_6H_{24}N_6CrBr_3$: Cr, 11.1; Br, 51.2. Found: Cr, 11.1; Br, 51.4%.

Properties

Freshly prepared $[Cr(en)_3] Br_3$ is a yellow solid which slowly turns violet-blue, probably because of some photochemically induced reaction (cf. Reference 9). The compound is sparingly soluble in water, from which it can be recrystallized as the hydrate. The resolution of $[Cr(en)_3]^{3+}$ salts by means of tartrate has been described in a previous volume of *Inorganic Syntheses.*[11]

C. HEXAKIS(DIMETHYL SULFOXIDE)CHROMIUM(III) BROMIDE, [Cr-(DMSO)₆] Br₃

$$CrBr_3 + 6dmso \longrightarrow [Cr(dmso)_6] Br_3$$

$$[dmso = (CH_3)_2 SO]$$

Dimethyl sulfoxide is now widely used both as a solvent and a ligand; the triperchlorate salt $(Cr(dmso)_6](ClO_4)_3$ was first prepared by Cotton and Francis[12] by the reaction sequence

$$Cr(OH)_3 \xrightarrow{\text{HClO}_2} Cr(ClO_4)_3 \xrightarrow{\text{dmso}} [Cr(dmso)_6] (ClO_4)_3$$

This compound, described as crystallizing as emerald-green needles, has been the subject of infrared study of both $\nu_{S=O}$[13,14] and ν_{M-O}[15] vibrations. The bromide salt of the $[Cr(dmso)_6]^{3+}$ cation has now been prepared directly from $CrBr_3$.

Procedure

Approximately 0.3 g (1.05 mmole) of solid $CrBr_3$ is dissolved at 45° in 20 mL of dimethyl sulfoxide which has been dried over molecular sieves and distilled before use. The solid dissolves immediately. When the solution has cooled to room temperature, the pale-green crystals which form are collected, washed with diethyl ether (two 25-mL portions), and dried *in vacuo* (1 hr, room temperature). Yield 0.74 g (0.97 mmole, 94%). *Anal.* Calcd. for $C_{12}H_{36}S_6O_6Cr-Br_3$: Cr, 6.6; Br, 34.2. Found: Cr, 6.5; Br 34.4%.

Properties

The infrared spectrum shows the characteristic absorption due to $\nu_{S=O}$ at 936 cm⁻¹, shifted from the value in free dimethyl sulfoxide,[13] and ν_{Cr-O}[15] at 520 (s) cm⁻¹.

D. POTASSIUM TRIS(OXALATO)CHROMATE(III) TRIHYDRATE

$$CrBr_3 + 3K_2C_2O_4 \longrightarrow K_3[Cr(C_2O_4)_3] + 3KBr$$

The classical preparation[16] of this salt involves the reaction of oxalic acid, dipotassium oxalate, and potassium dichromate, in which the reduction Cr(VI) $\longrightarrow$ Cr(III) is accompanied by complexation. The method described below has chromium in the +3 state in the highly reactive starting material.

Procedure

Approximately 0.5 g (1.7 mmole) of $CrBr_3$ as the green oil is added to an aqueous solution of dipotassium oxalate [1 g (6.0 mmole) in 10 mL of water]. Green crystals of $K_3[Cr(C_2O_4)_3] \cdot 3H_2O$ are immediately precipitated and are collected and washed with dry ethanol. Yield 0.7 g (1.43 mmole, 81%). *Anal.* Calcd. for $C_6H_6O_{15}K_3Cr$: K, 24.0; Cr, 10.7. Found: K(atomic absorption), 24.3; Cr, 10.6%.

Properties

The salt $K_3[Cr(C_2O_4)] \cdot 3H_2O$ is soluble in water. The crystals are dichroic, being green-blue by transmitted light and magenta by reflection. The infrared spectra of this and other tris(oxalato) complexes have been recorded and analyzed.[17]

References

1. R. J. Sime and N. W. Gregory, *J. Am. Chem. Soc.,* 82, 93 (1960).
2. G. Brauer, *Handbook of Preparative Inorganic Chemistry,* Vol. 2, 2nd ed., Academic Press Inc., New York, 1965, p. 1341.
3. J. J. Habeeb and D. G. Tuck, *Chem. Commun.,* 1975, 808.
4. J. J. Habeeb, L. Neilson, and D. G. Tuck, *Synth. React. Inorg. Metal-org. Chem.,* 6, 105 (1976).
5. J. J. Habeeb, A. Osman, and D. G. Tuck, *Chem. Commun.,* 1976, 379.
6. J. J. Habeeb and D. G. Tuck, *Inorg. Synth.,* 19, 257 (1979).
7. C. L. Rollinson and J. C. Bailar, *Inorg. Synth.,* 2, 196 (1946).
8. W. N. Shepard, *Inorg. Synth.,* 13, 233 (1972).
9. C. L. Rollinson and J. C. Bailar, *J. Am. Chem. Soc.,* 65, 250 (1943).
10. R. D. Gillard and P. R. Mitchell, *Inorg. Synth.,* 13, 184 (1972).
11. F. Galsbol, *Inorg. Synth.,* 12, 274 (1970).
12. F. A. Cotton and R. Francis, *J. Am. Chem. Soc.,* 82, 2986 (1960).
13. F. A. Cotton, R. Francis, and W. D. Horrocks, *J. Phys. Chem.,* 64, 1534 (1960).
14. R. S. Drago and D. W. Meek, *J. Phys. Chem.,* 65, 1446 (1961).

15. C. V. Berney and J. H. Weber, *Inorg. Chem.*, 7, 282 (1968).
16. J. C. Bailar and E. M. Jones, *Inorg. Synth.*, 1, 37 (1939).
17. J. Fujita, A. E. Martell, and K. Nakamoto, *J. Chem. Phys.*, 36, 324, 331 (1962).

29. DINUCLEAR MOLYBDENUM COMPLEXES

Submitted by J. SAN FILIPPO, JR.,* and H. J. SNIADOCH*
Checked by M. E. CLAY† and T. M. BROWN†

The structural diversity and chemical reactivity of dinuclear complexes that contain strong metal-to-metal bonds have made them a focus of interest.[1] Such bonds are extensive in the chemistry of molybdenum(II), and a considerable chemistry of the $[Mo_2]^{4+}$ unit has been developed. Many of the chloro complexes of dimolybdenum(II) can be prepared by direct reaction of ligand and a salt of the $[Mo_2Cl_8]^{4-}$ ion. Others can be synthesized by ligand replacement reactions employing the 1,2-bis(methylthio)ethane complex $Mo_2Cl_4(CH_3SCH_2CH_2SCH_3)_2$ as a convenient starting complex. Salts of the dinuclear molybdenum(III) ion $[Mo_2Br_8H]^{3-}$ serve as useful precursors for the preparation of $Mo_2Br_4(C_5H_5N)_4$, from which, in turn, a series of bromo complexes of dimolybdenum(II) can be obtained by ligand replacement. Convenient, high-yield preparations for these key starting reagents are described here. In addition, the preparation of a family of mixed halo-μ-(arylcarboxylato) complexes, $Mo_2X_2(O_2CAr)_2L_2$, is presented along with a more general procedure for the synthesis of tetra-μ-(carboxylato) complexes, $Mo_2(O_2CR)_4$.

General Procedure

(■**Caution.** *All reactions involving noxious reagents [pyridine, tributylphosphine and 1,2-bis(methylthio)ethane, i.e. 2,5-dithiahexane] or corrosive substances, particularly hydrogen chloride, are carried out in a good fume hood.*) Tributylphosphine‡ is distilled under nitrogen (bp 150°, 50 torr) prior to use. 1,2-Bis(methylthio)ethane (2,5-dithiahexane)§ is distilled (bp 183°) before use. All solvents are deoxygenated by purging with dry, high-purity nitrogen for 30 minutes prior to use.

*Wright and Rieman Chemistry Laboratories, Rutgers, The State University of New Jersey, New Brunswick, NJ 08903.
†Department of Chemistry, Arizona State University, Tempe, AZ 85281.
‡Available from Aldrich Chemical Co., 940 W. St. Paul Ave., Milwaukee, WI 53233.
§Available from Columbia Organics, Inc., P. O. Box 5273 Columbia, SC 29209.

A. OCTACHLORODIMOLYBDATE(4-), [Mo$_2$Cl$_8$]$^{4-}$, DI-μ-CHLORO-HEX-
ACHLORO-μ-HYDRIDO-DIMOLYBDATE(3-), [Mo$_2$Cl$_8$H]$^{3-}$, AND DI-μ-
BROMO-HEXABROMO-μ-HYDRIDO-DIMOLYBDATE(3-), [Mo$_2$Br$_8$H]$^{3-}$
ANIONS

$$Mo_2(O_2CCH_3)_4 \ + \ 8HCl \xrightarrow{0°} [Mo_2Cl_8]^{4-} \ + \ 4H^+ \ + \ 4CH_3CO_2H$$

$$Mo_2(O_2CCH_3)_4 \ + \ 8HX \xrightarrow{60°} [Mo_2X_8H]^{3-} \ + \ 3H^+ \ + \ 4CH_3CO_2H$$

$$(X = Cl, Br)$$

Anions of the type [Mo$_2$Cl$_8$]$^{4-}$ and [Mo$_2$Cl$_8$H]$^{3-}$ were first prepared by Cotton
and coworkers[2,3] by procedures similar to those outlined below. The latter
species, previously[3] purported to be [Mo$_2$Cl$_8$]$^{3-}$, has been recently determined
to be [Mo$_2$Cl$_8$H]$^{3-}$, that is, a mixed halohydrido dinuclear complex of moly-
bdenum(III).[4]

Procedure

1. *Preparation of pentaammonium nonachlorodimolybdate(5-)
monohydrate, (NH$_4$)$_5$[Mo$_2$Cl$_9$]·H$_2$O.* Into a 500-mL flask equipped with a
Teflon-coated stirrer bar is placed 250 mL of 12 M hydrochloric acid. The flask
is placed in an ice bath, and hydrogen chloride gas is bubbled into it for 1 hour.
Molybdenum(II) acetate[5] (10.0 g, 23.4 mmole) and ammonium chloride (10.0
g, 187 mmole) are added and the vessel is stoppered with a rubber septum. The
flask is removed from the ice bath and the resulting mixture is allowed to warm
to room temperature under nitrogen with stirring. After 1 hour the wine-red
solid is collected by suction filtration using a medium-porosity fritted-glass
funnel. The product is rinsed with two 100-mL portions of cold (- 15°) absolute
ethanol, and vacuum dried for 2 hours. The average yield is 11.6 g (80%). *Anal.*
Calcd. for H$_{22}$Cl$_9$Mo$_2$N$_5$O: H, 3.50; Cl, 51.12; N, 12.02. Found: H, 3.21; Cl,
51.05; N, 11.93.

2. *Preparation of tricesium di-μ-chloro-hexachloro-μ-hydrido-dimolybdate-
(3-), Cs$_3$[Mo$_2$Cl$_8$H].* Molybdenum(II) acetate (10.0 g, 23.4 mmole) is
placed in a three-necked, 500-mL flask containing a Teflon-coated stirrer
bar. The flask is capped with three rubber septums, and a thermometer is
inserted through one of the septums. After the flask and its contents have

been flushed with nitrogen for 10 minutes, a deoxygenated solution of hydrochloric acid (250 mL, 12 M) is added and the mixture is stirred under nitrogen at 60° for 1 hour. Cesium chloride (15.0 g, 89.1 mmole) is added to the warm solution with stirring and the mixture is allowed to cool to room temperature. The yellow precipitate is isolated by suction filtration, washed with two 100-mL portions each of absolute ethanol and diethyl ether, and then dried *in vacuo.* Yields of 18.4 g (90%) or better are obtained. *Anal.* Calcd. for $HCl_8Cs_3Mo_2$: Cl, 32.41; Mo, 21.91. Found: Cl, 32.27; Mo, 21.79.

3. *Preparation of tricesium di-μ-bromohexabromo-μ-hydridodimoiybdate-(3-), Cs_3[Mo_2Br_8H]*. Molybdenum(II) acetate (12.0 g, 28.0 mmole) is placed in a three-necked, 2000-mL flask equipped with a Teflon-coated stirrer bar. Each neck is stoppered with a rubber septum, and a thermometer is inserted through one of the septums. The entire system is flushed with nitrogen and 600 mL of deoxygenated 48% hydrobromic acid is then added.* This mixture is heated with stirring at 60° for 1 hour under a nitrogen atmosphere. The resulting solution is allowed to cool to ambient temperature and is then treated with a deoxygenated solution of cesium bromide (30.0 g in 300 mL of 48% hydrogen bromide). A brown precipitate forms immediately. The reaction mixture is chilled at 0° for 1 hour and the solid is collected by suction filtration, washed with 50 mL of absolute ethanol and 50 mL of absolute diethyl ether, and dried *in vacuo.* The yield averages 30.7 g (89%). *Anal.* Calcd. for HBr_8Cs_3-Mo_2: Br, 51.98; Mo, 51.98. Found: Br, 51.47; Mo, 51.77.

Properties

Salts of the anion $[Mo_2Cl_8]^{4-}$ show varying degrees of stability and sensitivity to air. However, all can be stored essentially indefinitely *in vacuo.* Their dissolution in neutral aqueous solution is accompanied by rapid decomposition. The cesium salts of $[Mo_2Cl_8H]^{3-}$ and $[Mo_2Br_8H]^{3-}$ are relatively stable in dry air and appear to be stable indefinitely when stored *in vacuo.* They are essentially insoluble in neutral water and in concentrated aqueous hydrogen halide. Salts of $[Mo_2Cl_8]^{4-}$ exhibit a characteristic Raman spectrum that is distinguished by an intense band around 340 cm^{-1} corresponding to ν_{Mo-Mo}.[5] The low-energy Raman spectra of $Cs_3[Mo_2Cl_8H]$ and $Cs_3[Mo_2Br_8H]$ are rich in detail, although not uniquely characterized by any single feature.[6]

---

*The submitters have observed that the history of the hydrobromic acid is frequently critical to the success of this preparation. Best results are obtained with freshly opened bottles that have not been stored for an extended period.

B. TETRAHALODIMOLYBDENUM(II) COMPLEXES

$$[Mo_2Cl_8]^{4-} + 2CH_3SCH_2CH_2SCH_3 \rightarrow Mo_2Cl_4(CH_3SCH_2CH_2SCH_3)_2 + 4Cl^-$$

$$[Mo_2Br_8H]^{3-} + C_5H_5N \rightarrow Mo_2Br_4(C_5H_5N)_4$$

$$Mo_2Br_4(C_5H_5N)_4 + 4P(n\text{-}C_4H_9)_3 \rightarrow Mo_2Br_4[P(n\text{-}C_4H_9)_3]_4 + 4C_5H_5N$$

Procedure

1. *Preparation of tetrachlorobis[1,2-bis(methylthio)ethane]dimolybdenum-(II), $Mo_2Cl_4(CH_3SCH_2CH_2SCH_3)_2$.* Pentaammonium nonachlorodimolybdate-(II) monohydrate (4.00 g, 6.5 mmole) is placed in a 500-mL flask containing a Teflon-coated stirring bar. The flask is capped with a rubber septum and the vessel is flushed with nitrogen; a deoxygenated solution of 1,2-bis(methylthio)ethane (8 mL) in methanol (200 mL) is then injected. After the contents of the flask have been stirred for 30 minutes, the precipitated product is collected by suction filtration and rinsed with three 10-mL portions of absolute ethanol, followed by three 10-mL portions of diethyl ether. It is finally dried *in vacuo*. The yield of blue-green $Mo_2Cl_4(CH_3SCH_2CH_2SCH_3)_2$ is 3.4 g (90%). *Anal.* Calcd. for $C_8H_{20}Cl_4Mo_2S_4$: C, 16.62; H, 3.49; Cl, 24.52; S, 22.19. Found: C, 16.57; H, 3.67; Cl, 24.21; S, 21.90.

2. *Preparation of Tetrabromotetrakis(pyridine)dimolybdenum(II), Mo_2Br_4-$(C_5H_5N)_4$.* Freshly prepared $Cs_3[Mo_2Br_8H]$ (5.0 g, 4.1 mmole) is placed in a 250-mL flask equipped with a condenser, a Teflon-coated stirrer bar, an addition funnel, and a nitrogen inlet. The system is flushed with nitrogen, and pyridine (100 mL, freshly distilled from calcium hydride under nitrogen) is added. The resulting mixture is heated at reflux for 50 minutes with efficient stirring and then chilled ($-10°$) for 1 hour. The green, crystalline solid that precipitates is collected by suction filtration and washed, first with distilled water (15 mL), then with absolute ethanol (30 mL) and finally with absolute diethyl ether (30 mL), before it is dried *in vacuo*. The yield averages 1.9 g (55-60%). *Anal.* Calcd. for $C_{20}H_{20}Br_4Mo_2N_4$: C, 29.02; H, 2.43; Br, 38.60; N, 6.77. Found: C, 28.83; H, 2.32; Br, 38.49; N, 6.70.

3. *Preparation of Tetrabromotetrakis(tributylphosphine)dimolybdenum(II), $Mo_2Br_4[P(C_4H_9)_3]_4$.* $Mo_2Br_4(C_5H_5N)_4$ (1.73 g, 2.09 mmole) is placed in a 100-mL flask fitted with a nitrogen inlet and a condenser. The vessel is flushed with nitrogen and then a deoxygenated solution of tributylphosphine (3 mL) in methanol (60 mL) is introduced by syringe. The resulting mixture is heated at

reflux for 1 hour and allowed to cool to room temperature. The flask is then placed in a Dry Ice-acetone bath for 15 minutes. The blue crystals of Mo_2Br_4-$[P(n\text{-}C_4H_9)_3]_4$ are collected by suction filtration, rinsed with two 10-mL portions of cold ($-78°$), deoxygenated methanol, and dried *in vacuo.* The yield averages 2.48 g (85-90%). *Anal.* Calcd. for $C_{48}H_{108}Br_4Mo_2P_4$: C, 43.81; H, 8.21; Br 24.29. Found: C, 43.56; H, 8.67; Br, 24.18.

Properties

Tetrahalodimolybdenum(II) complexes are brightly colored solids. Their solubilities in organic solvents vary considerably with the nature of the coordinating ligand. The stability of these substances in air is limited, although in most cases the solids can be handled briefly in air. Solutions of these complexes are much more air sensitive. The tetrahalo complexes of dimolybdenum(II) exhibit a characteristic visible absorption between 500 and 700 nm ($\epsilon \approx$ 200-5000), the position of which is strongly dependent on the nature of the coordinating ligands: $Mo_2Cl_4[P(C_4H_9)_3]_4$, 588 (1.3×10^3); Mo_2Cl_4-$(CH_3SCH_2^-CH_2SCH_3)_2$, 629 (KBr); $Mo_2Br_4(C_5H_5N)_4$, 656 (9.3×10^2); Mo_2Br_4-$[P(C_4H_9)_3]$, 600 (1.5×10^3).[7] Like the octahalo complexes, these compounds also display an intense Raman band between ~330 and 380 cm^{-1} assigned as ν_{Mo-Mo}.[7]

C.   BIS-   AND   TETRAKIS(CARBOXYLATO)DIMOLYBDENUM(II) COM-PLEXES

$$Mo_2X_4[P(n\text{-}C_4H_9)_3]_4 + 2ArCO_2H \rightarrow Mo_2X_2(O_2CAr)_2[P(n\text{-}C_4H_9)_3]_2$$

$$+ 2HX + 2P(n\text{-}C_4H_9)_3$$

$$Mo_2X_4[P(n\text{-}C_4H_9)_3]_4 + 4RCO_2H \rightarrow Mo_2(O_2CR)_4 + 4P(n\text{-}C_4H_9)_3 + 4HX$$

Tetrakis-μ-(carboxylato)-dimolybdenum(II) complexes have been obtained by only one general route, namely by the direct interaction of carboxylic acids with molybdenum hexacarbonyl.[8] This reaction requires elevated reaction temperatures and prolonged reaction times. These same compounds are obtained in comparable or better yields by the brief reaction of tetrachloro- or tetrabromotetrakis(tributylphosphine)dimolybdenum(II) with alkyl- or aryl-carboxylic acids in refluxing benzene. The bis-μ-(arylcarboxylato) complexes

$Mo_2X_2(O_2CAr)_2[P(n\text{-}C_4H_9)_3]_2$ can be obtained by a similar reaction using an adjusted stoichiometry.

Procedure

1. *Preparation of Tetrakis-μ-(butyrato)-dimolybdenum(II),* $Mo_2(O_2CCH_2\text{-}CH_2CH_3)_4$. Into a 100-mL flask, equipped with a Teflon-coated stirrer bar, is placed 1.00 g (0.76 mmole) of $Mo_2Br_4[P(C_4H_9)_3]_4$. A distillation head is attached and the system is thoroughly flushed with nitrogen, and a deoxygenated solution of butyric acid (0.50 mL, 5.4 mmole) in benzene (10.0 mL) is injected by syringe. The resulting mixture is heated gently at reflux under nitrogen with stirring for 2 hours and is then concentrated by distillation under nitrogen to a volume of about ∼3 mL. The pale-yellow product that precipitates upon cooling is collected by filtration under nitrogen, washed with two 10-mL portions of deoxygenated petroleum ether, and finally dried in a stream of nitrogen. Yield 0.27 g (66%). Tetrakis-μ-(butyrato)-dimolybdenum(II) is extremely air sensitive and must be handled exclusively under nitrogen.

2. *Preparation of bis-μ-(benzoato)-dibromobis(tributylphosphine)dimolybdenum(II),* $MO_2Br_2(O_2CC_6H_5)_2[P(C_4H_9)_3]_2$. $Mo_2Br_4[P(C_4H_9)_3]_4$ (0.50 g, 0.38 mmole) and benzoic acid (0.093 g, 0.76 mmole) are placed in a 50-mL, round-bottomed flask equipped with a Teflon-coated stirrer bar and a condenser capped with a rubber septum. The system is flushed with dry nitrogen and then 10 mL of deoxygenated benzene is injected by syringe. This mixture is then heated at reflux for 3 hours. The resulting solution is concentrated to about 2 mL by distillation under nitrogen, and chilled (−15°), oxygen-free methanol (10 mL) is added by syringe. This mixtures is allowed to stand for 1 hour at −15°. The product that precipitates is collected by suction filtration, washed with two 10-mL portions each of methanol and diethyl ether, and dried *in vacuo*. The yield is 0.27 g (70-75%). *Anal.* Calcd. for $C_{38}H_{64}Br_2Mo_2O_4P_2$: C, 45.71; H, 6.46; Br, 16.01; P, 6.20. Found: C, 44.76; H, 6.18; Br, 15.68; P, 6.15.

Properties

Tetrakis-μ-(butyrato)-dimolybdenum(II) is a bright-yellow complex that is soluble in a spectrum of organic solvents, including THF and benzene (but not petroleum ether). It is very air sensitive in both the solid state and in solution and must be handled under an inert atmosphere at all times. It can be stored under nitrogen indefinitely.

Dibromobis-μ-(benzoato)-bis(tributylphosphine)-dimolybdenum(II) is a bright orange-red compound. It is soluble in a broad spectrum of organic solvents including benzene. Although noticeably less air sensitive than $Mo_2(O_2CCH_2CH_2CH_3)_4$, it, nonetheless, should be handled under an inert atmosphere and can be stored indefinitely under nitrogen.

References

1. F. A. Cotton, *Chem. Soc. Rev.,* **4**, 1, (1975) and references therein.
2. J. Brencic and F. A. Cotton, *Inorg. Chem.,* 8, 7, 2698 (1969); 9, 346 (1970).
3. M. J. Bennett, J. V. Brencic, and F. A. Cotton, *Inorg. Chem.,* 8, 1060 (1969).
4. F. A. Cotton and B. J. Kalbacher, *Inorg. Chem.,* 15, 522 (1975).
5. J. San Filippo, Jr. and H. J. Sniadoch, *Inorg. Chem.,* 12, 2326 (1973).
6. J. San Filippo, Jr. and M. A. King, *Inorg. Chem.,* 15, 1228 (1976).
7. J. San Filippo, Jr., H. J. Sniadoch, and R. L. Grayson, *ibid.,* 13, 2121 (1974).
8. A. B. Brignole and F. A. Cotton, *Inorg. Synth.* 13, 81 (1972).

30.   (2,2′-BIPYRIDINE)TRICHLOROOXOMOLYBDENUM(V)

$$[bpyH_2][MoOCl_5] \xrightarrow[\substack{\text{boiling} \\ \text{98\% ethanol}}]{-2HCl} [MoOCl_3(bpy)] \quad \text{(red form)}$$

$$[bpyH_2][MoOCl_5] \xrightarrow[\substack{\text{boiling} \\ \text{acetonitrile}}]{-2HCl} [MoOCl_3(bpy)] \quad \text{(green form)}$$

Submitted by H. K. SAHA* and M. C. HALDAR*
Checked by T. M. BROWN[†] and ROBERT WILEY[†]

Isomers of (2,2′-bipyridine)trichlorooxomolybdenum(V) were first isolated and characterized by Saha and Halder.[1] These materials are interesting both as starting materials for the syntheses of other compounds of quinquevalent molybdenum and also for studying molybdenum-oxygen bonding interactions in oxomolybdenum compounds. Synthesis of [MoOCl$_3$(bpy)] involves two stages of reaction, that is, (1) isolation of the salt 2,2′-bipyridinium pentachlorooxomolybdate(V), [bpyH$_2$][MoOCl$_5$] and (2) dehydrohalogenation of [bpyH$_2$][MoOCl$_5$] in a solvent to yield the desired stereoisomer of [MoOCl$_3$(bpy)].

*Inorganic Chemistry Laboratory, Pure Chemistry Department, University College of Science, 92, Acharya Prafulla Chandra Road, Calcutta-9, India.
†Department of Chemistry, Arizona State University, Tempe, AZ 85281.

A. 2,2'-BIPYRIDINIUM PENTACHLOROOXOMOLYBDATE(V)   [bpyH₂]-[MoOCl₅]

Procedure

A sample of 2.0 g (14 mmole) of molybdenum trioxide is dissolved in 15 mL of hot concentrated hydrochloric acid (12 M) with stirring. Then 2 mL of hydriodic acid (D_{20} = 1.70) is added, and the mixture is gently boiled to drive off the liberated iodine. The concentrated small mass is again boiled with a 10-mL volume of fresh 12 M hydrochloric acid to ensure the complete removal of iodine. Finally, the dark-brown, moist residue is taken up with a minimum quantity of 12 M hydrochloric acid to give a bright-green solution. A sample of 2.7 g (17 mmole) of 2,2'-bipyridine, dissolved in 10 mL of 12 M hydrochloric acid, is added to this bright-green solution, and the mixture is cooled and vigorously stirred. There is immediate precipitation of a yellowish-green crystalline solid, which is filtered through a sintered-glass Gooch crucible and dried in a vacuum desiccator over solid KOH. Yield about 6.0 g (96%). The filtrate on slow evaporation deposits larger dark-green crystals. *Anal.* Calcd. for $C_{10}H_{10}N_2MoOCl_5$: Mo, 21.46; Cl, 39.64; N, 6.26. Found: Mo, 21.12; Cl, 39.50; N, 6.24. The oxidation state of the metal is 5.0 (by ceric sulfate method).

Properties

Magnetic susceptibility measurements by Guoy balance show μ_{eff} = 1.73 BM (30°). Infrared spectra show $\nu_{Mo=O}$ at 985 cm^{-1}. The salt rapidly decomposes on exposure to moist air.

B. [2,2'-BIPYRIDINE)TRICHLOROOXOMOLYBDENUM(V)] – RED FORM, [MoOCl₃(bpy)]

Procedure

A sample of 2.5 g (5.6 mmole) of 2,2'-bipyridinium pentachlorooxomolybdate-(V) is placed in a 100-mL, round-bottomed flask fitted with reflux condenser and a bent guard tube (filled with fused calcium chloride) to avoid entrance of moisture. About 20 mL of ethanol (98%) is added, and the materials are shaken. There is an immediate reddish-pink coloration. The mixture is heated at reflux for about 1 hour, until there is no more evolution of hydrogen chloride (tested with blue litmus paper). The mixture is then filtered through a dry sintered-glass Gooch crucible and washed twice with small quantities of ethanol. The product, a reddish-pink solid, is dried in a vacuum desiccator. Yield 1.5 g (71%). *Anal.*

Calcd. for $MoOCl_3 \cdot C_{10}H_8N_2$: Mo, 25.64; Cl, 28.41; N, 7.48. Found: Mo, 25.68; Cl, 27.85; N, 7.44. The oxidation state of the metal is 5.0 (found by ceric sulfate method in the presence of saturated silver sulfate solution).

Properties

Magnetic susceptibility measurements show μ_{eff} = 1.60 BM at 32°. Infrared spectra show $\nu_{Mo=0}$ at 980 cm^{-1}. It hydrolyses slightly in warm water. This compound can also be prepared by heating solid $[bpyH_2][MoOCl_5]$ in an inert atmosphere at 150°.

C. [(2,2'-BIPYRIDINE)TRICHLOROOXOMOLYBDENUM (V)] –GREEN FORM, $[MoOCl_3(bpy)]$

Procedure

A sample of 1.5 g (3.4 mmole) 2,2'-bipyridinium pentachlorooxomolybdate(V) is heated at reflux with 20 mL of acetonitrile (CH_3CN) for 6-7 hours in an inert atmosphere to avoid access to air and moisture as far as possible. Hydrogen chloride gas is evolved during this reaction (tested with moist blue litmus paper), and a bright-green crystalline solid separates out at the bottom. The solid is removed by filtration using a dry sintered Gooch crucible. It is washed twice with small quantities of acetonitrile and dried in a vacuum desiccator. Yield 1.0 g (79%). *Anal.* Calcd. for $MoOCl_3 \cdot C_{10}H_8N_2$: Mo, 25.64; Cl, 28.41; N, 7.48. Found: Mo, 25.50; Cl, 28.34; N, 7.40. The oxidation state (found by ceric sulfate method in the presence of saturated silver sulfate solution) is 5.0.

Properties

Magnetic susceptibility measurement show μ_{eff} = 1.76 BM at 30°. Infrared spectra show $\nu_{Mo=0}$ at 980 cm^{-1} It is exceptionally stable and does not hydrolyze even in warm water. It is isomorphous with the red form.

Analysis

Molybdenum[1] is estimated gravimetrically as MoO_3. The compound is decomposed with Na_2O_2 and then treated with sodium sulfide solution. The resulting thiomolybdate solution is acidified with dilute sulfuric acid to precipitate molybdenum sulfide, which is ignited at 540° to constant weight as MoO_3. The filtrate and washing are used for the estimation of chloride as silver salts.

Reference

1. H. K. Saha and M. C. Halder, *J. Inorg. Nucl. Chem.*, **33**, 3719 (1971).

31. METAL PENTAFLUORIDES

$$MF_6 + 1/4\,Si \xrightarrow{\text{HF } (\ell)} MF_5 + 1/4\,SiF_4$$

$$2UF_6 + CO \xrightarrow{h\nu} 2UF_5 + COF_2$$

$$(M = Mo, Re, Os, U)$$

Submitted by R. T. PAINE* and L. B. ASPREY†
Checked by L. GRAHAM‡ and N. BARTLETT‡

The transition metal and actinide pentafluorides are both structurally intriguing and synthetically useful reagents. A number of synthetic routes to these reactive compounds have been described, including direct fluorination of the metal,[1] halogen exchange,[2] and decomposition or chemical reduction of a higher oxidation state metal fluoride.[1-7] We have not found any of the previously described methods to be general or suitable for preparation of large quantities of pure compounds.

The syntheses described here involve simple one-electron reduction reactions of metal hexafluorides. The procedures have been found to be conveniently applicable to the preparation of MoF_5,[5] ReF_5,[6] OsF_5,[6] and UF_5.[7] High-purity hexafluorides can be obtained relatively easily by known methods,[2] and the reducing agents are ordinary reagents.

It should be pointed out that the procedures and conditions described here do *not* result in the reduction of SF_6 or WF_6.

■**Caution.** *Metal hexafluorides are volatile, toxic, corrosive, and highly hygroscopic materials. They must be handled in a very dry, clean, fluorine-preconditioned metal vacuum system. The vacuum system should be constructed from stainless steel or Monel materials (glass is not acceptable if pure products are desired). Hydrogen fluoride also is quite toxic and volatile. If only one metal*

*Department of Chemistry, The University of New Mexico, Albuquerque, NM 87131.
†CNC-4, Los Alamos Scientific Laboratory, Los Alamos, NM 87545.
‡Department of Chemistry, The University of California, Berkeley, CA 94720.

vacuum system is available, it should be throughly cleaned when it is changed from one hexafluoride to another to minimize cross contamination.

Procedure A

The metal vacuum system used here is essentially identical to those used at the Argonne National Laboratory.[8] A simpler vacuum system described in an earlier volume of this series is also satisfactory.[9] The metal hexafluorides are stored in Kel-F storage tubes sealed with a medium-pressure stainless steel valve (Series 30 VM Autoclave Engineers).* Anhydrous hydrogen fluoride† may be used directly from the tank, but we have found that it is often sufficiently wet so that it promotes hydrolysis of the metal hexafluoride. We have found that the following procedure provides pure, dry HF. Crude HF is subjected first to trap-to-trap distillation through -78 and $-196°$ baths. The HF retained at $-78°$ is treated with fluorine at $25°$ and 4 atm for 12 hours or more in a stainless steel tank, and this HF is distilled again. The HF retained at $-78°$ is treated with K_2NiF_6 for 2-6 hours, and the HF is distilled directly into a Kel-F storage tube fitted with a Kel-F valve.‡

In a typical experiment the metal vacuum line and connections to the hexa-fluoride and HF storage tubes are pumped to at least 10^{-4} torr over several hours. A 30-mL Kel-F reaction tube[9] fitted with a metal valve closure is charged with silicon powder and a Teflon-covered stirring bar. The reaction tube is attached to the vacuum system and evacuated.

The remainder of the experiments are described in terms of preparing UF_5 from UF_6; however, the MoF_6, ReF_6, and OsF_6 reductions proceed in exactly the same fashion. Ten milliliters of anhydrous HF is condensed with liquid nitrogen onto 30.8 mg (1.1 mmole) of silicon powder. This mixture provides a slurry of silicon powder when stirred. Onto the frozen slurry is condensed 1.7 g (4.9 mmole) of UF_6. The mixture is warmed to room temperature and stirred. After about 1 hour, the reaction is complete, as evidenced by the disappearance of the gray silicon powder and the cessation of gas evolution (SiF_4). In this case, the UF_5 product is insoluble in HF, and a fine yellow-green powder settles in the tube. The other metals yield HF-soluble pentafluorides.

■**Caution.** *Metal hexafluorides should not be condensed directly onto dry silicon powder. Explosions can occur.*

The pentafluoride products are easily recovered by vacuum evaporation of the HF, SiF_4, and excess MF_6. These volatile substances are passed through soda-lime and charcoal traps to scrub the HF and MF_6. For each hexafluoride a

---

*Available from Autoclave Engineers, 2930 W. 22nd St., Erie, PA 16512.

†Available from Matheson Gas Products, P.O. Box 85, E. Rutherford, NJ 07073.

‡Container identical to that shown in Reference 8, p. 132.

100% yield of MF_5 is obtained based on the reducing agent, which is the limiting reagent. The products are quite pure at this stage; however, MoF_5, ReF_5, and OsF_5 can be sublimed to a cold finger with continuous pumping. Sublimations are best performed in metal or Kel-F sublimation apparatus.

Procedure B

In a typical gas-phase carbon monoxide-UF_6 reduction, 1.7 g (4.9 mmole) of UF_6 is condensed at $-196°$ into an evacuated 30-ml Kel-F tube. Carbon monoxide (2.3 mmole) is expanded into the tube, and the valve of the reactor is closed. A Hanovia, 6515-34, 450-W Hg lamp is placed about 30 cm from the Kel-F tube and the gaseous sample is irradiated for 2-4 hours. (■**Caution.** *The lamp should be shielded by a heavy black screen from laboratory personnel.*) The pentafluoride product collects on the walls and the remaining volatile products can be vacuum evaporated. The yield is in excess of 90%. *Anal.* Calcd. for UF_5: U, 71.47; F, 28.52. Found: U, 71.33; F, 28.45.

Larger scale preparations using either Procedure A or B are possible; however, the scales described here are most convenient for normal laboratory uses. Procedure B is the preferred synthesis for UF_5.

Properties

The pentafluorides MoF_5 (yellow), ReF_5 (green), OsF_5 (blue), and UF_5 (pale yellow-green) are extremely moisture sensitive and must be handled and stored in a dry box. The samples can be stored in Kel-F bottles. With the exception of UF_5 these compounds have sufficient vapor pressure so that they can be sublimed. Uranium pentafluoride, on the other hand, is a nonvolatile solid at room temperature. The melting points for these compounds are : MoF_5, 65°; ReF_5, 47°; OsF_5, 70°. The infrared spectra (Nujol mull) show the following broad bands: MoF_5, 740, 693, 653, 520 cm^{-1}; ReF_5, 720, 691, 660, 530 cm^{-1}; OsF_5, 710, 690, 655, 530 cm^{-1}; UF_5, 620, 565, 510, 405 cm^{-1}. More detailed spectroscopic and powder diffraction data have been summarized elsewhere.[5-7]

References

1. H. H. Canterford and R. Colton, *Halides of the Transition Elements, Halides of the Second and Third Row Transition Metals*, J. Wiley & Sons, Inc., New York, 1968; N. Bartlett, *Prep. Inorg. React.*, **2**, 301 (1968); R. D. Peacock, *Adv. Fluorine Chem.*, **7**, 114 (1973).
2. R. D. Peacock, *Proc. Chem. Soc. Lond.*, **1957**, 59.
3. T. A. O'Donnell and D. F. Stewart, *J. Inorg. Nucl. Chem.*, **24**, 309 (1962).
4. W. E. Falconer, G. R. Jones, W. A. Sunder, I. Haigh, and R. D. Peacock, *J. Inorg. Nucl. Chem.*, **35**, 751 (1973).

5. R. T. Paine and L. B. Asprey, *Inorg. Chem.*, **13**, 1529 (1974).
6. R. T. Paine and L. B. Asprey, *Inorg. Chem.*, **14**, 1111 (1975).
7. L. B. Asprey and R. T. Paine, *Chem. Commun.*, **1973**, 920.
8. D. F. Shriver, *The Manipulation of Air-Sensitive Compounds*, McGraw Hill Book Co., New York, 1969, p. 136.
9. T. J. Ouellette, C. T. Ratcliff, D. W. A. Sharp, and A. M. Steven, *Inorg. Synth.*, **13**, 146 (1972).

32. SODIUM AND AMMONIUM DECAVANADATES(V)

$$10Na_3[VO_4] \cdot nH_2O \ + \ 24HOAc \ \rightarrow \ Na_6[V_{10}O_{28}] \cdot 18H_2O \ + \ 24NaOAc$$

$$6NH_3 \ + \ 5V_2O_5 \ + \ 9H_2O \ \rightarrow \ (NH_4)_6[V_{10}O_{28}] \cdot 6H_2O$$

Submitted by GARRETT K. JOHNSON* and R. KENT MURMANN*
Checked by R. DEAVIN† and W. P. GRIFFITH†

The aqueous chemistry of vanadium(V) is not very well understood. One reason for this may be the lack of well-defined compounds of known composition and structure.

In the very alkaline region $[VO_4]^{3-}$ is said to be the main ion, gradually giving dimers and trimers (or tetramers) as the basicity is decreased.[1] Equilibrium measurements in the pH region 4-7 suggest oligomerization predominantly to the decavanadate ion, $[V_{10}O_{28}]^{6-}$, which may be partially protonated.[2] A region of electroneutrality and minimum solubility lies at pH 2-3, and in very acidic solutions the $[VO_2]^+$ ion is thought to predominate.

The existence of the $[V_{10}O_{28}]^{6-}$ ion has been unequivocally demonstrated in the compounds $K_2Zn_2[V_{10}O_{28}] \cdot 16H_2O$ and $Ca_3[V_{10}O_{28}] \cdot 17H_2O$ by X-ray crystal structure determinations.[3] Raman spectra of these solids are very similar to the spectra of the solutions from which they crystallize, suggesting that the same highly symmetric conformation (D_{2h}) persists in both solutions and salts.[4]

Recent studies[5] of the ^{18}O isotropic exchange of $[V_{10}O_{28}]^{6-}$ and the ortho-vanadate ion, $[VO_4]^{3-}$, have shown slow kinetic exchange ($t_{1/2}$, minutes to hours) and have proven the identity of discrete decavanadate ions in solution.

At this time the decavanadate ion is better characterized than any other aqueous vanadium(V) species and may be the starting material for future structural advances in vanadium(V) chemistry. Presented here are simple

*Department of Chemistry, University of Missouri, Columbia, MO, 65201.
†Department of Chemistry, Imperial College, London SW 7, England.

syntheses of the sodium and ammonium salts of the decavanadate(V) ion and some guidelines for the preparation of others.

Two general approaches to syntheses seem to exist. In one method, exemplified in Section A for the sodium salt, a solution of a simple vanadate is acidified in a controlled way to produce the condensed $[V_{10}O_{28}]^{6-}$ ion, which is then crystallized as the salt of the original cation. This method has somewhat limited applicability inasmuch as the solubilities of many simple vanadates (e.g., of calcium and ammonium) are too low.

The second method, illustrated in Section B for the ammonium salt, relies on the direct stoichiometric acid-base reaction between V_2O_5 and the oxide, hydroxide, carbonate, or hydrogen carbonate (bicarbonate) of the desired positive ion. Surprisingly, this method works well for quite a variety of ions, even in cases where both reactants are only slightly soluble in water (e.g., CaO + V_2O_5).

In summary, the method in Section A is adaptable to K^+ and Li^+, as well as Na^+, and that in Section B has been found to be applicable to all Group I ions, as well as NH_4^+, and to Ca^{2+}, Sr^{2+}, and Ba^{2+}.

It should be noted that both methods strictly avoid the introduction of foreign positive ions (except H^+). This is necessary because of the very large variety of stoichiometric mixed-cation decavanadates known, which are generally less soluble than the pure end members and thus contaminate the latter if their formation is possible.

The typical side products in decavanadate preparations are the metavanadate and/or hexavanadate, as well as unreacted V_2O_5. These impurities are generally less soluble than the decavanadate salt and can be removed by filtration. X-Ray powder diffraction is felt to be the best criterion of phase purity unless macroscopic crystals are obtained.

Although many salts of the decavanadate ion can be made, their preparations in many cases depend implicitly on the relative kinetic inertness of the ion, rather than on true thermodynamic stability. This is certainly the case in the one system for which an order of stabilities can presently be assigned, namely, NH_3-V_2O_5-H_2O. Thus the following reaction is observed in the damp salt and in concentrated solutions:

$$(NH_4)_6[V_{10}O_{28}]\,(\text{solution}) \;\rightarrow\; 4NH_4[VO_3]\,(s) \;+\; (NH_4)_2[V_6O_{16}]\,(s)$$

$$\text{(orange)} \qquad\qquad \text{(white)} \qquad\qquad \text{(yellow-orange)}$$

Corroborating this, only the products of the above reaction appear in the equilibrium phase diagram of the system at 30°.[6] Decavanadate salts nevertheless occur naturally, and synthesized samples may be stored for years without change (see further under Properties).

A. SODIUM DECAVANADATE(V) OCTADECAHYDRATE, $Na_6[V_{10}O_{28}]\cdot$ 18H$_2$O

Procedure

A mixture of 50 g (0.182 mole) of commercial sodium (ortho)vanadate hydrate $(Na_3VO_4\cdot nH_2O)$* and 100 g of water is stirred with a magnetic bar for 15 minutes, or until nearly all of the solid is dissolved, at about 30°. Glacial acetic acid (50 mL total, 1.142 moles) is added with vigorous stirring over a period of 15-20 minutes, keeping the temperature in the 30-35° region. The solution becomes orange, and toward the end an orange crystalline product is deposited. The solution is stirred for 1 hour at 25° and the crystalline product is collected on a filter, washed four times with 50-mL portions of acetone, and air dried. The yield of crude product is approximately 25 g.

Purification is accomplished by crystallization from a solution prepared from the crude product and 100 mL of water at a maximum temperature of 50°. (The time at this temperature should be kept minimal.) Filtration of the nearly saturated solution through coarse fiber-glass paper (Whatman GF/A) removes impurities of low solubility. At 0° the solid product crystallizes slowly; 12 hours are required for complete deposition. The orange solid is collected on a filter and dried under atmospheric conditions (not vacuum). Yield: 17.0 g, 66% of theory.* Water of hydration is lost under low humidity conditions or under vacuum. *Anal.* Calcd. for $Na_6V_{10}O_{28}\cdot 18H_2O$: V, 35.89; H$_2$O, 22.84. Found: V, 35.51; H$_2$O, 22.65%.

Properties

The formula weight of this compound is 1419.6. The yellow-orange crystals are stable for an indefinite period in closed vessels at room temperature. They dehydrate to a yellow powder at low humidities or upon gentle heating. The density determined by flotation is 2.35 g/mL and the pH of a 5% aqueous solution is 6.7. Concentrated aqueous solutions show noticeable decomposition in 1 day at 25°, while dilute solutions $\simeq$ 0.01 M show little change in weeks. At 25° the compound has a solubility of 0.284 g/ml H$_2$O.

The principal X-ray powder diffraction lines of $Na_6[V_{10}O_{28}]\cdot 18H_2O$ are given on card number 20-1176 of the *Powder Diffraction File.*[7] Patterns reported for other hydrates of this salt are also in the file.

---

*The value for n in the commercial Na_3VO_4 is variable but usually is in the range of 3-6. The yield is based on $n = 5$.

B. AMMONIUM DECAVANADATE(V) HEXAHYDRATE, $(NH_4)_6[V_{10}O_{28}]\cdot 6H_2O$

Procedure

The starting material is high-grade powdered V_2O_5 having appreciable surface area (at least a few m^2/g). Materials that have proven satisfactory are: V_2O_5 produced by pyrolyzing NH_4VO_3 in air at $400°C$ for 2-4 hours; Spex Industries high-purity V_2O_5*; and Fisher Company certified reagent V_2O_5†. Fused V_2O_5 (nonporous) must not be used. V_2O_5 may contain varying amounts of V(IV); calcination in air or oxygen for a few hours at $400°$ ensures an adequate degree of purity with respect to oxidation state.

Minor departures from the following deceptively simple procedure may drastically reduce the yield.

A 20-g sample of V_2O_5 (0.22 mole V) is mixed with 60 mL of H_2O in a 200-250 mL glass container. The mixture is magnetically stirred continuously for 2 hours, or until the oxide is thoroughly slurried and suspended. (For larger preparations, ball-milling is preferable.)

A 10-mL sample of concentrated aqueous NH_3 (29%) (0.15 mole NH_3) is separately diluted with H_2O to a volume of 40 ml.

The V_2O_5 slurry is heated to $35°$, the source of heat is removed, and rapid dropwise addition of the NH_3 solution is begun. The complete addition should take 10-15 minutes, during which period the oxide must remain entirely suspended, through adequate stirring. The temperature must remain in the 30-$35°$ range and the pH should not exceed 6.5. (If the preparation is considerably scaled up, addition of small amounts of cracked ice is necessary to prevent overheating.)

Immediately after the addition of NH_3 is complete, the mixture is cooled as rapidly as practicable to $15-20°$. Complete reaction of V_2O_5 should not be awaited. The mixture is filtered with suction through glass or paper of medium porosity. The clear orange filtrate of pH 5.5-6 is returned to a clean vessel, and 75 mL of reagent grade acetone is added gradually with constant stirring. Stirring is continued for 5-10 minutes after addition is complete to allow maximum precipitation.

The orange crystalline product is collected on a coarse filter and sucked as dry as possible. It is washed once with a 100-mL portion of 50:50 acetone-water. (Neat acetone causes dehyration.)

The damp product is spread in thin layers on glass or filter paper and dried, with

---

*Available from Spex Industries Inc., 3880 Park Ave., Metuchen, NJ 08840.
†Available from Fischer Scientific Co., P.O. Box 171, Itasca, IL 60143.

frequent turning, at room temperature. Some decomposition occurs if complete drying is not effected within a few hours. This is indicated by lack of total solubility in H_2O.

The yield is 20.7 g (80%). Since no foreign species are introduced into the synthesis, the product may be considered essentially pure at this point. If recrystallization is desirable, the best method is to refrigerate for several days $(0°)$ an aqueous solution initially saturated around $25°$. *Anal.*: Calcd. for $(NH_4)_6V_{10}O_{28} \cdot 6H_2O$: V, 43.4; NH_3, 8.71; V_2O_5, 77.5; H_2O, 9.21. Found: V, 43.5; NH_3, 8.71; V_2O_5, 77.7; H_2O, 9.45%.

Properties

The formula weight of this compound is 1173.7. The crystals are red-orange equiaxed triclinic prisms, stable indefinitely in the dry state in air. The streak is yellow. The flotation density is 2.37 g/mL and the pH of 5% aqueous solution $(24°)$ is 7.2. The hydrate vapor pressure is 0.3 torr at $25°$. The solid may be dehydrated without further decomposition at $80°$ *in vacuo* in 2 hours. Loss of NH_3 is rapid above $180°$ in air. The salt may be pyrolyzed to pure V_2O_5 in air at $350\text{-}400°$ in about 2 hours. The solubilites are given below.

H_2O		H_2O-Acetone, $26°$	
Temp. ($°C$)	Conc. (M)	Ratio	Conc. (M)
40	0.5 (unstable)	10:0	0.32
30	0.37	9:1	0.10
20	0.23	8:2	0.04
10	0.10	7:3	0.02
0	0.08	6:4	0.004

The solution stabilities (room temperature) are: 0.1 M, decomposition apparent in about 1 day; 0.01 M, stable for months.

Principal X-ray powder lines are: 8.32 (10), 8.75 (7), 8.01 (4), 6.19 (2), 6.24 (1), 4.14 (1), 3.16 (1)Å.

Solutions of the NH_4^+ and the Na^+ salts at acidities about pH 7 contain only one vanadium oxy ion, $[V_{10}O_{28}]^{6-}$. Isotopic ^{18}O exchange with the solvent is slow,[5] the $t_{1/2}$ being of the order of 15 hours at $25°$, showing that the ion does not reversibly dissociate to smaller fragments at a rapid rate. The ion decomposes to V_2O_5 in acidic media and slowly to $(VO_3^-)_n$ and HVO_4^{2-} in strongly alkaline media. It behaves reversibly as a base having $pK_{a1} = 3.6 \pm 0.3$ and $pK_{a2} = 5.8 \pm 0.10$ at $25°C$; $\mu = 1.0$ with $NaClO_4$.[2] At $0°$, 10^{-2} M solutions of $Na_6V_{10}O_{28} \cdot 18H_2O$ give apparent $pK_{a1} = 4.7$ and $pK_{a2} = 7.2$.[5]

References

1. O. W. Howarth and R. E. Richards, *J. Chem. Soc. A*, **1965**, 864.
2. F. J. C. Rossotti and H. Rossotti, *Acta Chem. Scand.*, **10**, 957 (1956).
3. H. T. Evans, Jr., A. G. Swallow, and W. H. Barnes, *J. Am. Chem. Soc.*, **86**, 4209 (1964); H. T. Evans, Jr., *Inorg. Chem.*, **5**, 967 (1966); A. G. Swallow, F. R. Ahmed and W. H. Barnes, *Acta Cryst.*, **21**, 397 (1966).
4. W. P. Griffith and P. J. B. Lesniak, *J. Chem. Soc., A*, **1969**, 1066.
5. R. Kent Murmann, *J. Am. Chem. Soc.*, **96**, 7836 (1974).
6. A. D. Kelmers, *J. Inorg. Nucl. Chem.*, **17**, 168 (1961).
7. Powder Diffraction File, Joint Committee on Powder Diffraction Standards, Swarthmore, Pennsylvania.

33. *cis*-[DIHALOBIS(2,4-PENTAEDIONATO)TITANIUM(IV)] COMPLEXES

$$TiX_4 + 2C_5H_8O_2 \longrightarrow Ti(C_5H_7O_2)_2X_2 + 2HX$$

Submitted by C. A. WILKIE,* G. LIN,* and D. T. HAWORTH*
Checked by DAVID W. THOMPSON†

In this procedure *cis*-[dihalobis(2,4-pentanedionato)titanium(IV)] complexes (halo = F, Cl, Br) are prepared in high yields through the reaction of a titanium tetrahalide with 2,4-pentanedione in dichloromethane.

Procedure

1. *cis-[Difluorobis(2,4-pentanedionato)titanium(IV)]*. A 250-mL, three-necked flask is equipped with a magnetic stirring bar, 50-mL dropping funnel, and nitrogen inlet (vented through a safety bubbler). The apparatus is flushed with dry nitrogen and charged with 4.19 g (0.034 mole) of titanium(IV) fluoride and 150 mL of dichloromethane that has been dried over calcium hydride. A solution of redistilled 2,4-pentanedione (acetylacetone) (7.24 g, 0.0723 mole) in 25 mL of dichloromethane is slowly added from the dropping funnel to the flask while the contents of the latter are stirred. The dropping funnel is then replaced with a condenser, topped with a calcium chloride drying tube. The solution is heated at reflux for about 12 hours with a slow stream

*Department of Chemistry, Marquette University, Milwaukee, WI 53233.
†Department of Chemistry, The College of William and Mary, Williamsburg, VA 23185.

of nitrogen flowing from the condenser. The yellow solution is filtered under nitrogen using appropriate inert-atmosphere apparatus such as that described by Shriver[1] and Holah.[2] The filtrate is reduced in volume to about 50-60 mL by boiling off solvent under a nitrogen flow or removing the solvent under reduced pressure. Hexane, which has been dried over calcium hydride, is added to the cooled solution until the crystals begin to precipitate. The solution is allowed to stand until crystallization is complete (about 10 hr). The product is filtered, washed with 3 10-mL portions of hexane, and dried at room temperature *in vacuo*. The product (8.2 g, 84% yield) is recrystallized by dissolving it in dichloromethane and then adding hexane until precipitation begins; mp 165-166°, lit. 165-166°.[3] *Anal.* Calcd. for $C_{10}H_{14}O_4F_2Ti$: C, 42.27; H, 4.97; Ti, 16.8. Found: C, 41.55; H, 4.68; Ti, 16.7.

   2. cis-*[Dichlorobis(2,4-pentanedionato)titanium(IV)]*. A similarly equipped flask as used in the difluoro synthesis is flushed with dry nitrogen and charged with 150 mL of dichloromethane to which 4.24 g (0.022 mole) of titanium(IV) chloride is added from a preweighed syringe. A solution of redistilled acetylacetone (5.46 g, 0.055 mole) in 25 mL of dichloromethane is slowly added through the dropping funnel to the flask while the contents of the latter are stirred. The red solution is stirred for 30 minutes as it is purged with nitrogen. The solution is evaporated to about 50-60 mL under a nitrogen flow. Hexane is added to the cooled solution with stirring until crystallization begins. The solution is allowed to stand for about 6 hours and it then filtered; the precipitate is washed with 3 10-mL portions of hexane. The product (6.0 g, 90% yield) is redissolved in dichloromethane (about 10 mL) and recrystallized by adding hexane; mp 191-192°, lit. 191-192°.[3] *Anal.* Calcd. for $C_{10}H_{14}O_4Cl_2Ti$: C, 37.89; H, 4.45; Ti, 15.1. Found: C, 37.80; H, 4.24; Ti. 14.9.

   3. cis-*[Dibromobis(2,4-pentanedionato)titanium(IV)]*. The preparation of the dibromo derivative is similar to that of the dichloro complex. Since titanium(IV) bromide fumes in moist air, a pulverized sample of titanium(IV) bromide is prepared in a nitrogen-filled glove bag and placed in a preweighed glass-stoppered bottle (about 5 mL). The titanium(IV) bromide (5.32 g, 0.145 mole) is quickly poured from the weighing bottle into 150 mL of nitrogen-purged dichloromethane. Acetylacetone (6.24 g, 0.062 mole) in 25 mL of dichloromethane is slowly added from the dropping funnel to the flask while the contents of the latter are stirred. The dark-red solution is purged with nitrogen for 30 minutes, filtered, and then evaporated to about 50-60 mL under a nitrogen flow. The product (5.91 g, 88% yield) is crystallized from dichloro-methane-hexane; mp 242-244°, lit. 243-244°.[3] *Anal.* Calcd. for $C_{10}H_{14}O_4Br_2Ti$: C, 29.59; N, 3.48; Ti, 11.8. Found: C, 29.28; H, 3.15; Ti, 11.6.

Properties

Difluorobis(2,4-pentanedionato)titanium(IV) is a bright yellow solid, dichlorobis(2,4-pentanedionato)titanium(IV) is an orange-red solid, and dibromobis-(2,4-pentanedionato)titanium(IV) is a dark-red solid. They all exhibit parallel extinction and pleochroism.[3] Under atmospheric conditions the dibromo complex is easily hydrolyzed (1 hr), whereas the dichloro and difluoro complexes are not easily hydrolyzed (1-3 days).[3] All three complexes are soluble in nitrobenzene, dichloromethane, and chloroform, with the solubility increasing from the bromo to the fluoro derivative, and they are insoluble in saturated hydrocarbon solvents.

Previous ^{1}H NMR work has shown these *cis*-disubstituted (2,4-pentanedio-nato)titanium(IV) complexes exist in solution as nonrigid molecules.[4] Fay and coworkers[3,5-6] have shown that dihalobis(2,4-pentanedionato)titanium(IV) compounds rearrange by way of a unimolecular mechanism. In addition, the dihalo- and also the dialkoxybis(2,4-pentanedionato)titanium(IV) complexes undergo rapid ligand-exchange reactions that scramble both monodentate and bidentate ligands.[7] Except for the diiodobis(2,4-pentanedionato)titanium,[8] which exists in a dichloromethane solution in both the cis and trans forms, the other compounds of this type all have a cis-octahedral structure.[9] On reaction of these complexes with a titanium tetrahalide the apparent five-coordinate trihalo(diketonato)titanium(IV) can be synthesized. The trichloro-(diketonato)titanium(IV) complex has been shown to exist in solution as a monomer and as a dimer in the solid state.[10-11]

The IR spectra of these complexes are all very similar in the 100-3100 cm^{-1} range, which is indicative of acetylacetone as a bidentate ligand.[12] The KBr pellet technique gives the following major absorption bands below 1100 cm^{-1}. Ti(dik)$_2$F$_2$: 1025, 935, 807, 616, 547, 466, 411, 355, 312, 240, and 219 cm^{-1}; Ti(dik)$_2$Cl$_2$; 1029, 952, 932, 805, 667, 541, 471, 372, 319, 265, and 220 cm^{-1}; Ti(dik)$_2$Br$_2$; 1031, 931, 811, 674, 667, 510, 410, 361, 320, 239 and 210 cm^{-1}. The low-frequency spectra on Nujol mulls between polyethylene sheets give the following major absorption bands. Ti(dik)$_2$F$_2$; 408, 352, 309, 257, and 220 cm^{-1}; Ti(dik)$_2$Cl$_2$; 372, 315, 265, 218; Ti(dik)$_2$Br$_2$; 406, 358, 305, and 215 cm^{-1}.

References

1. D. F. Shriver, *The Manipulation of Air-Sensitive Compounds*, McGraw-Hill Book Co., New York, 1969, Chap. 7.
2. D. G. Holah, *J. Chem. Educ.*, **42**, 561 (1965).
3. R. C. Fay and R. N. Lowry, *Inorg. Chem.*, **6**, 1512 (1967).
4. D. C. Bradley and C. E. Holloway, *Chem. Commun.*, 1965, 284; *J. Chem. Soc. Dalton Trans.*, 1969, 282.
5. N. Serpone and R. C. Fay, *Inorg. Chem.*, **6**, 1835 (1967).
6. R. C. Fay and R. N. Lowry, *Inorg. Nucl. Chem. Lett.*, **3**, 117 (1967).
7. R. C. Fay and R. N. Lowry, *Inorg. Chem.*, **13**, 1309 (1974).
8. R. C. Fay and R. N. Lowry, *Inorg. Chem.*, **9**, 2048 (1970).
9. N. Serpone and R. C. Fay, *Inorg. Chem.*, **8**, 2379 (1969).

10. N. Serpone, P. H. Bird, D. G. Bickley, and D. W. Thompson, *Chem. Commun.,* **1972,** 217.

11. N. Serpone, P. H. Bird, A. Somogyvari, and D. G. Bickley, *Inorg. Chem.,* **16,** 2381 (1977).

12. K. Nakamoto, *Infrared Spectrum of Inorganic and Coordination Compounds,* 3rd ed., John Wiley & Sons, New York, 1978, 251.

Chapter Four

TRANSITION METAL ORGANOMETALLIC COMPOUNDS

34. BIS(η^8-CYCLOOCTATETRAENE)URANIUM(IV)

$$2K + C_8H_8 \longrightarrow K_2[C_8H_8]$$

$$2K_2[C_8H_8] + UCl_4 \longrightarrow U(C_8H_8)_2 + 4KCl$$

Submitted by ANDREW STREITWIESER, JR.,* U. MUELLER-WESTERHOFF,*
FRANTISEK MARES*, CHARLES B. GRANT,* and DENNIS G. MORRELL*
Checked by TOBIN J. MARKS† and STEVEN S. MILLER†

The synthesis of bis(η^8-cyclooctatetraene)uranium(IV) (uranocene)‡ from uranium tetrachloride and (cyclooctatetraene)dipotassium was first published in 1968.[1] The method reported here is a modification of that procedure and is suitable for a large variety of cyclooctatetraene complexes.[2-4] Bis(η^8-cyclooctatetraene)uranium(IV) has also been prepared by the reaction of uranium tetrafluoride with (cyclooctatetraene)magnesium in the absence of solvent.[5] Direct reaction of finely divided uranium metal with cyclooctatetraene vapors at 150° also gives some uranocene.[5,6] However, both methods give low yields.

*Department of Chemistry, University of California, Berkeley, CA 94720; work supported in part by grants from the National Science Foundation.
†Department of Chemistry, Northwestern University, Evanston, IL 60201.
‡The suffix "ocene" is formally reserved for bis(η-cyclopentadienyl) metal complexes. The name "uranocene" does not fit this convention, but has been used as a *trivial* name in the literature.

Bis(η^8-cyclooctatetraene)uranium(IV) and its precursor, dipotassium cyclooctatrienediide, are air and water sensitive. A knowledge of inert-atmosphere techniques is essential to conduct this synthesis successfully.[7] (■**Caution.** *Dry dipotassium cyclooctatrienediide may react explosively with air; therefore it always should be handled under an inert atmosphere or in vacuum.*)

Procedure

■**Caution.** *Tetrahydrofuran must be free of peroxides before it is distilled from LiAlH$_4$.*[8]

Dry, degassed tetrahydrofuran (THF) is first prepared by distillation from LiAlH$_4$ or sodium-benzophenone under nitrogen to ensure the absence of water and peroxides. Degassing is accomplished by two or three freeze-pump-thaw cycles. Approximately 100 mL of this THF is distilled by vacuum transfer into the reaction vessel shown in Fig. 1. The condenser coils are checked for leaks, both visually and by ability to hold a vacuum (10^{-3} torr). Potassium metal is cut under dry toluene. (■**Caution.** *Potassium metal is highly reactive with water and with the skin of potassium oxides. The oxides should be scraped off of each chunk before it is cut.*) The chunks are then weighed [4.16 g, 0.106 mole] in a tared beaker of dry toluene and then washed with hexane. The metal is added to the reactor vessel through the side arm while the reactor is flushed with prepurified argon. Since the THF attacks the vacuum grease, a small amount of fresh grease should be applied to the side-arm joint each time it is opened. The cap is carefully flushed with argon before tightening.

Over a safety pan, the THF is heated to reflux with a heat gun and stirred vigorously. When the potassium is finely divided, the stirring is stopped. Heating is continued until the circular motion of the liquid has stopped. The potassium sand should be finely divided to ensure short reaction times. The mixture is allowed to cool to room temperature without external cooling. Usually, the THF may be reheated with stirring if an unsatisfactory sand results. After the potassium has hardened, the reaction mixture is cooled to $-30°$, and cyclooctatetraene* (5.70 g, 0.053 mole; used as supplied) is added with a syringe or pipe through the side arm. The reaction mixture is maintained at $-30°$ until all of the potassium has dissolved.[9] Cooling of the reaction mixture is unnecessary when preparing alkyl- or aryl-substituted cyclooctatetraene uranium complexes. The corresponding substituted dianions appear to be more thermally stable in THF than the parent. A small excess of potassium metal has little effect on the yield or quality of the product.

Uranium tetrachloride for the next step is prepared from UO$_3$ and hexachloropropene according to the method of Hermann and Suttle[10] and may be stored

---

*Available from Aldrich Chemical Co., 940 W. St. Paul Ave., Milwaukee, WI 53233.

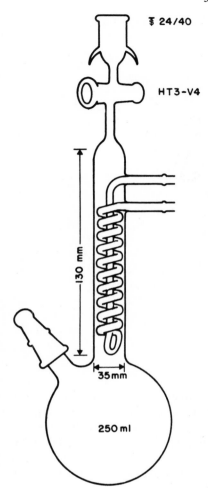

Fig. 1. Reaction vessel.

in a glove box. Commercially available UCl_4 contains enough impurities to reduce the yield significantly. A solution of UCl_4 [10.0 g, 0.0264 mole] is prepared in about 100 mL of THF that has been dried and degassed as above. (■**Caution.** *Uranium tetrachloride evolves heat when dissolving in THF and can form a dark crust that only slowly dissolves. A safe, convenient way to prepare this solution is to place about 50 ml of THF in the addition flask shown in Fig. 2 and add the uranium salt slowly with stirring in the glove box.*) When all the salt has been added, excess UCl_4 is washed down with the remaining 50 mL of solvent. The stopper is replaced and tied down with rubber bands. The stopcock is closed, and a small cork is placed in the open end of the delivery

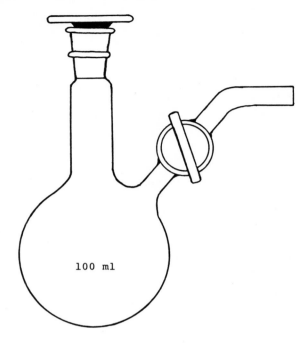

Fig. 2. Addition flask.

tube. This THF solution is added to the solution of dipotassium cyclooctatriene-diide by removing the cork and inserting the tube into the reactor side-arm. The stopcock is opened, and the addition flask is tilted to allow the solution to run out. Heating with a heat gun speeds delivery. The solution immediately turns emerald green. The reaction is quite rapid, and may be over within a few minutes; however, our usual procedure is to allow at least 3 hours of reaction time. The reaction mixture is then frozen and degassed, and the solvent is removed by vacuum transfer. The crude solid is dried under vacuum for at least 4 hours.

Purification of bis(η^8-cyclooctatetraene)uranium(IV) is achieved through Soxhlet extraction of the crude solid by THF in the extractor shown in Fig. 3. In the glove box, the green, powdery crude material is placed in a thoroughly degassed paper thimble. The thimble is placed in the extractor, and the solvent is added. The bis(η^8-cyclooctatetraene)uranium(IV) is extracted under argon until the solution surrounding the thimble is no longer dark green. Depending on the rate of reflux, extraction can take as long as 9 days. Many alkyl- and aryl-substituted cyclooctatetraene uranium complexes are highly soluble in THF. To minimize washing losses, other solvents are recommended.[2] When toluene is used, the extractor should have a round bottom; thawing of frozen toluene can crack cylindrical extractor bottoms.

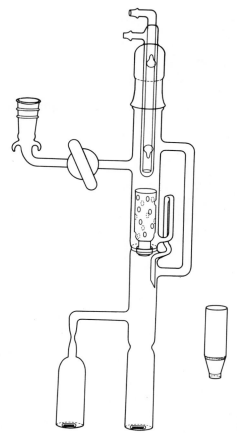

Fig. 3. Soxhlet extractor for highly air sensitive compounds.

When the solution cools to room temperature, a green solid collects in the bottom of the extractor. Cooling to $-70°$ is recommended at this point for the more soluble uranium compounds, such as bis(butyl-η^8-cyclooctatetraene)-uranium(IV). The extractor stopcock is closed, and the extractor is detached from the vacuum manifold. The supernatant liquid is poured into the side arm. The solid remaining is cooled with liquid nitrogen, and portions of solvent are distilled in from the side arm using a heat gun. After it is thawed, the solid is mixed thoroughly with the solvent and allowed to settle. The supernatant liquid is poured into the side arm. This washing process is then repeated. The solution in the side arm and the solid are both frozen with liquid nitrogen, and the liquid is degassed once to remove condensed argon. The stopcock to the extractor is opened carefully and slowly. Enough solvent remains in the thimble to spray the unextracted solids onto the pure product. The solution in the side arm is refrozen, and the side arm is sealed off at the constriction with a flame under full vacuum (10^{-3} torr). The solid bis(η^8-cyclooctatetraene)uranium(IV) is warmed to room temperature, and the residual solvent is carefully pumped off with stirring. When all traces of solvent have been removed from the product, the extractor is opened in the glove box. The thimble and plug are removed, and the stub end of the side arm is scored and broken off. The bis(η^8-cyclooctate-traene)uranium(IV) is scraped out through the resultant hole with a long-handled spatula. Yield 9.6 g (82%).

The bis(η^8-cyclooctatetraene)uranium(IV) can be purified further by extraction with benzene or by sublimation at $140°/10^{-3}$ torr.

Properties

The green crystals of bis(η^8-cyclooctatetraene)uranium(IV) are air and water sensitive and only slightly soluble in aromatic and polar organic solvents. The visible spectrum is characteristic. λ_{max}^{THF}: 615 (1850); 641 (890); 660 (600); and 680 nm (350 L/mole-cm). The IR spectrum shows only four strong bands at 3000, 1430, 900, and 741 cm^{-1}.[4]

References

1. A. Streitwieser, Jr. and U. Mueller-Westerhoff, *J. Am. Chem. Soc.*, **90**, 7364 (1968).
2. A. Streitwieser, Jr. and C. A. Harmon, *Inorg. Chem.*, **12**, 1102 (1973).
3. K. O. Hodgson, F. Mares, D. F. Starks, and A. Streitwieser, Jr., *J. Am. Chem. Soc.*, **95**, 8650 (1973).
4. A. Streitwieser, Jr., U. Mueller-Westerhoff, G. Sonnichsen, F. Mares, D. G. Morrell, K. O. Hodgson, and C. A. Harmon, *J. Am. Chem. Soc.*, **95**, 8644 (1973).
5. D. F. Starks, Ph.D. Dissertation, University of California, Berkeley, 1974.
6. D. F. Starks and A. Streitwieser, Jr., *J. Am. Chem. Soc.*, **95**, 3423 (1973).
7. D. F. Shriver, *The Manipulation of Air-Sensitive Compounds*, McGraw-Hill Book Co., New York, 1969.
8. *Inorg. Synth.*, **12**, 317 (1970).
9. T. J. Katz, *J. Am. Chem. Soc.*, **82**, 3784 (1960).
10. J. A. Hermann and J. F. Suttle, *Inorg. Synth.*, **5**, 143 (1957).

35. (η^6-ARENE)TRICARBONYLCHROMIUM COMPLEXES

Submitted by C. A. L. MAHAFFY* and P. L. PAUSON*
Checked by M. D. RAUSCH[†] and W. LEE[†]

Most simple arenes react smoothly with Cr(CO)$_6$ to give hexahapto complexes

*Deparement of Pure and Applied Chemistry, University of Strathclyde, Thomas Graham Building, Cathedral Street, Glasgow, G1 1XL.
[†]Department of Chemistry, University of Massachusetts, Amherst, MA 01002.

$(\eta^6\text{-ArH})\text{Cr(CO)}_3$. The reactions have been conducted under a wide variety of conditions,[1] and the chief problems encountered arise from (a) the volatility of Cr(CO)_6, (b) slowness of the reaction, and (c) difficulty in removing high-boiling solvents or excess arene from the product. Sublimation of Cr(CO)_6 from the reaction vessel has been overcome by the use of rather complex apparatus,[2] but the following procedure shows this to be unnecessary. In inert solvents [decahydronaphthalene (decalin) has been widely used], reaction is excessively slow. Donor solvents (D) lead to appreciably more rapid reaction, probably by way of intermediates $\text{Cr(CO)}_{6-n}\text{D}_n$ (where n = 1-3); but if the donor is too good (e.g., $\text{C}_6\text{H}_5\text{CN}$), it may compete, especially with the less reactive arenes (e.g., $\text{C}_6\text{H}_5\text{CN}$), leading to incomplete complex formation. Alkylpyridines have been recommended,[3,4] but ethers have been much more widely used. Tetrahydrofuran (THF), a good donor, allows reaction to proceed cleanly[5] but too slowly because of its low boiling point. Dibutyl ether also leads to rather slow reaction, probably because of its weak donor properties, whereas bis(2-methoxyethyl) ether (diglyme), which is better in this respect, is relatively difficult to remove. The following procedure therefore uses a mixture of dibutyl ether with sufficient THF to "catalyze" the reaction and to wash back most of the Cr(CO)_6 that sublimes into the condenser, but not enough to lower the boiling point too much.

A. (η⁶-ANISOLE)TRICARBONYLCHROMIUM

Procedure

■**Caution.** *The reaction should be carried out in a well ventilated hood, as hexacarbonylchromium is toxic and carbon monoxide is evolved during the reaction. Both ether solvents peroxidize; they should be carefully freed from peroxide and dried (conveniently by distilling from lithium tetrahydridoaluminate or from sodium) before use. Benzene is toxic; contact with the liquid or vapor should be avoided.*

In a 250-mL, round-bottomed flask fitted with a gas inlet and a simple reflux condenser [not spiral or similar type from which subliming Cr(CO)_6 is washed back less efficiently] are placed hexacarbonylchromium (4 g, 18 mmole), anisole (25 mL), dibutyl ether (120 mL) and tetrahydrofuran (10 mL). A bubbler is placed at the top of the condenser to prevent access of air. The apparatus is thoroughly purged with nitrogen. The nitrogen stream is stopped and the mixture is then heated at reflux for 24 hours (the checkers found stirring beneficial). The yellow solution is cooled and filtered through kieselguhr (diatomaceous earth) or a similar material (the checkers used Celite or, preferably, a small pad of anhydrous silica gel) on a sintered-glass filter, which is then washed with a little additional solvent. The solvents are distilled off on a rotary evaporator from a water bath held at 60° (an oil pump may be required to remove the solvents completely); a deep-yellow oil remains to which dry light petroleum ether (bp. 40-60°) or hexane (20 mL) is added. Crystalline (η^6-anisole)tricarbonylchromium [4.1 g (92%), mp 83-84°] separates. A small amount of unreacted $Cr(CO)_6$ may be recoverable from the condenser; the remainder distills off with the solvent. If a very pure product is required the compound may be recrystallized by dissolving it in benzene or in diethyl ether and adding light petroleum ether to give 3.53 g (80%), mp 84-85°; lit.[6] mp 84-85°. *Anal.* Calcd. for $C_{10}H_8CrO_4$: C, 49.2; H, 3.3. Found: C, 49.5; H, 3.4.

Solutions of this compound and other arenetricarbonylchromiums can only be handled for very brief periods in air. The workup procedures are therefore best carried out in an inert atmosphere throughout, but they can be conducted in air if done rapidly and efficiently.

Properties

(η^6-Anisole)tricarbonylchromium is a yellow crystalline solid. Its NMR spectrum[8] ($CDCl_3$ solution) shows a singlet at τ 6.4 for the CH_3O group and three well-resolved signals for the metal-bound ring at τ 5.23 (1H, t) 4.97 (2H,d) and 4.6 (2H, t). Its solutions are air sensitive, forming greenish precipitates, but the pure solid is stable. However, refrigeration is recommended for prolonged storage.

B. OTHER ARENETRICARBONYLCHROMIUM COMPLEXES

The general applicability of the above method is illustrated by the examples in Table I. With the lowest boiling arenes, subliming $Cr(CO)_6$ is completely washed back, and none is recovered at the end of the reaction. However, the reduced boiling point of the mixture necessitates a longer reaction time. For arenes of higher boiling points, it is preferable not to use a large excess, since it would be difficult to remove at the end of the reaction. This leads to a somewhat less complete reaction, but since much of the unchanged $Cr(CO)_6$ is readily recovered from the condenser and the distillate, yields in the table are based on unrecovered carbonyl. The result with anisole when only a small excess is used is included for comparison. The arenes for which results are tabulated are liquids, but several solid arenes have been used similarly. Thus the checkers obtained an 81% yield of (triphenylene)tricarbonyl chromium using a 1:1 molar ratio of the hydrocarbon to $Cr(CO)_6$.

TABLE I Reaction of Arenes with 4 g of $Cr(CO)_6$ to Give the Corresponding Complexes $(\eta^6\text{-Arene})Cr(CO)_3$[a]

Arene	Volume (mL)	Reflux Time (hr)	$Cr(CO)_6$ Recovered (g)	Yield g	Yield %	Product mp[b] (°C)	ν_{CO} peaks in C_6H_{12}[c] (cm^{-1})
$C_6H_5(OCH_3)$	5	42	–	3.77	85	83-84 (84-85, Ref. 6)	1980, 1908
C_6H_6	20	40	–	3.44	89[d]	159-160 (161.5-163, Ref. 6)	1982, 1915
C_6H_5F	16	42	–	3.79	90	108[e] (116-117, Ref. 7)	1990, 1929, 1926
C_6H_5Cl	20	20	0.35	2.6	64[f]	101-102 (102-103, Ref. 6)	1991, 1929, 1925
$C_6H_5(NMe_2)$	5[g]	19	0.79	3.2	85	144 (145-146, Ref. 6)	1969, 1894, 1888
C_6H_5COOMe	10	20	0.70	3.63	89	92-93 (95.5-96, Ref. 6)	1990, 1927

[a]The procedure given for $[(\eta^6\text{-}C_6H_5(OCH_3)]Cr(CO)_3$ was used, except for the conditions noted.

[b]Melting point of the product as initially isolated, before recrystallization; literature melting point of the pure compound in parentheses.

[c]Values from Reference 9.

[d]A 19-hr run [0.21 g recovered $Cr(CO)_6$] gave 2.3 g (62%).

[e]This mp is exceptionally sensitive to the rate of heating: an analytically pure sample of mp 116° was obtained after two crystallizations.

[f]Slight decomposition during the reaction indicated by formation of grey-green precipitates was found difficult to avoid with chlorobenzene and some other arenes. This appears to become progressive and reaction should be stopped and solutions filtered when such precipitation is observed.

[g]Solvent: dibutyl ether (60 mL) + tetrahydrofuran (5 mL); most of the product from this reaction crystallized from the solution when cooled in ice.

Reactivity

The chemistry of arenetricarbonylchromium complexes has been extensively studied and reviewed.[10,11] Compared to the uncomplexed arenes they show greatly enhanced reactivity towards nucleophiles.[12]

References

1. Comprehensive references may be found in *Gmelin's Handbuch der anorganischen Chemie,* Supplement to the 8th ed., Vol. 3, **1971**, pp. 181-289).
2. W. Strohmeier, *Chem. Ber.,* **94**, 2490 (1961), A. T. T. Hsieh, W. C. Matchan, H. van den Bergen, and B. B. West, *Chem. and Ind.* (London), **1974**, 114.
3. R. L. Pruett, J. E. Wyman, D. R. Rink, and L. Parts. U. S. Patent 3378569 (1968); *Chem. Abstr.,* **69**, 77512 (1968); U. S. Patent 3382263 (1968); *Chem. Abstr.,* **69**, 59376 (1968).
4. M. D. Rausch, *J. Org. Chem.,* **39**, 1787 (1974).
5. W. P. Anderson, N. Hsu, C. W. Stanger, and B. Munson, *J. Organomet. Chem.,* **69**, 249 (1974).
6. W. R. Jackson, B. Nicholls, and M. C. Whiting, *J. Chem. Soc.,* **1960**, 469.
7. J. F. Bunnett and H. Hermann, *J. Org. Chem.* **36**, 4081 (1971); V. S. Khandkarova, S. P. Gubin, and B. A. Kvasov, *J. Organomet. Chem.,* **23**, 509 (1970).
8. W. McFarlane and S. O. Grim, *J. Organomet. Chem.,* **5**, 147 (1966); A. Mangini and F. Taddei, *Inorg. Chim. Acta.,* **2**, 8 (1968).
9. R. D. Fischer, *Chem. Ber.,* **93**, 165 (1960); D. A. Brown and H. Sloan, *J. Chem. Soc.,* **1963**, 4389; D. A. Brown and J. R. Raju, *J. Chem. Soc., A,* **1966**, 1617; D. M. Adams and A. Squire, *J. Chem. Soc., Dalton Trans.,* **1974**, 558.
10. R. P. A. Sneeden, *Organochromium Compounds,* Academic Press Inc., New York, 1975.
11. W. E. Silverthorn, *Adv. Organomet. Chem.,* **13**, 47 (1975).
12. M. F. Semmelhack, in *New Applications of Organometallic Reagents in Organic Synthesis,* D. Seyferth, ed., Elsevier Publishing Co., Amsterdam, 1976, p. 361.

36. PENTACARBONYLMANGANESE HALIDES

The pentacarbonylmanganese halides have been known for some time[1,2] but are still the subject of intense interest. Numerous vibrational analyses have been carried out,[3] as well as synthetic and kinetic studies of carbonyl substitution reactions.[4] Halide substitution, which requires elevated temperatures,[5] leads to charged species. Diazocyclopentadienes insert into the manganese-halogen bond.[6]

The best method for preparation of pentacarbonylchloromanganese is the reaction of chlorine with dimanganese decacarbonyl in carbon tetrachloride solution. Earlier versions of this procedure[2] often give low yields; however, the present version is reliable and produces a good yield.

The action of CCl_4 alone on $Mn_2(CO)_{10}$ also gives $MnCl(CO)_5$,[7] but the reaction is too slow to be of synthetic utility. Recent reports[8] suggest that photolysis enhances both the rate and yield of this reaction.

A commonly used preparation of $MnBr(CO)_5$ is the reaction of $Mn_2(CO)_{10}$ with Br_2 in CCl_4.[2,9] However, the product is frequently contaminated with $MnCl(CO)_5$ owing to reaction of the decacarbonyl with the solvent.[7] A substantial increase in the yield and purity of this compound can be achieved by the use of CS_2 as the reaction solvent, as in the procedure given here.

The iodo analogue has been prepared by thermal[1,7] and photochemical[8] reactions of $Mn_2(CO)_{10}$ with I_2. The former method, which is simple to perform, is described in Section D. The reaction of $MnH(CO)_5$ and I_2 also has been reported to give $MnI(CO)_5$.[10] Another procedure utilizes the nucleophilic character of the anion $Mn(CO)_5^-$. This reaction, which forms the basis of the procedure in Section C, was mentioned briefly some years ago,[11] but no experimental detail was provided. Advantages of the method are speed, high yield, and absence of product contamination by unreacted $Mn_2(CO)_{10}$.

Commercial $Mn_2(CO)_{10}$ should be purified by sublimation (70°, 0.05 torr) before use in these preparations. Solutions of $MnX(CO)_5$ must not be heated during evaporations, owing to the facile decarbonylation of these compounds to $[MnX(CO)_4]_2$[2,12] (especially for X = Cl). Evaporations can be assisted by immersion of the flask in a *room temperature* water bath.

A. PENTACARBONYLCHLOROMANGANESE, MnCl(CO)$_5$

$$Mn_2(CO)_{10} + Cl_2 \longrightarrow 2MnCl(CO)_5$$

Submitted by KENNETH J. REIMER* and ALAN SHAVER[†]
Checked by MICHAEL H. QUICK[‡] and ROBERT J. ANGELICI[‡]

The reaction is conducted in a 100-mL three-necked round-bottom flask fitted with a nitrogen inlet (attached to a nitrogen line equipped with a mercury bubbler),[13] a magnetic stirrer, and an equipressure dropping funnel. Finely ground $Mn_2(CO)_{10}$(4.0 g, 0.01 mole) is dissolved in a minimum of degassed carbon tetrachloride at 0° in the nitrogen-filled flask. Another sample of carbon tetrachloride (25 mL) in a dropping funnel is saturated by means of a stream of chlorine and the yellow solution is added dropwise over 30 minutes to the cooled, stirred solution of $Mn_2(CO)_{10}$. As the addition proceeds, some

*Department of Chemistry, University of British Columbia, Vancouver, B.C., Canada.
[†] Department of Chemistry, McGill University, Montreal, Quebec, Canada.
[‡] Department of Chemistry, Iowa State University, Ames, IA, 50011.

MnCl(CO)$_5$ begins to precipitate. After complete addition, the reaction mixture is allowed to warm to room temperature and is then stirred for 4 hours. The yellow precipitate is filtered in air and washed several times with carbon tetrachloride. The yield of crude product is 4.04 g (86%). The compound is contaminated with only small amounts of [MnCl(CO)$_4$]$_2$ and white insoluble material. These impurities are negligible and usually no further purification is undertaken. If pure MnCl(CO)$_5$ is required, it is easily obtained by sublimation (40°/0.1 torr, yield 65%) [based on Mn$_2$(CO)$_{10}$]. *Anal.* Calcd. for MnCl(CO)$_5$: C, 26.06; Cl, 15.39. Found: C, 26.08; Cl, 15.44.

B. PENTACARBONYLBROMOMANGANESE, MnBr(CO)$_5$

$$Mn_2(CO)_{10} + Br_2 \longrightarrow 2MnBr(CO)_5$$

Submitted by MICHAEL H. QUICK* and ROBERT J. ANGELICI*
Checked by KENNETH J. REIMER[†] and ALAN SHAVER[‡]

A Schlenk tube of about 100-mL capacity, equipped with a magnetic stirring bar, is flushed with N$_2$. Dimanganese decacarbonyl (2.00 g, 5.13 mmole) and CS$_2$ (50 mL) are then added, under a nitrogen flush, and the mixture is stirred for about 10 minutes under nitrogen. (■**Caution.** *Carbon disulfide is highly flammable and toxic and should be vented through an efficient fume hood.*) Some of the Mn$_2$(CO)$_{10}$ remains undissolved, but this does not affect the reaction. A solution of Br$_2$ (0.35 mL, 1.0 g, 6.3 mmole) in 20 mL of CS$_2$ is then added under nitrogen from an equipressure dropping funnel over a period of 20-30 minutes; some of the product begins to precipitate during this time. Stirring is continued for an additional 1 hour. The dark-red mixture is then evaporated under reduced pressure to give an orange powder.

The crude product is stirred with 150 mL of CH$_2$Cl$_2$, and the orange solution is filtered. Hexane (60 mL) is added, and the solution is slowly evaporated under reduced pressure (50-60 torr) to about 30 mL. The yellow-orange, micro-crystalline MnBr(CO)$_5$ is filtered off, washed with cold (0°) pentane, and dried *in vacuo*. Yield 2.53 g (90%). The product thus obtained is sufficiently pure for most purposes, but it can be sublimed (50°, 0.05 torr) with only a

*Department of Chemistry, Iowa State University, Ames, IA 50011.
†Department of Chemistry, University of British Columbia Vancouver, B.C., Canada.
‡Department of Chemistry, McGill University, Montreal, Quebec, Canada.

slight decrease in yield if further purification is desired.

The reaction may also be carried out in dichloromethane or cyclohexane. However, the yield and purity of the product are generally superior in CS_2. *Anal.* Calcd. for $MnBr(CO)_5$: C, 21.8; Br, 29.1. Found: C, 21.7; Br, 28.7.

C. PENTACARBONYLIODOMANGANESE, MnI(CO)₅

$$Mn_2(CO)_{10} + 2Na \longrightarrow 2Na[Mn(CO)_5]$$

$$Na[Mn(CO)_5] + I_2 \longrightarrow MnI(CO)_5 + NaI$$

Submitted by MICHAEL H. QUICK* and ROBERT J. ANGELICI*
Checked by KENNETH J. REIMER[†] and ALAN SHAVER[‡]

The reaction is conducted in a 100-mL three-neck amalgam reduction flask, which is described elsewhere.[14] The mixture is agitated with a paddle stirrer driven by a small electric motor. (The checkers found a Teflon-coated stirring bar and magnetic stirrer adequate.) The apparatus is flushed with nitrogen, after which 6 mL of 1% Na(Hg) and a solution of $Mn_2(CO)_{10}$ (2.00 g, 5.13 mmole) in 40 mL of tetrahydrofuran (distilled from $LiAlH_4$; see Reference 15 for the distillation procedure) are added. The mixture is stirred vigorously for 30-40 minutes, after which the excess amalgam is drained off. The flask and solution are then "washed" for a few minutes by vigorous stirring with 3-4 mL of fresh mercury, which is then removed as before. A somewhat cloudy greenish-gray solution of $Na[Mn(CO)_5]$ is obtained.[14]

A solution of I_2 (2.65 g, 10.4 mmole) in 20 mL of THF is then added from an equipressure dropping funnel to the stirred solution over a period of 15-20 minutes. The solution, now clear and deep red-orange, is stirred for an additional 5 minutes and then evaporated to dryness under reduced pressure. The residue is extracted with 120 mL of 1:1 hexane-dichloromethane, the mixture is filtered, and the filtrate is evaporated to dryness. Sublimation (50°, 0.05 torr) or recrystallization from hexane (about 40 mL per gram of crude product) at $-20°$ gives red-orange crystals of pure $MnI(CO)_5$. Yield 2.50-2.72 g (75-82%). *Anal.* Calcd. for $MnI(CO)_5$: C, 18.6; I, 39.4. Found: C, 18.4; I, 40.3.

*Department of Chemistry, Iowa State University, Ames, IA 50011.
†Department of Chemistry, University of British Columbia, Vancouver, B.C., Canada.
‡Department of Chemistry, McGill University, Montreal, Quebec, Canada.

D. PENTACARBONYLIODOMANGANESE, ALTERNATE PROCEDURE

$$Mn_2(CO)_{10} + I_2 \longrightarrow 2MnI(CO)_5$$

Submitted by KENNETH J. REIMER* and ALAN SHAVER[†]
Checked by MICHAEL H. QUICK[‡] and ROBERT J. ANGELICI[‡]

Stoichiometric quantities of $Mn_2(CO)_{10}$ and I_2 are introduced into a thick-walled Carius tube (the checkers used a capable glass pressure bottle[16]), mixed thoroughly, and cooled to $0°$ and the tube is sealed *in vacuo*. The mixture is uniformly heated in an oven at $90°$ until the color is uniform (12 hr is convenient for a reaction employing about 1 g of $Mn_2(CO)_{10}$). (■**Caution.** *Care should be taken when handling sealed reaction vessels.*[1] *Face and hands should be protected from the possibility of flying glass and chemicals.*) The residue is placed in a sublimation apparatus fitted with a water-cooled probe. Sublimation ($40°/0.1$ torr) separates the product from a nonvolatile powder, but the sublimate is usually contaminated with $Mn_2(CO)_{10}$. Purification may be accomplished by fractional sublimation or, more conveniently, by recrystallization from pentane to give an overall 50% yield of ruby-red crystals.

Properties of MnX(CO)₅, X = Cl, Br, I

The yellow $MnCl(CO)_5$ is slightly to moderately soluble in nonpolar organic solvents. Although stable in air, prolonged exposure to light results in the formation of $[MnCl(CO)_4]_2$. In a closed vessel an equilibrium exists between the monomer and dimer in benzene[17] and pentane.[6b] Unstoppered reaction vessels allow loss of CO and subsequent dimer formation. Heating solutions or solid samples of $MnCl(CO)_5$ accelerates dimer formation.

Pentacarbonylbromomanganese is an air-stable yellow-orange crystalline solid. Unlike the chloro analogue, $MnBr(CO)_5$ is not particularly light sensitive. Its properties, such as solubility in organic solvents and ease of decarbonylation,[12] are intermediate between those of the chloro and iodo compounds.

The red crystalline (orange powder) $MnI(CO)_5$ is more soluble in common organic solvents thatn the chloro analogue. It dimerizes slowly in solution. The

*Department of Chemistry, University of British Columbia, Vancouver, B.C., Canada.
[†]Department of Chemistry, McGill University, Montreal, Quebec, Canada.
[‡]Department of Chemistry, Iowa State University, Ames, IA 50011.

compound is best stored at 5°, since iodine is liberated on prolonged standing at room temperature.

The infrared spectrum in the carbonyl stretching region is very useful in characterizing these complexes (Table I). Three infrared active bands are predicted;[18] however, limited solubility may preclude observation of the weaker bands. Dimer formation is easily detected by the presence of characteristic bands.[12]

TABLE I

Compound	νCO (cm^{-1}) in CCl_4 [a]			
$MnCl(CO)_5$	2140 (w)	2054 (s)	2021 (vw)	1998 (m)
$MnBr(CO)_5$	2134 (w)	2051 (s)	2020 (vw)	2000 (m)
$MnI(CO)_5$	2127 (w)	2045 (s)	2016 (vw)(sh)	2005 (m)

[a]Referenced to 2147 cm^{-1} band of CO gas.

References

1. E. O. Brimm, M. A. Lynch, Jr., and W. J. Sesny, *J. Am. Chem. Soc.*, **76**, 3831 (1954).
2. E. W. Abel and G. Wilkinson, *J. Chem. Soc.*, **1959**, 1501.
3. L. M. Haines and M. H. B. Stiddard, *Adv. Inorg. Radiochem.*, **12**, 53 (1969) and references therein.
4. (a) G. R. Dobson, *Acc. Chem. Res.*, **9**, 300 (1976); (b) D. A. Brown, *Inorg. Chim. Acta,* **1**, 35 (1967); (c) R. J. Angelici, *Organomet. Chem. Rev.*, **3**, 173 (1968).
5. R. H. Reimann and E. Singleton, *J. Organomet. Chem.*, **59**, C24 (1973).
6. (a) K. J. Reimer and A. Shaver, *Inorg. Synth.*, **19**, 000 (0000); (b) K. J. Reimer and A. Shaver, *Inorg. Chem.*, **14**, 2707 (1975), *J. Organomet. Chem.*, **93**, 239 (1975).
7. J. C. Hilemann, D. K. Huggins, and H. D. Kaesz, *Inorg. Chem.*, **1**, 933 (1962).
8. (a) S. A. Hallock and A. Wojcicki, *J. Organomet. Chem.*, **54**, C 27 (1973); (b) M. S. Wrighton and D. S. Ginley, *J. Am. Chem. Soc.*, **97**, 2065 (1975).
9. R. B. King, *Organomet. Synth.*, **1**, 174 (1965).
10. I. G. DeJong, S. C. Srinivasan, and D. R. Wiles, *J. Organomet. Chem.*, **26**, 119 (1971).
11. M. F. Farona and L. M. Frazee, *J. Inorg. Nucl. Chem.*, **29**, 1814 (1967).
12. F. Zingales and U. Satorelli, *Inorg. Chem.*, **6**, 1243 (1967).
13. D. F. Shriver, *The Manipulation of Air-Sensitive Compounds,* McGraw-Hill Book Co., New York, 1969.
14. (a) R. B. King, *Organomet. Synth.*, **1**, 149 (1965); (b) R. B. King and F. G. A. Stone, *Inorg. Synth.*, **7**, 198 (1963).
15. *Inorg. Synth.*, **12**, 111, 317 (1970).
16. See Ref. 13, p. 157.
17. C. H. Bamford, J. W. Burley, and M. Coldbeck, *J. Chem. Soc. Dalton Trans.*, **1972**, 1846.
18. F. A. Cotton, *Inorg. Chem.*, **3**, 702 (1964) and references therein.

37.   METHYLENE (CARBENE) COMPLEXES OF TRANSITION METALS

A. CARBENE COMPLEXES OBTAINED FROM ACYL- AND CARBAMOYL METALLATES

Submitted by E. O. FISCHER,* U. SCHUBERT,* W. KLEINE,* and H. FISCHER*
Checked by KEVIN P. DARST,[†] C. M. LUKEHART,[†] and L. T. WARFIELD[†] D. J.
DARENSBOURG,[‡] R. R. BURCH, Jr.,[‡] and J. A. FROELICH[‡]

A variety of different methods for preparing carbene complexes of transition metals have been developed in recent years and reviewed in some detail.[1-5] The basic idea behind the oldest and best established method is the stepwise synthesis of the carbene ligand at the metal by sequential treatment of a transition metal carbonyl with a nucleophile and an alkylating agent. Metal carbonyl compounds of Cr, Mo, W, Mn, Tc, Re, Fe, and Ni were employed in this synthesis,[1] as well as a variety of bases, such as organolithium compounds,[1,6] Grignard reagents,[7] and potassium alkoxides.[8] Reaction of the metal carbonyl complexes with bases leads to formation of acylmetallates or related compounds that can be protonated[9,10] or alkylated[1-5] to give hydroxy or alkoxy carbene complexes. Trialkyloxonium salts are usually used as alkylating reagents.[1-5] Other reagents, such as methyl fluorosulfonate[11] or iodomethane in the presence of macrocyclic crown ethers[12] offer no advantages. The intermediate acylmetallates can also be made to react with acyl halides,[2,13] trialkylhalosilanes[2,14] and dichlorodicyclopentadienyltitanium[15] to give the corresponding acyl-, trialkylsiloxy-, and titanoxy-substituted carbene complexes, respectively.

Alkoxy carbene complexes are useful starting compounds for other organometallic complexes,[1-5] particularly methylidyne (carbyne) complexes.[16] By modification of the coordinated (noncarbene) ligands or of the carbene ligand, other carbene complexes can be synthesized. The use of carbene complexes in organic syntheses has been reviewed recently.[17,18]

The preparations of $(CO)_5W[C(OCH_3)C_6H_5]$,[19] $(CO)_5W[C[OSi(CH_3)_3]$-$C_6H_5]$[14] and $(CO)_5Cr[C(OC_2H_5)N(C_2H_5)_2]$[20] are described here. The scale can

---

*Anorganisch-Chemisches Institut der Technischen Universität, 8 München 2, West Germany.

[†]Department of Chemistry, Vanderbilt University, Nashville, TN 37235; checked Sections A.1 and A.3.

[‡]Department of Chemistry, Tulane University, New Orleans, LA 70118; checked Section A.2.

be increased provided that the increased sizes of the apparatus and the increased volumes of solvents can be handled without difficulties. Large amounts of crude carbene complexes are better divided into several smaller portions for chromatography.

■**Caution.** *Volatile metal carbonyls are highly toxic. Contact of phenyllithium and lithium diethylamide with the skin must be avoided. Lithium diethylamide is pyrophoric on contact with air.*

1. **Pentacarbonyl(methoxyphenylmethylene)tungsten(0),** ˙˙
   [Pentacarbonyl[methoxy(phenyl)carbene] tungsten(0)]

$$W(CO)_6 + LiC_6H_5 \longrightarrow [(CO)_5W[C(O)C_6H_5]] Li,*$$

$$[(CO)_5W[C(O)C_6H_5]] Li + [(CH_3)_3O][BF_4] \longrightarrow$$

$$(CO)_5W[C(OCH_3)C_6H_5] + LiBF_4 + (CH_3)_2O$$

Procedure

All operations are carried out under nitrogen free from oxygen and moisture, using nitrogen-saturated solvents. Organic solvents must be dried by the usual methods. The products obtained are oxidized by air and must be handled in an inert atmosphere. A suspension of 15.0 g (0.0427 mole) of powdered $W(CO)_6$ in 600 mL of diethyl ether is prepared in a two-necked, 1-L flask equipped with an additional nitrogen inlet and a dropping funnel. To the well-stirred suspension, 0.0427 mole of LiC_6H_5[†21] in 75 mL of diethyl ether is added over a period of 3 hours. To obtain a good yield of product, local excesses of LiC_6H_5 should be avoided. During addition of the base, the color of the solution changes to orange-red and the $W(CO)_6$ dissolves completely. After addition of all the LiC_6H_5 the solvent is removed *in vacuo* (15 torr) without heating the solution above room temperature. The residue is dissolved in 100 mL of water saturated with N_2, the solution is filtered under a nitrogen atmosphere through a G3-frit (medium porosity), and 200 mL of pentane is added. A sample of $[(CH_3)_3O][BF_4]$[22†] (6.35 g, 0.0427 mole) is added in small portions to this mixture, which results in an immediate change in color to dark red. After addition of each portion of $[(CH_3)_3O][BF_4]$, the flask must be shaken well to extract the carbene complex into the organic phase. After addition of the last portion of the oxonium salt, the aqueous layer should be *slightly* acidic. Otherwise more $[(CH_3)_3O][BF_4]$ must be added. The organic layer is separated and the aqueous

---

*The fact that diethyl ether is coordinated to solid lithium acylmetallates is neglected here and throughout Section 37.

†Available from Alfa Products, Ventron Corp., P.O. Box 299, Danvers, MA 01923.

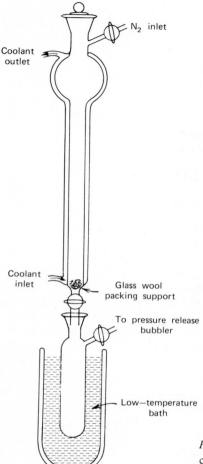

N₂ inlet

Coolant
outlet

Coolant
inlet

Glass wool
packing support

To pressure release
bubbler

Low—temperature
bath

Fig. 1. Low-temperature, inert-atmosphere
chromatography column.

layer is extracted under a nitrogen atmosphere with successive 50-mL portions
of deoxygenated pentane until the organic layer no longer becomes dark red on
shaking. The combined organic extracts are dried with 50 g of anhydrous
Na_2SO_4 to remove moisture and concentrated to a volume of 75 mL *in vacuo.*
On cooling to $-15°$ (ice-salt bath) the complex crystallizes to give 16.3-17.4 g of
product [86-92% based on $W(CO)_6$], $(CO)_5W[C(OCH_3)C_6H_5]$, that is
sufficiently pure for most purposes. Analytically pure samples are obtained by
column chromatography on silica gel with hexane. A water-cooled column
40 × 2.5 cm equipped with a nitrogen inlet is appropriate (Fig. 1). Prior to use,
the silica gel in the column is held at high vacuum for 10 hours and flushed with

N_2. The subsequent separation is performed under a nitrogen atmosphere with deoxygenated hexane. Approximately 2 g of carbene complex can be purified at one time on this column. After the portion of the eluant containing the deep-red carbene complex is isolated, this solution is reduced under vacuum to a volume of 10 mL and crystallization is induced by cooling to $-15°$. *Anal.* Calcd. for $C_{13}H_8O_6W$: C, 35.16; H, 1.82; O, 21.62; W, 41.40. Found: C, 35.12; H, 1.67; O, 21.50; W, 41.60.

Properties

The red crystalline complex can be stored under N_2 at $-20°$ for months. It is soluble in common organic solvents. Mp $59°$; subl. 45-$50°/10^{-4}$ torr, IR: ν_{CO} at 2083 (s), 1988 (s), 1946 (vs) cm^{-1} in Nujol; ^{1}H NMR (TMS, C_6D_6, 60 MHz): τ_{OCH_3} at 6.04 ppm (singlet).

2. Pentacarbonyl[*p*-tolyl(trimethylsiloxy)methylene]tungsten(0), [Pentacarbonyl[*p*-tolyl(trimethylsiloxy)carbene]tungsten(0)]

$$W(CO)_6 + LiC_6H_4CH_3 \longrightarrow [(CO)_5W[C(O)C_6H_4CH_3]]\,Li$$

$$[(CO)_5W[C(O)C_6H_4CH_3]]\,Li + (CH_3)_3SiCl \longrightarrow$$

$$(CO)_5W[C[OSi(CH_3)_3]C_6H_4CH_3] + LiCl$$

Procedure

■**Caution.** *Since metal carbonyls and chlorotrimethylsilane are volatile toxic substances, the operations described here should be performed in a fume hood.*

All operations are carried out under nitrogen free from oxygen and moisture, using nitrogen-saturated solvents. Organic solvents must be dried by the usual methods. The products obtained are oxidized by air and must be handled in an inert atmosphere. In a 500-mL, two-necked flask with a nitrogen inlet and a dropping funnel, a suspension is prepared of 3.2 g (0.010 mole) of powdered $W(CO)_6$ in 200 mL of diethyl ether. *p*-Tolyllithium ($LiC_6H_4CH_3$) (0.010 mole)[34] in 50 mL of diethyl ether is added to the well-stirred suspension over a period of 2 hours. The orange-red solution is concentrated *in vacuo* to a volume of 50 mL, and filtered through a G3-frit (medium porosity). At $-50°$ an excess of cold pentane is added to precipitate $[(CO)_5W[C(O)C_6H_4CH_3]]\,Li$. The light-yellow powder is collected on a G3-frit (medium porosity) and redissolved in 150 mL of diethyl ether. To the magnetically stirred solution in a 250-mL, two-necked flask with nitrogen supply 2 mL (0.0155 mole) of freshly distilled

$(CH_3)_3SiCl$ in 30 mL diethyl ether is added slowly at $-20°$. After 10 minutes of stirring at $0°$ the solvent and excess $(CH_3)_3SiCl$ are removed at $-10°/10^{-4}$ torr (ice-salt bath). The brown-red residue is extracted with pentane and the solution is filtered through a G3-frit and concentrated to a volume of 15 mL. On cooling to $-50°$ and two more recrystallizations from pentane, dark-red crystals are obtained which are dried at $-5°/10^{-4}$ torr for 5 hours. Yield 2.9 g (56% based on $W(CO)_6$).* *Anal.* Calcd. for $C_{16}H_{16}OSiW$: C, 37.22; H, 3.12; Si, 5.43; W, 35.61. Found: C, 36.99; H, 3.10; Si, 5.60; W, 36.20.

Properties

The red thermally unstable crystals are soluble in organic solvents. Since the complex is extremely sensitive to nucleophiles, traces of moisture and alcohols have to be removed from all solvents before use. The complex must be stored under nitrogen below $-20°$. IR: ν_{CO} at 2070 (m), 1952 (vs, br), 1942 (vs, br) cm^{-1} in pentane, ^{1}H NMR (TMS, C_6D_6, 60 MHz): τ_{SiCH_3} at 9.93 ppm (singlet), τ_{CH_3} at 8.11 ppm (singlet), $\tau_{C_6H_4}$ at 2.68 ppm (doublet) + 3.08 ppm (doublet).

3. Pentacarbonyl[ethoxy(diethylamino)methylene]chromium(0).
{Pentacarbonyl[ethoxy(diethylamino)carbene]chromium(0)}

$$Cr(CO)_6 \ + \ LiN(C_2H_5)_2 \ \longrightarrow \ [(CO)_5Cr[C(O)N(C_2H_5)_2]] \, Li$$

$$[(CO)_5Cr[C(O)N(C_2H_5)_2]] \, Li \ + \ [(C_2H_5)_3O] \, [BF_4] \ \longrightarrow$$

$$(CO)_5Cr[C(OC_2H_5)N(C_2H_5)_2] \ + \ LiBF_4 \ + \ (C_2H_5)_2O$$

Procedure

All operations are carried out under nitrogen free from oxygen and moisture, using nitrogen-saturated solvents. Organic solvents must be dried by the usual methods. The products obtained are oxidized by air and must be handled in an inert atmosphere. A suspension of powdered $Cr(CO)_6$ (4.4 g, 0.020 mole) in 200 mL of tetrahydrofuran (THF) is prepared in a two-necked, 500-mL flask with an additional nitrogen inlet and a dropping funnel. The well-stirred suspension is cooled to $0°$, and 1.58 g (0.020 mole) of very pure $Li[N(C_2H_5)_2]$ [†] in 100 mL of THF is added dropwise within 1 hour. Stirring is continued at $0°$ for an additional hour. By concentration to a volume of 50 mL at high vacuum (without warming above $0°$) and addition of 200 mL of pentane, $[(CO)_5Cr-[C(O)N(C_2H_5)_2]]\,Li$ is precipitated. To complete precipitation, the mixture

---

*The checkers obtained an oil that yields an infrared spectrum of the desired material contaminated with $W(CO)_6$.

†Available from Alfa Products, Ventron Corp., P.O. Box 299, Danvers, MA 01923.

is allowed to stand overnight at $-20°$. Then the solvents are decanted and the residue is dissolved in 30 mL of CH_2Cl_2. A solution of 3.61 g (0.019 mole) $[(C_2H_5)_3O][BF_4]$,[22] in 30 mL CH_2Cl_2 is then added in several portions at $-10°$ and with vigorous stirring. Excess $[(C_2H_5)_3O][BF_4]$ must be avoided; therefore if the yield of $[(CO)_5Cr[C(O)N(C_2H_5)_2]]Li$ is low, the amount of oxonium salt used here must be reduced proportionately. The solution is stirred at $0°$ for an additional 10 minutes, filtered through silica gel at $-78°$, and concentrated at high vacuum without warming above $0°$. The brown oily residue is twice extracted with 50 mL of pentane. The combined pentane layers are concentrated to a volume of 20 mL, filtered (medium-porosity frit), and purified by chromatography at $-10°$ (Fig. 1) on silica gel with pentane-diethyl ether (7:1) (column 25×2.5 cm, silica gel held at high vacuum for 10 hr and flushed with N_2 before use). The brown nonmigrating side products are retained on the top of the column and can be removed by a pipette. The carbene complex then is eluted by diethyl ether. The eluate is concentrated in high vacuum and recrystallized from pentane to give 1.92 g of carbene complex [30% yield based on $Cr(CO)_6$]. *Anal.* Calcd. for $C_{12}H_{15}NO_6Cr$: C, 44.86; H, 4.70; N, 4.26; O, 29.88; Cr, 16.18. Found: C, 44.57; H, 4.75; N, 4.22; O, 30.55; Cr, 15.98.

Properties

The yellow solid can be stored under N_2 at $-20°$ for months. It is soluble in common organic solvents. Mp $29°$; subl $30°/10^{-4}$ torr. IR: ν_{CO} = 2057 (m), 1927 (vs) cm^{-1} in hexane, ^1H NMR (TMS, C_6D_6, 60 MHz): τ_{OCH_2} at 5.49 (quartet); $\tau_{OCH_2CH_3}$ at 9.07 (triplet); $\tau_{NCH_2CH_3}$ at 9.25 (triplet), + 8.96 (triplet); τ_{NCH_2} at 7.10 (quartet) + 6.29 (quartet).

B. CARBENE COMPLEXES OBTAINED BY NUCLEOPHILIC SUBSTITUTION.
PENTACARBONYL[(DIMETHYLAMINO)PHENYLMETHYLENE]TUNGSTEN(0)

$$(CO)_5W[C(OCH_3)C_6H_5] + (CH_3)_2NH \longrightarrow$$

$$(CO)_5W[C[N(CH_3)_2]C_6H_5] + CH_3OH$$

Submitted by E. O. FISCHER,* U. SCHUBERT* and H. FISCHER*
Checked by D. T. HOBBS† and C. M. LUKEHART†

*Anorganisch-Chemisches Institut der Technischen Universität, 8 München 2, West Germany.
†Department of Chemistry, Vanderbilt University, Nashville, TN 37235.

Substitution of the alkoxy group of alkoxy carbene complexes is an important method of preparing carbene complexes with different substituents. Although reactions with primary and secondary amines,[1-4] thiols,[1-4] and selenols[23,24] proceed spontaneously to give the corresponding amino-, thio-, or seleno-carbene complexes, exchange of the methoxy group for aryl,[25,26] 2-thienyl, or 2-furyl groups[26] must be carried out in two steps: (1) addition of the heteroaryllithium compound to the carbene carbon atom at low temperatures and (2) subsequent removal of the methoxy group by acidification.

The preparation of $(CO)_5W[C[N(CH_3)_2]C_6H_5]$[27] by aminolysis of $(CO)_5$-$W[C(OCH_3)C_6H_5]$[19,28] is reported here (the scale may be increased to any size). Without major changes, the procedure is applicable to any known alkoxy carbene complexes and to all kinds of primary aryl or alkyl amines,[27,29] secondary aliphatic amines,[27,29] diamines,[30] and even esters of amino acids,[31] as well as ammonia.[29] Elimination reactions at the nitrogen atom may occur during the preparation of aminocarbene complexes with certain secondary amines.[32] The kinetics of the reaction of $(CO)_5W[C(OCH_3)C_6H_5]$ with primary amines have been investigated.[33]

Aminocarbene complexes usually are less sensitive towards oxygen than the corresponding alkoxycarbene complexes, but nevertheless should be handled in an inert atmosphere. Their preparation is carried out under nitrogen free from oxygen and moisture, using dry, nitrogen-saturated solvents.

Procedure

■**Caution.** *The following synthesis involves noxious compounds (dimethylamine and metal carbonyls), and therefore the operations should be performed in a fume hood.*

In a two-necked, 500-mL flask with an additional nitrogen inlet, 3 mL (0.04 mole) of dimethylamine is added to a cooled ($-20°$, ice-salt bath) solution of 2.5 g (0.0056 mole) of $(CO)_5W[C(OCH_3)C_6H_5]$[19,28] in 200 mL of pentane. The color of the solution changes immediately from dark red to yellow. After a short time, $(CO)_5W[C[N(CH_3)_2]C_6H_5]$ starts to precipitate. Precipitation is completed by cooling to $-70°$ (Dry Ice-acetone). After the crystallization is complete (about 4 hr) the solvent is decanted, and the crystals are dried at $25°/10^{-4}$ torr to give 2.26 g of the aminocarbene complex (88% yield based on $(CO)_5W[C(OCH_3)C_6H_5]$). *Anal.* Calcd. for $C_{14}H_{11}NO_5W$: C, 36.75; H, 2.43; N, 3.06. Found: C, 36.84; H, 2.50; N, 2.99.

Properties

The yellow complex can be stored under nitrogen for months. It is easily dissolved in benzene, ethers, and halogenated hydrocarbons and less easily in aliphatic hydrocarbons. Mp 93-94°. ^{1}H NMR (TMS, CDCl$_3$, 60 MHz): τ_{NCH_3}

at 5.97 (singlet) + 6.93 (singlet); $\tau_{C_6H_5}$ 2.12-3.28 (multiplet). IR (in benzene): ν_{CO} 2090 (m), 1975 (w), 1935 (vs), and 1930 (sh) cm^{-1}.

References

1. E. O. Fischer, *Angew. Chem.,* **86**, 651 (1974); *Adv. Organomet. Chem.,* **14**, 1 (1976); *Pure Appl..Chem.,* **24**, 407 (1970); **30**, 353 (1972).
2. D. J. Cardin, B. Cetinkaya, and M. F. Lappert, *Chem. Rev.,* **72**, 545 (1972).
3. F. A. Cotton and C. M. Lukehart, *Prog. Inorg. Chem.,* **16**, 487 (1972).
4. D. J. Cardin, B. Cetinkaya, M. J. Doyle, and M. F. Lappert, *Chem. Soc. Rev.,* **2**, 99 (1973).
5. R. P. A. Sneeden, *Organochromium Compounds,* Academic Press Inc., New York, 1975.
6. E. O. Fischer, H. Hollfelder, F. R. Kreissl, and W. Uedelhoven, *J. Organomet. Chem.,* **113**, C31 (1976).
7. M. Y. Darensbourg, H. L. Conder, D. J. Darensbourg, and C. Hasday, *J. Am. Chem. Soc.,* **95**, 5919 (1973).
8. E. O. Fischer, K. Scherzer, and F. R. Kreissl, *J. Organomet. Chem.,* **118**, C33 (1976).
9. E. O. Fischer, F. Kreis, and F. R. Kreissl, *J. Organomet. Chem.,* **56**, C37 (1973).
10. K. Weiss and E. O. Fischer, *Chem. Ber.,* **109**, 1120 (1976).
11. C. P. Casey and C. R. Cyr, *J. Organomet. Chem.,* **57**, C69 (1973).
12. M. Y. Darensbourg, D. Burns, and D. Drew, Abstract of Papers, 1st Chemistry Congress of the North American Continent, Mexico City, 1975.
13. E. O. Fischer, T. Selmayr, and F. R. Kreissl, *Chem. Ber.,* **110**, 2947 (1977).
14. E. O. Fischer, T. Selmayr, F. R. Kreissl, and U. Schubert, *Chem. Ber.,* **110**, 2574 (1977).
15. (a) E. O. Fischer and S. Fontana, *J. Organomet. Chem.,* **40**, 159 (1972); (b) H. G. Raubenheimer and E. O. Fischer, *J. Organomet. Chem.,* **91**, C23 (1975).
16. E. O. Fischer and U. Schubert, *J. Organomet. Chem.,* **100**, 59 (1974).
17. K. H. Dötz, *Naturwissenschaften,* **62**, 365 (1975).
18. C. P. Casey, *J. Organomet. Chem. Libr.,* **1**, (1976).
19. E. O. Fischer and A. Maasböl, *Chem. Ber.,* **100**, 2445 (1967).
20. E. O. Fischer and H. J. Kollmeier, *Angew. Chem.,* **82**, 325 (1970); *Angew. Chem., Int. Ed.,* **9**, 309 (1970).
21. G. Wittig, *Angew. Chem.,* **53**, 243 (1940); G. Wittig, *Newer Methods of Preparative Organic Chemistry,* Interscience Publishers, New York, 1968, p. 575; L. A. Walter, *Org. Synth.,* Coll. Vol. 3, 757 (1955).
22. (a) H. Meerwein, G. Hinz, P. Hofmann, E. Kronig, and E. Pfeil, *J. Prakt. Chem.,* **147**, No. 2, 257 (1937). (b) H. Meerwein, E. Battenberg, A. Gold, E. Pfeil, and G. Willfang, *J. Prakt. Chem.,* **154**, No. 2, 83 (1940); (c) H. Meerwein, *Org. Synth.,* **46**, 120 (1966).
23. E. O. Fischer, G. Kreis, F. R. Kreissl, C. G. Kreiter, and J. Müller, *Chem. Ber.,* **106**, 3910 (1973).
24. C. T. Lam, C. V. Senoff, and J. E. H. Ward, *J. Organomet. Chem.,* **70**, 273 (1974).
25. C. P. Casey and T. J. Burkhardt, *J. Am. Chem. Soc.,* **95**, 5833 (1973).
26. E. O. Fischer, W. Held, F. R Kreissl, A. Frank, and G. Huttner, *Chem. Ber.,* **110**, 656 (1977).
27. E. O. Fischer, K. R. Schmid, W. Kalbfus, and C. G. Kreiter, *Chem. Ber.,* **106**, 3893 (1973).
28. E. O. Fischer, U. Schubert, W. Kleine, and H. Fischer, *Inorg. Synth.,* **19**, 165 (1979).

29. E. O. Fischer and M. Leupold, *Chem. Ber.,* **105**, 599 (1972) and references cited therein.

30. E. O. Fischer and S. Fontana, *J. Organomet. Chem.,* **40**, 367 (1972).

31. K. Weiss and E. O. Fischer, *Chem. Ber.,* **109**, 1868 (1976).

32. J. A. Connor and E. O. Fischer, *J. Chem. Soc., A,* **1969**, 578.

33. H. Werner, E. O. Fischer, B. Heckl, and C. G. Kreiter, *J. Organomet. Chem.,* **28**, 367 (1971).

34. H. Gilman, W. Langham, and F. W. Moore, *J. Am. Chem. Soc.,* **62**, 2327 (1940).

38. METHYLIDYNE (CARBYNE) COMPLEXES. *trans*-[BROMO-TETRACARBONYL(PHENYLMETHYLIDYNE)TUNGSTEN]

$$(CO)_5W[C(OCH_3)C_6H_5] \ + \ BBr_3 \ \longrightarrow$$

$$trans\text{-}Br(CO)_4W(CC_6H_5) \ + \ CO \ + \ \{Br_2BOCH_3\}$$

Submitted by E. O. FISCHER.* U. SCHUBERT* and H. FISCHER*
Checked by C. M. LUKEHART† and G. PAULL TORRENCE†

Most carbyne complexes synthesized to date have been derived from chromium and manganese group carbene complexes in which the carbene carbon atom is bound to a substituent such as an alkoxy, hydroxy, alkylthio or amino acid ester.[1,2] Trihalides of boron, aluminium, or gallium cleave this substituent, generating the carbyne moiety. Depending on the σ-donor/π-acceptor relation of the ligand which is trans to the carbene ligand, neutral or cationic carbyne complexes can be isolated.[2-5] Other methods of preparing carbyne complexes have been reviewed recently.[2]

The preparation of *trans*-[Br(CO)₄W(CC₆H₅)] is reported here. The substituent bound to the carbyne carbon atom can be varied considerably[2,7] without major changes in the procedure. Although BCl₃ or BI₃ may be applied with equal success to give the corresponding *trans*-chloro or *trans*-iodo complexes, the reaction with BF₃ is slightly different.[8,9] Both carbene and carbyne complexes are sensitive to oxidation and therefore must be handled in an inert gas atmosphere.

■**Caution.** *Boron tribromide is very corrosive. Inhalation of the vapors and*

*Anorganisch-Chemisches Institut, der Technischen Universität, 8 München 2, West Germany.

†Department of Chemistry, Vanderbilt University, Nashville, TN 37235.

contact with the skin is dangerous. Reaction with moisture may be vigorous.
Therefore exposure of BBr₃ to the atmosphere must be minimized and
manipulations must be performed in a fume hood.

Procedure

All operations are carried out under nitrogen free from oxygen and moisture
using nitrogen-saturated dry solvents. In a round-bottomed, two-necked, 250-mL
flask with an additional nitrogen inlet a suspension of 4.4 g (0.010 mole) $(CO)_5$-
$W[C(OCH_3)C_6H_5]$ [10] in 80 mL of pentane is prepared at $-15°$ (ice-salt bath).
To the well-stirred suspension, 2.5 g (0.010 mole) of BBr_3* is added slowly in
several portions by syringe. A nearly colorless solid starts to precipitate
immediately, while the carbene complex dissolves completely. After addition of
all the BBr_3, the solution is stirred at $-10°$ (ice-salt bath) for some minutes and
then cooled to $-78°$ (Dry Ice-acetone) to complete precipitation of the carbyne
complex. The solvent is decanted and the crude solid is dried at $-20°/10^{-4}$ torr
(ice-salt bath) for some hours. The unreacted carbene complex and $W(CO)_6$
(first band) are separated from the ivory-colored second band, containing the
carbyne complex, by column chromatography under N_2 on silica gel with cold
pentane-CH_2Cl_2 (6:1). The column (see Fig. 1, Sec. 37) is 20 × 2.5 cm and is
filled with silica gel, which is held at high vacuum for 10 hours and then flushed
with nitrogen before use. The separation is performed at $-25°$. After elution of
the first band and removal of nonmigrating by-products at the top of the column
by a pipette, the carbyne complex is eluted with CH_2Cl_2. The eluate is
concentrated to a volume of 2 mL at $-20°/10^{-4}$ torr and 20 mL of cold pentane
is added. On cooling to $-100°$, (methanol-slush bath) *trans*-$Br(CO)_4W(CC_6H_5)$
crystallizes as fine needles. The mother liquid is decanted, and the pale-yellow
crystals are washed with three portions of 5 mL of cold ($-100°$) pentane each
and dried at $-25°/10^{-4}$ torr for 24 hours. Yield 3.8 g [82% based on $(CO)_5$-
$WC(OCH_3)C_6H_5$]. *Anal.* Calcd. for $C_{11}H_5BrO_4W$: C, 28.42; H, 1.08; Br, 17.19;
O, 13.77; W, 39.54. Found: C, 28.63, H, 1.18; Br, 17.80; O, 13.80; W, 38.60.

Properties

The thermally unstable crystalline complex must be stored under nitrogen below
$-20°$. It is easily dissolved in benzene, diethyl ether, or CH_2Cl_2 but is only
slightly soluble in aliphatic hydrocarbons. IR: ν_{CO} at 2125 (m), 2040 (vs) cm^{-1}
in hexane.

*Available from Alfa Products, Ventron Corp., P.O. Box 299, Danvers, MA
01923.

References

1. E. O. Fischer, *Angew. Chem.,* 86, 651 (1974); *Adv. Organomet. Chem.,* 14, 1 (1976).
2. E. O. Fischer and U. Schubert, *J. Organomet. Chem.,* 100, 59 (1975).
3. E. O. Fischer and K. Richter, *Chem. Ber.,* 109, 3079 (1976).
4. E. O. Fischer, P. Stückler, H. -J. Beck, and F. R. Kreissl, *Chem. Ber.,* 109, 3089 (1976).
5. E. O. Fischer, E. W. Meineke, and F. R. Kreissl, *Chem. Ber.,* 110, 1140 (1977).
6. E. O. Fischer and G. Kreis, *Chem. Ber.,* 109, 1673 (1976).
7. E. O. Fischer, A. Schwanzer, H. Fischer, D. Neugebauer, and G. Huttner, *Chem. Ber.,* 110, 53 (1977).
8. K. Richter, E. O. Fischer, and C. G. Kreiter, *J. Organomet. Chem.,* 122, 187 (1976).
9. E. O. Fischer, W. Kleine, and F. R. Kreissl, *Angew. Chem.,* 88, 646 (1976); *Angew. Chem. Int. Ed. Engl.,* 15, 616 (1976).
10. (a) E. O. Fischer and A. Maasböl, *Chem. Ber.,* 100, 2445 (1976); (b) E. O. Fischer, U. Schubert, W. Kleine, and H. Fischer, *Inorg. Synth.,* 19, 165 (1979).

39. ISOCYANIDE AND METHYLENE (CARBENE) COMPLEXES OF PLATINUM(II)

Submitted by R. L. RICHARDS*
Checked by P. M. TREICHEL[†]

When coordinated to metal ions in their normal or higher oxidation states, an isocyanide is rendered susceptible to attack at the ligating carbon atom by nucleophilic reagents.[1] When alcohols or amines are the nucleophiles, carbene complexes that may be prepared for a variety of metals and substituent groups are obtained. The first fully characterized compounds were of platinum(II),[2] and general methods of their preparation, with particular examples, are given below.

A. *cis*-[DICHLORO(PHENYL ISOCYANIDE)(TRIETHYLPHOSPHINE)-PLATINUM(II)]

$$\text{\textit{trans-}}[Pt_2Cl_4(PEt_3)_2] + 2RNC \xrightarrow{\text{benzene}} 2\text{\textit{cis-}}[PtCl_2(RNC)(PEt_3)]$$

■**Caution.** *Isocyanides are toxic and nasal anaesthetics and must be handled*

*School of Molecular Sciences, University of Sussex, Brighton Sussex, England, BN1 9QJ.
[†]Department of Chemistry, University of Wisconsin, Madison, WI 53706.

only in an efficient hood. Benzene is a suspected mild carcinogen and must be handled in a hood. Isocyanide residues may be destroyed by treatment with commercial bleach (calcium hypochlorite).

Procedure

Isocyanides are prepared by the methods of Ugi et al.,[3] except for MeNC.[4]

Phenyl isocyanide (0.4 mL, 3.9 mmole) is added dropwise under nitrogen with stirring to *trans*-[Pt$_2$Cl$_4$(PEt$_3$)$_2$] (1 g)[5] in suspension in 30 mL of dry benzene. Initially a yellow oil is formed, but after about 1 hour a colorless oil is obtained that solidifies to give white crystals of the product. These white crystals can be recrystallized from benzene as colorless needles, mp 162-163°. Yield 0.9 g (70%). *Anal.* Calcd. for C$_{13}$H$_{20}$Cl$_2$NPPt: C, 32.0; H, 4.1; N, 2.9; Cl, 14.6. Found: C, 32.2; H, 4.3; N, 3.0; Cl, 14.5.

A further quantity of product may be obtained from the mother liquor of the original reaction by removing the solvent *in vacuo* and then recrystallizing the residue from ethanol.

These and succeeding preparations may occasionally be contaminated by small quantities of platinum metal, which may be removed by dissolving the product in the recrystallizing solvent, treating the hot solution with charcoal, filtering it, and allowing it to crystallize as usual.

The *trans*-[chloro(isocyanide)bis(triethylphosphine)platinum(II)] perchlorate complexes used below were prepared by the method of Church and Mays.[6]

■**Caution.** *These compounds should be treated with the extreme caution normally afforded organometallic perchlorate salts, which are potentially explosive.*

B. *cis*-[(ANILINOETHOXYMETHYLENE)DICHLORO(TRIETHYL-PHOSPHINE)PLATINUM(II)]

$$\text{cis-}[PtCl_2(RNC)(PEt_3)] \ + \ R'OH \ \rightarrow \ \text{cis-}[PtCl_2\{C(OR')NHR\}\,(PEt_3)]$$

$$(R = \text{aryl}, \ R' = \text{alkyl})$$

Procedure

cis-[PtCl$_2$(PhNC)(PEt$_3$)] (1 g, 2.05 mmole) is heated under reflux in 50 mL of dry ethanol for 3 hours. The solution is concentrated to about one-third the

original volume under vacuum and white crystals separate; the crystals are filtered and recrystallized from ethanol or an acetone-diethyl ether mixture as colorless needles, mp 209-211°. Yield 1 g (91%). *Anal*. Calcd. for $C_{15}H_{26}Cl_2NOPPt$;
C, 33.8; H, 4.9; N, 2.6; Cl, 13.3. Found: C, 34.0; H, 4.9; N, 2.6; Cl, 12.9.

C. *cis*-[(DIANILINOMETHYLENE)DICHLORO(TRIETHYLPHOSPHINE) PLATINUM(II)]

$$cis\text{-}[PtCl_2(RNC)(PEt_3)] + R'NH_2 \rightarrow cis\text{-}[PtCl_2\{C(NHR')NHR\}(PEt_3)]$$

$$(R = R' = \text{alkyl or aryl})$$

Procedure

cis-$[PtCl_2(PhNC)(PEt_3)]$ (1 g, 2.05 mmole) is dissolved in 30 mL of analytical
grade aniline, and the solution is stirred for 1 hour at 20°. The aniline is then
removed in *vacuo* at 50° to give a brown oil, which solidifies on trituration with
dry ethanol. The product, which is sparingly soluble in ethanol, may be recrystallized from a large volume of this solvent (solubility ~0.001 g/mL) as white
needles, mp 235-236°. Yield 0.5 g (42%). *Anal*. Calcd. for $C_{19}H_{27}Cl_2N_2PPt$: C,
39.3; H, 4.7; N, 4.8; Cl, 12.2. Found: C, 39.4; H, 4.7; N, 4.8; Cl, 12.3.

D. *trans*-[[ANILINO(ETHYLAMINO)METHYLENE]CHLOROBIS(TRI-ETHYLPHOSPHINE)PLATINUM(II)] PERCHLORATE

$$trans\text{-}[PtCl(PhNC)(PEt_3)_2]ClO_4 \xrightarrow{\text{EtNH}_2}$$

$$trans\text{-}\{PtCl[C(NHPh)(NHEt)](PEt_3)_2\}ClO_4$$

Procedure

trans-$[PtCl(CNPh)(PEt_3)_2]ClO_4$ (1 g, 2.05 mmole) is dissolved in 50 mL of
ethylamine and the solution is stirred for 30 minutes and is then taken to dryness *in vacuo* at 20°, leaving a white solid, which is recrystallized from ethanol
to give colorless needles. Yield 0.7 g (66%); mp 154-156°. *Anal*. Calcd. for
$C_{21}H_{42}Cl_2N_2O_4P_2Pt$; C, 35.3; H, 5.9; N, 3.9; Cl, 9.9. Found: C, 35.6; H, 5.9; N,

3.9; Cl, 10.1. Conductivity = 24.7 ohms^{-1} cm^2 mole^{-1} in approximately 10^{-3} M PhNO$_2$ solution.

The methylamino derivative is prepared from an excess of methylamine, passed as gas into a benzene suspension of the isocyanide complex, which dissolves as the reaction proceeds. Subsequent work-up is same as in Section C.

Properties

The isocyanide and carbene complexes are all colorless or white, air-stable, crystalline solids. The molecular structures of the cis isomers have been established by X-ray crystallography.[2] The isocyanide complexes have a strong IR absorption [(N—C) stretching] in the range 2150-2280 cm^{-1}; for example, *cis*-PtCl$_2$(PhNC)(PEt$_3$) has $\nu_{(NC)}$ at 2190 cm^{-1} (Nujol mull). The carbene complexes have IR bands [(N—H) stretching] in the range 3100-3450 cm^{-1}; for example, *cis*-PtCl$_2$[C(OEt)NHPh](PEt$_3$) has $\nu_{(NH)}$ at 3150 and 3110 cm^{-1}; *cis*-PtCl$_2$[C(NHPh)$_2$](PEt$_3$) has $\nu_{(NH)}$ at 3250 m, 3210 cm^{-1}; *cis*-PtCl$_2$[C-(NHEt)NHPh](PEt$_3$) has $\nu_{(NH)}$ at 3265, 3100 cm^{-1} (Nujol mulls). A variety of phosphine coligands may be employed. In particular, cis complexes of PPr$_3$ or PBu$_3$ are sufficiently soluble for NMR studies of them (and the soluble cationic PEt$_3$ complexes) to be made.[7] On treatment with base, the cationic bis(amino)carbene complexes give uncharged *trans*-halo(amidino) complexes, which revert to the parent carbene complexes when treated with acid.[8] The carbene complexes react with chlorine, without cleavage of the platinum-carbon bond, to give complexes of platinum(IV); those having *N*-phenyl substituents are metallated at the 2-phenyl position.[9]

References

1. J. Chatt, R. L. Richards and G. D. H. Royston, *J. Chem. Soc. Dalton Trans.*, **1973**, 1433, and references therein.
2. E. M. Badley, J. Chatt, and R. L. Richards, *J. Chem. Soc., A*, **1971**, 21.
3. I. Ugi, U. Fetzer, U. Eholzer, H. Kaupfer, and K. Offerman, *Angew. Chem. Int. Ed.*, **4**, 472 (1965).
4. R. E. Schuster, J. E. Scott, and J. Casanova, *Org. Synth.*, **46**, 75 (1966).
5. Prepared according to A. C. Smithies, P. Schmidt, and M. Orchin, *Inorg. Synth.*, **12**, 240 (1970).
6. M. J. Church and M. J. Mays, *J. Chem. Soc., A*, **1968**, 3074.
7. B. Crociani and R. L. Richards, *J. Chem. Soc. Dalton Trans.*, **1974**, 693.
8. E. M. Badley, U. Belluco, L. Busetto, B. Crociani, B. J. L. Kilby, A. Palazzi, and R. L. Richards, *J. Chem. Soc. Dalton Trans.*, **1972**, 1800.
9. J. Chatt, G. H. D. Royston, and R. L. Richards, *J Chem. Soc. Dalton Trans.*, **1976**, 599.

40.   PENTACARBONYL(DIHYDRO-2(3*H*)-FURANYLIDENE) CHROMIUM(0), (CO)$_5$Cr(C$_4$H$_6$O)

Submitted by C. P. CASEY,* R. L. ANDERSON,[†] S. M. NEUMANN,* and D. M. SCHECK*

Checked by E. O. FISCHER[‡] and W. HELD[‡]

Dihydro-2(3*H*)-furanylidene (1-oxacyclopent-2-ylidene) complexes of transition metals have previously been prepared by novel procedures involving either the reaction of metal carbonyl anions with 1,3-dibromopropanes[1,2] or the reaction of acetylenic alcohols with platinum complexes.[3] The reaction of the conjugate base of pentacarbonyl(1-methoxyethylidene)chromium(0) with epoxides[4] provides a convenient method for obtaining several members of this class of compounds. Pentacarbonyl(1-oxacyclopent-2-ylidene)chromium(0) has been shown to be easily elaborated via its conjugate base,[5] affording new carbene complexes and, by oxidation, substituted γ-butyrolactones.[6,7]

The following procedure can be extended to prepare 1-oxacyclopent-2-ylidene chromium complexes substituted in the 3 or 5 position.[4] Thus the reaction of the anion of pentacarbonyl(1-methoxypropylidene)chromium(0) with ethylene oxide gives pentacarbonyl[dihydro-3-methyl-2(3*H*)-furylidene] chromium(0) [pentacarbonyl(3-methyl-1-oxacyclopent-2-ylidene)chromium(0)], and the reaction of the anion of pentacarbonyl(1-methoxyethylidene)chromium(0) with propylene oxide gives pentacarbonyl(dihydro-5-methyl-2(3*H*)-furylidene)-chromium(0) [pentacarbonyl(5-methyl-1-oxacyclopent-2-ylidene)chromium(0)].

The synthesis described below is a minor modification of our original procedure,[4] and requires less than 1 day to complete.

*Department of Chemistry, University of Wisconsin, Madison, WI 53706.

[†]Amoco Research Center, P.O. Box 400, Naperville, IL 60540.

[‡]Anorganisch-Chemisches Institut der Technischen Universität, Arcisstr. 21, D 8000 München 2, West Germany.

Procedure

All manipulations are carried out under a nitrogen atmosphere using standard septum, syringe, and hypodermic tubing techniques.[8] ■**Caution.** *Since ethylene oxide is a colorless, extremely flammable, moderately toxic gas at room temperature and atmospheric pressure, all procedures except the final column chromatography are carried out in a well-ventilated hood.*

Pentacarbonyl(1-methoxyethylidene)chromium(0)[9] (2.14 g, 8.54 mmole) is placed in a flame-dried 250-mL round-bottomed flask fitted with a rubber septum, dissolved in 50 mL of diethyl ether, and cooled to -78°. Butyllithium in hexane* (5.80 mL, 1.63 M, 9.45 mmole, 1.73 M total base, 10.03 mmole) is added dropwise to the stirred solution. The solution becomes dark brown after it is stirred for 5 minutes at -78°. Double titration of the lithium reagent using a 1,2-dibromoethane quench is recommended.[10] Ethylene oxide[†] (1 mL, 20.0 mmole) is condensed at 0° into a graduated tube and is then added as a liquid by means of hypodermic tubing to the reaction mixture. The mixture is warmed from -78° to ambient temperature and is stirred for ½ hour. Distilled water (100 mL) is added to the mixture, and the deep red-orange diethyl ether layer is carefully decanted through hypodermic tubing under a slight positive nitrogen pressure. The aqueous layer is extracted with 50 mL of diethyl ether, and the combined ether layers are dried over Na_2SO_4. The red-orange ether solution is decanted from the Na_2SO_4 using hypodermic tubing and a positive nitrogen pressure, 2 g of silica gel[‡] is added, and the solvent is immediately removed on a rotary evaporator. The rotary evaporator is vented to nitrogen, the flask is removed, and the material adsorbed on the silica gel is subjected to column chromatography (silica gel, 45 g) with hexane as the eluent. A water-jacketed chromatography column [15 mm id × 56 cm] is employed to maintain the temperature at about 15°.[11] The column and receiver are kept under nitrogen throughout the elution. Hexane is distilled from CaH_2 before use. Two bands are eluted; the fast-moving yellow band contains unreacted starting material (0.13 g, 6%) and the second, slower-moving, orange band contains the desired product. Removal of solvent from the orange solution by rotary evaporation gives pure pentacarbonyl(1-oxacyclopent-2-ylidene)chromium(0) (1.10 g, 49% yield), mp 65-66°, which can be recrystallized by dissolving in hexane (0.025 g/ml) at room temperature and cooling to -22°. *Anal.* Calcd. for $C_9H_6O_6Cr$: C, 41.24; H, 2.31; Cr, 19.84. Found: C, 41.40; H, 2.30; Cr, 19.64.

---

*Available from Foote Mineral and Chemicals, Route 100, Exton, PA 19341.

[†]Available from Matheson Gas Products, P.O. Box 85, 932 Paterson Plank Rd., E. Rutherford, NJ 07073.

[‡]Available from Grace, Davison Chemical Co., Baltimore, MD 21226; Grade 62, 60-200 mesh.

Properties

Pentacarbonyl(1-oxacyclopent-2-ylidene)chromium(0) is an air-stable, bright yellow compound that dissolves readily in most organic solvents. The infrared spectrum of a heptane solution shows bands at 2066 (s), 1983 (m), 1958 (s), and 1944 (s) cm^{-1}. The proton NMR spectrum in CS_2 shows a two-proton triplet (J = 8 Hz) at δ 4.90, a two-proton triplet (J = 8 Hz) at δ 3.67, and a two-proton quintet (J = 8 Hz) at δ 1.96 all relative to internal tetramethylsilane.

References

1. C. P. Casey, *Chem. Commun.*, **1970**, 1220.
2. F. A. Cotton and C. M. Lukehart, *J. Am. Chem. Soc.*, **93**, 2672 (1971); **95**, 3552 (1973).
3. M. H. Chisholm and H. C. Clark, *Inorg. Chem.*, **10**, 1711 (1971); *Acc. Chem. Res.*, **6**, 202 (1973).
4. C. P. Casey and R. L. Anderson, *J. Organomet. Chem.*, **73**, C28 (1974).
5. C. P. Casey and W. R. Brunsvold, *J. Organomet. Chem.*, **118**, 309 (1976).
6. C. P. Casey and W. R. Brunsvold, *J. Organomet. Chem.*, **102**, 175 (1975).
7. C. P. Casey in *New Applications of Organometallic Reagents in Organic Synthesis, J. Organomet. Chem. Lib.*, **1** (1976).
8. D. F. Shriver, *The Manipulation of Air-Sensitive Compounds*, McGraw-Hill Book Co., New York, 1969.
9. E. O. Fischer and A. Maasböl, *Angew. Chem.*, **76**, 645 (1964); C. V. Senoff, C. T. Lam and C. D. Malkiewich, *Inorg. Synth.*, **17**, 95 (1978).
10. H. Gilman and F. K. Cartledge, *J. Organomet. Chem.*, **2**, 447 (1964).
11. E. O. Fischer, U. Schubert, W. Klein, and H. Fischer, *Inorg. Synth.*, **19**, 164 (1979). Fig. 1.

41. PENTACARBONYL(DIPHENYLMETHYLENE)-TUNGSTEN(0)

Submitted by C. P. CASEY,* T. J. BURKHARDT,[†] S. M. NEUMANN,*
D. M. SCHECK,* and H. E. TUINSTRA*
Checked by E. O. FISCHER[‡] and W. HELD[‡]

*Department of Chemistry, University of Wisconsin, Madison, WI 53706.
[†] Elastomers Division, E. I. DuPont de Nemours and Company, Wilmington, DE 19898.
[‡] Anorganisch-Chemisches, Institut der Technischen Universität, Arcisstr. 21, 8000 Munchen 2, West Germany.

Both $W(CO)_5[C(C_6H_5)_2]$[1] and the analogous di-*p*-tolylmethylene complex have been used in model studies of the olefin metathesis reaction.[2,3] In contrast to heteroatom-stabilized carbene complexes such as $W(CO)_5[C(OCH_3)(C_6H_5)]$, pentacarbonyl(diphenylmethylene)tungsten(0) reacts with alkenes to give cyclopropanes and 1,1-diphenylalkenes.[2] The compound $W(CO)_5[C(C_6H_5)_2]$ is the best reported catalyst for the metathetical polymerization of 1-methylcyclobutene.[4]

The preparation described here is a composite of the published procedures of Casey et al.[1] and of Fischer et al.[5] and can be completed in 1 day. Similar synthetic procedures have been employed for the preparation of $W(CO)_5[C(p\text{-}CH_3C_6H_4)_2]$[3] and of $Cr(CO)_5[C(C_6H_5)_2]$.[5]

Procedure

Solutions of $W(CO)_5[C(C_6H_5)_2]$ can be handled for short times in the air, but it is recommended that all manipulations be carried out under a nitrogen atmosphere. Standard septum, syringe, hypodermic tubing techniques[6] are used in this procedure. Since this synthesis involves the use of phenyllithium, it should be carried out in a hood.

Pentacarbonyl(α-methoxybenzylidene)tungsten(0)[7] (0.89 g, 2.0 mmole) and a magnetic stirring bar are placed in a flame-dried, 100-mL, round-bottomed flask fitted with a rubber septum. The flask is flushed with nitrogen, 50 mL of diethyl ether (freshly distilled from sodium and benzophenone) is added, and the resulting red solution is cooled to $-78°$. A double-titrated[8] solution of phenyllithium* in 70:30 benzene/diethyl ether (1.25 mL, 1.85 M, 2.31 mmole, 2.10 M total base, 2.63 mmole) is added by syringe to the well-stirred solution. The anionic intermediate formed is thermally unstable at room temperature and is kept at $-78°$ (Dry Ice-acetone) to avoid decomposition.

After 1 hour at $-78°$, dry HCl in diethyl ether (4.0 mL, 0.73 M, 2.92 mmole, prepared by bubbling anhydrous HCl through dry diethyl ether) is added. At this point, the red-brown suspension becomes wine red and homogeneous. The cold solution is filtered under nitrogen through a fritted funnel containing a 1-cm layer of silica gel into a Schlenk tube.[6] The red solution is transferred to a 250-mL round-bottomed flask by means of hypodermic tubing and solvent is removed by distillation on a rotary evaporator at $0°$ to give a red-purple oil.

The oil is dissolved in several milliliters of hexane† and the solution is added to the top of a 27.5 × 2.0 cm silica gel‡ chromatography column, which is main-

---

*Available from Alfa Products, Ventron Corp., 8 Congress St., Beverly, MA 01915.

†Olefin-free hexane (obtained by stirring over H_2SO_4 for several days, washing with bicarbonate and water, drying over $MgSO_4$, and distilling from CaH_2) is used for chromatography.

‡Available from Grace, Davison Chemical Co., Baltimore, MD 21226. Grade 62, 60-200 mesh.

tained at approximately $-40°C$ by circulation of cold methanol through the column jacket.[7] [Column chromatography also can be carried out with a water-cooled column at $15°$, since $W(CO)_5[C(C_6H_5)_2]$ is stable under these conditions. However, a 55-cm column is required to attain adequate separation, and recrystallization is required to remove small amounts of $W(CO)_5[C(OCH_3)(C_6H_5)]$. While chromatography at $-40°$ is more time consuming (4.5 vs. 1.5 hr) purer material is obtained.] After elution with 200 mL of hexane, a red-purple band begins to elute and is collected with 500 mL of hexane in a receiver cooled to $-78°$. Solvent is evaporated at $0°$ on a rotary evaporator and the residual red-purple oil is dissolved in 60 mL of hexane and filtered through a sintered-glass filter. Solvent is evaporated and the red oil is dissolved in 3 mL of pentane and cooled to $-78°$ to give crystalline material. Solvent is removed by evaporation under vacuum at $-78°$ and the solid is dried under vacuum for several minutes at room temperature to give $W(CO)_5[C(C_6H_5)_2]$ (0.73 g, 75% yield), mp 65-66°, which is homogeneous as judged by thin-layer chromatography on silica gel (10% diethyl ether-90% hexane). The crystals may also be isolated by rapid filtration of the cold solution. However, if the solution becomes warm during filtration some material is lost.

Recrystallization from pentane at $-22°$ gives analytically pure material, mp 65-66°. *Anal.* Calcd. for $C_{18}H_{10}O_5W$: C, 44.11; H, 2.05; W. 37.51. Found: C, 43.94; H, 2.11; W, 37.69.

Properties

Pentacarbonyl(diphenylmethylene)tungsten(0) is a moderately air-stable soild that is readily soluble in most organic solvents. The resulting solutions are air and light sensitive and decomposed thermally at about 50°. The infrared spectrum of a heptane solution shows bands in the metal carbonyl region at 2070 (m), 1971 (s), and 1963 (s) cm^{-1}, characteristic of a group VI pentacarbonyl species. The proton NMR spectrum in CS_2 or acetone-d_6 shows a complex multiplet at δ 7.2 relative to internal tetramethylsilane.

References

1. (a) C. P. Casey and T. J. Burkhardt, *J. Am. Chem. Soc.,* **95,** 5833 (1973); (b) C. P. Casey, T. J. Burkhardt, C. A. Bunnell, and J. C. Calabrese, *J. Am. Chem. Soc.,* **99,** 2127 (1977).
2. C. P. Casey and T. J. Burkhardt, *J. Am. Chem. Soc.,* **96,** 7808 (1974).
3. C. P. Casey, H. E. Tuinstra, and M. C. Saeman, *J. Am. Chem. Soc.,* **98,** 608 (1976).
4. T. J. Katz, J. McGinnis, and C. Alters, *J. Am. Chem. Soc.,* **98,** 606 (1976).
5. E. O. Fischer, W. Held, S. Riedmuller, and F. Kohler, unpublished results referred to by E. O. Fischer, *Angew. Chem.,* **86,** 651 (1974); E. O. Fischer, W. Held, F. R. Kreissl, A. Frank, and G. Huttner, *Chem. Ber.,* **110,** 656 (1977).

6. D. F. Shriver, *The Manipulation of Air-Sensitive Compounds,* McGraw Hill Book Co., New York, 1969, pp. 154-158.
7. E. O. Fischer, U. Schubert, W. Kleine, and H. Fischer, *Inorg. Synth.,* **19**, 164 (1979).
8. H. Gilman and F. K. Cartledge, *J. Organomet. Chem.,* **2**, 447 (1964).

42. THIOCARBONYL COMPLEXES OF TUNGSTEN(0)

Submitted by B. DUANE DOMBEK* and ROBERT J. ANGELICI*
Checked by P. TELLIER[†] and C. FRANK SHAW III[†]

The thiocarbonyl (CS) ligand of certain thiocarbonyl complexes of tungsten exhibits a chemical reactivity that is substantially different from that of CO in its analogous complexes.[1-4] The parent complex in this series, $W(CO)_5(CS)$, is prepared by reducing $W(CO)_6$ in tetrahydrofuran with sodium amalgam (a reaction that gives mainly $[W_2(CO)_{10}{}^{2-}]$)[5] and allowing the resulting solution to react with thiophosgene (thiocarbonyl dichloride), Cl_2CS. The $W(CO)_5(CS)$ product may be converted to numerous thiocarbonyl-containing derivatives, one of the most useful of which is $[Bu_4N]$ [*trans*-$W(CO)_4(CS)I$]. The preparation and purification of these two complexes are described here. Somewhat lower yields of $Cr(CO)_5(CS)$ are obtained in similar preparations using $Cr(CO)_6$.[1]

■**Caution.** *Because of the very toxic nature of carbon monoxide, all of the following reactions should be performed in an efficient hood.* Except where specified otherwise, the preparations are performed under an atmosphere of commercial prepurified nitrogen, in glassware previously flushed with nitrogen.

A. PENTACARBONYL(THIOCARBONYL)TUNGSTEN(0)

$$W(CO)_6 \xrightarrow[\text{(2) } Cl_2CS]{\text{(1) } Na/Hg} W(CO)_5(CS)$$

Procedure

An apparatus consisting of a 1000-mL, three-necked flask fitted with a mechanical stirrer and reflux condenser is flushed with nitrogen for several minutes. A positive nitrogen pressure is maintained thereafter by a mineral oil bubbler connected to a T-tube in the gas line between the nitrogen cylinder and the

*Department of Chemistry, Iowa State University, Ames, IA 50011.
[†]Department of Chemistry, University of Wisconsin-Milwaukee, Milwaukee. WI 53201.

flask. Mercury (about 120 mL) is added to the flask, and a sodium amalgam is prepared by stirring vigorously while adding, one at a time, pieces of sodium metal (7.0 g, 304 mmole, cut into about 10 pieces) through the third neck of the flask against a counter-current of nitrogen.

After the amalgam is cooled, 400 mL of tetrahydrofuran (THF) distilled from lithium tetrahydridoaluminate is added to the flask, followed by 50.0 g (142 mmole) of $W(CO)_6$.* ■**Caution.** *The THF should be checked for peroxide and high water content before $LiAlH_4$ is added. The distillation procedure is given in Reference 6.*

An electric heating mantle is then fitted under the flask, and the mixture is vigorously stirred and heated to reflux under nitrogen for 12-18 hours. After cooling to room temperature, a positive nitrogen pressure is maintained while the mechanical stirrer is replaced with a glass stopper.

The solution is then decanted under nitrogen flush through a 90° bent tube into a 500-mL pressure-equalizing addition funnel fitted with a small flask with a side arm (Fig. 1). A nitrogen source is then connected to this side arm to maintain positive nitrogen pressure over the solution while the bent tube at the top of the funnel is replaced with a nitrogen inlet. Any amalgam that has also been transferred is removed through the funnel stopcock. (The mercury in the amalgam may be recovered by washing first with ethanol and then with water.[7])

The funnel is then placed on a 2000-mL, single-necked, round bottomed flask

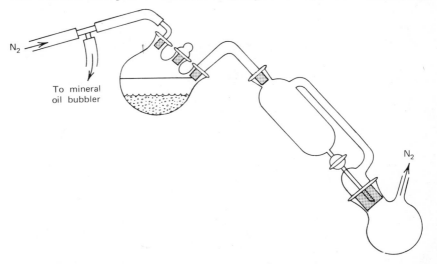

Fig. 1. Apparatus used in the transfer of THF solution from sodium amalgam to addition funnel.

*Available from Alfa Products, Ventron Corp., P.O. Box 299, Danvers, MA 01923.

containing 150 mL of dried THF, 14 mL (183 mmole) of thiophosgene,* and a magnetic stirring bar under nitrogen. (■**Caution.** *Thiophosgene is toxic and should be handled only in the hood.*) The THF-Cl_2CS solution is stirred vigorously with a magnetic stirrer, and the contents of the funnel are added rapidly (< 5 min) to the flask. (■**Caution.** *Large amounts of gas are liberated during this exothermic reaction, and frothing occurs.*) The black solution is then stirred until it has cooled to room temperature. It may be transferred in air to a 1000-mL flask; the flask is rinsed with THF. The solution is concentrated under water-aspirator vacuum on a rotary evaporator. The solution may be transferred to a 500-mL flask during the concentration, and evaporation is continued until the solvent has been removed.

A water-cooled sublimation probe with a vacuum inlet, similar to that shown in Reference 8, is then inserted into the flask, which is immersed in an oil bath at 50-60°. The contents are sublimed under a static vacuum (i.e., the apparatus is closed after being evacuated), with occasional removal of the sublimate from the probe. Approximately 10-15 g of yellow crystals, a mixture of $W(CO)_6$ and 3-6 g of $W(CO)_5(CS)$, is obtained. Much of the $W(CO)_6$ may be removed by dissolving the yellow solid in approximately 400 mL of warm hexane and allowing the solution to cool to 0° in a refrigerator. The solution is decanted from the pale-yellow crystals [mainly $W(CO)_6$] and evaporated to dryness to yield 4-7 g of a mixture that is generally 60-80% $W(CO)_5(CS)$.

The amount of $W(CO)_5(CS)$ in this mixture may be determined by preparing a solution of about 20 mg of the mixture in 50 mL of hexane and measuring the absorbance of the 420-nm peak of $W(CO)_5(CS)$. Since $W(CO)_6$ absorption is negligible at this wavelength, the concentration of $W(CO)_5(CS)$ may be calculated from its molar extinction coefficient, 7.52×10^2 L/cm mole.

Alternatively, the composition of this mixture may be estimated by gas chromatographic analysis of a sample in hexane solution. A 6.2 mm $\times$ 1.5 m 2% SE-30 column at temperatures of 70-150° and a helium flow rate of 25 mL/min gives excellent separations. Using a temperature program of 8°/min and an initial temperature of 70°, the retention time of $W(CO)_6$ is 4 minutes and that of $W(CO)_5(CS)$ is 10 minutes. The peak areas for these compounds are taken to be indicative of their relative molar quantities. Small amounts of pure $W(CO)_5(CS)$ may be collected from the gas chromatograph using oven and detector temperatures below 70°.

Further recrystallizations remove more of the $W(CO)_6$, but much $W(CO)_5(CS)$ is lost in the process. The thiocarbonyl complex in the $W(CO)_5(CS)$-$W(CO)_6$ mixture is more conveniently purified by first converting it to [Bu_4N] [*trans*-$W(CO)_4(CS)I$], which may then be allowed to react with Ag^+ and CO to yield pure $W(CO)_5(CS)$, as outlined in Sections B and C. *Anal.* Calcd.: C, 19.55; H, 0.00; S, 8.70. Found: C, 18.93; H, <0.01; S, 8.36.

*Available from Aldrich Chemical Co., 940 W. St. Paul Ave., Milwaukee, WI 53233.

Properties

The $W(CO)_5(CS)$ complex is an air- and moisture-stable yellow solid, soluble in polar and nonpolar organic solvents and insoluble in water. The infrared spectrum of $W(CO)_5)(CS)$ exhibits ν_{CO} bands in hexane solvent at 2096 (w), 2007 (m) and 1989 (vs) cm^{-1}, and a very strong ν_{CS} band in CS_2 solvent at 1258 cm^{-1}. (Tungsten hexacarbonyl exhibits one strong band at 1983 cm^{-1} in hexane.) The similarity of the physical properties of $W(CO)_5(CS)$ and $W(CO)_6$ makes their separation difficult. The thiocarbonyl ligand in this complex is susceptible to nucleophilic attack, reacting rapidly at room temperature with primary and secondary amines.[1,9] The complex undergoes carbonyl substitution reactions with various ligands at lower temperatures than does $W(CO)_6$.[1,2,9] This property is used to separate the thiocarbonyl complex from $W(CO)_6$, since the latter complex does not react with $[Bu_4N]I$ ($Bu = n-C_4H_9$) under the mild conditions that convert $W(CO)_5(CS)$ to $[Bu_4N][W(CO)_4(CS)I]$. After removal of the $W(CO)_6$, the halide complex may be reconverted to $W(CO)_5(CS)$ as described in Section C.

B.   TETRABUTYLAMMONIUM *trans*-[TETRACARBONYLIODO-(THIOCARBONYL)TUNGSTATE]

$$W(CO)_5(CS) + [Bu_4N]I \rightarrow [Bu_4N][W(CO)_4(CS)(I)] + CO$$

Procedure

Approximately equimolar amounts of $W(CO)_5(CS)$ and $[Bu_4N]I$ should be used in this preparation. The $W(CO)_5(CS)$ content of the mixture obtained in Section A may be determined by visible spectroscopy or gas chromatography, or it can be estimated from the infrared spectrum. Such a mixture (5.1 g) of $W(CO)_5(CS)$ and $W(CO)_6$ containing 3.3 g of $W(CO)_5(CS)$ (9.0 mmole) is placed in a 250-mL, single necked flask with 3.3 g (9.0 mmole) of $[Bu_4N]I$. The flask is flushed with nitrogen, 125 mL of dried THF is added, and the flask is connected to a nitrogen source and mineral oil bubbler. The solution is magnetically stirred at 50-55° in an oil bath for 1.5 hours and is then concentrated under water-aspirator vacuum to about 25 mL on a rotary evaporator. Precipitation of the crude product is completed by addition of 50-75 mL of diethyl ether. The brown solid (6.4 g, 99%) is washed well with diethyl ether and air dried. Unless a large excess of $[Bu_4N]I$ is employed in the preparation, this crude product may be used in further syntheses without purification. Recrystallization by dissolving the solid in a minimum volume of CH_2Cl_2 and adding 4 volumes of diethyl ether gives yellow crystals of the desired product. *Anal.* Calcd.: C, 35.54; H, 5.08; S, 4.52. Found: C, 35.05; H, 5.09; S, 4.67.

Properties

This air- and moisture-stable complex is soluble in many polar organic solvents. Its infrared spectrum in CH_2Cl_2 solution exhibits ν_{CO} bands at 2062 (w), and 1947 (vs) cm^{-1} and a ν_{CS} band at 1195 (vs) cm^{-1}. Strong alkylating and acylating agents add to the thiocarbonyl sulfur atom in the complex to yield neutral S-alkyl- and S-acylthiocarbonylium complexes.[3] The halide ion may be abstracted in acetone or THF solutions with $AgBF_4$ or $Ag[CF_3SO_3]$. Addition of ligands (L) to these solutions yields *trans*-$[W(CO)_4(CS)(L)]$ complexes.[4] One such reaction (L = CO) is outlined below, and other ligand additions may be performed similarly, in ground-joint glassware under a nitrogen atmosphere.

C. PENTACARBONYL(THIOCARBONYL)TUNGSTEN(0)

$$[Bu_4N] [W(CO)_4(CS)I] + Ag[CF_3SO_3] + CO \rightarrow$$

$$W(CO)_5(CS) + [Bu_4N] [CF_3SO_3] + AgI\downarrow$$

Procedure

■**Caution.** *This synthesis should be performed in a hood, and the pressure bottle should be adequately shielded.*

The crude $[Bu_4N] [W(CO)_4(CS)I]$ product from Section B is heated overnight at 50° under high vacuum to remove the last traces of $W(CO)_6$. Then 5.1 g of this product is placed in a 250-mL soft-glass pressure bottle (pop bottle)*[10] with a magnetic stirring bar and 150 mL of acetone. The solution is briefly flushed with CO (*in a hood!*) and fitted with a metal cap having a self-sealing Neoprene liner. A solution of 1.85 g (7.2 mmole) of $Ag[CF_3SO_3]$ [†] in 5 mL of acetone is added through a syringe with magnetic stirring, and the bottle is charged using a syringe needle with 30 psi of carbon monoxide. The mixture is stirred vigorously for 6 hours with occasional carbon monoxide repressuring. The carbon monoxide is then vented (*into a hood!*) through a syringe needle and the bottle cap is removed. The solution is filtered through Celite filter-aid and evaporated to dryness under reduced pressure. The residue is extracted with 25-mL portions of warm hexane until the extracts are nearly colorless. The combined extracts are concentrated to about 25 mL and passed through a 2 × 40 cm Florisil column with hexane eluant. Evaporation of the yellow band under reduced pressure gives 2.15 g (83%) of $W(CO)_5(CS)$. Use of ^{13}CO instead of

---

*The reaction may also be performed at atmospheric pressure with ground-glass equipment if carbon monoxide saturation of the solution is maintained. However, somewhat lower yields (about 65%) result under these conditions.

[†]Available from Alfa Products, Ventron Corp., P.O. Box 299, Danvers, MA 01923.

^{12}CO in this reaction gives *trans*-$[W(CO)_4(^{13}CO)(CS)]$.[4] $AgBF_4$ also has been used successfully.

References

1. B. D. Dombek and R. J. Angelici, *J. Am. Chem. Soc.*, **95**, 7516 (1973).
2. B. D. Dombek and R. J. Angelici, *J. Am. Chem. Soc.*, **96**, 7568 (1974). B. D. Dombek and R. J. Angelici, *Inorg. Chem.*, **15**, 2397 (1976).
3. B. D. Dombek and R. J. Angelici, *J. Am. Chem. Soc.*, **97**, 1261 (1975).
4. B. D. Dombek and R. J. Angelici, *J. Am. Chem. Soc.*, **98**, 4110 (1976).
5. J. E. Ellis, *J. Am. Chem. Soc.*, **96**, 7825 (1974); J. K. Ruff and W. J. Schlientz, *Inorg. Synth.*, **15**, 84 (1974).
6. *Inorg. Synth.*, **12**, 317 (1970).
7. R. B. King, *Organomet. Synth.*, **1**, 150 (1965).
8. *Ibid.*, p. 24.
9. B. D. Dombek and R. J. Angelici, *Inorg. Chem.*, **15**, 1089 (1976); S. S. Woodard, R. J. Angelici, and B. D. Dombeck, *Inorg. Chem.*, **17**, 1634 (1978). B. D. Dombek and R. J. Angelici, *Inorg. Chem.*, **15**, 2403 (1976); S. S. Woodard, R. A. Jacobson, and R. J. Angelici, *J. Organometal. Chem.*, **117**, C75 (1976).
10. D. F. Shriver, *The Manipulation of Air-Sensitive Compounds,* McGraw-Hill Book Co., New York, 1969, p. 157.

43. CYCLOPENTADIENYLMANGANESE THIOCARBONYLS

$$(\eta^5\text{-}C_5H_5)Mn(CO)_2(CS) + (C_6H_5)_3P \rightarrow$$

$$(\eta^5\text{-}C_5H_5)Mn(CO)(CS)[(C_6H_5)_3P] + CO \qquad (1)$$

$$(\eta^5\text{-}C_5H_5)Mn(CO)_2(CS) + (C_6H_5)_2PCH_2CH_2P(C_6H_5)_2 \rightarrow$$

$$(\eta^5\text{-}C_5H_5)Mn(CS)[(C_6H_5)_2PCH_2CH_2P(C_6H_5)_2] + 2CO \qquad (2)$$

Submitted by I. S. BUTLER,* N. J. COVILLE,* D. COZAK,*
S. R. DESJARDINS,* A. E. FENSTER,* and K. R. PLOWMAN*
Checked by D. A. SLACK[†] and M. C. BAIRD[†]

Since the discovery of the first transition metal thiocarbonyl complexes by Baird and Wilkinson in 1966,[1] over 100 of these complexes have been synthesized, encompassing essentially all the transition metals.[2,3] X-Ray data[4,5] reveal that

*Department of Chemistry, McGill University, 801 Sherbrooke St. West, Montreal, Quebec, Canada H3A 2K6.
[†]Department of Chemistry, Queen's University, Kingston, Ontario, Canada K7L 3N6.

metal-C(S) bond distances are about 0.07 Å shorter than metal-C(O) bond distances in analogous metal carbonyl derivatives. This fact, together with other physical[6-8] and chemical evidence,[2] indicates that CS is both a better σ-donor and π-acceptor ligand than CO and, consequently, metal-C(S) bonds should be stronger than metal-C(O) bonds.

In an earlier volume of *Inorganic Syntheses,* we described the synthesis of dicarbonyl(η^5-cyclopentadienyl)(thiocarbonyl)manganese(I),[9] and since then we have demonstrated the greater strength of the Mn—C(S) bond compared to the Mn—C(O) bonds in this complex with respect to photolytic substitution by preparing numerous (η^5-C$_5$H$_5$)Mn(CO)(CS)L,[10,12] (η^5-C$_5$H$_5$)Mn(CS)L$_2$,[12] and (η^5-C$_5$H$_5$)Mn(CS)(L-L)[13] derivatives, where L is a monodentate ligand and L-L is a bidentate ligand. Presented are the preparations of two typical derivatives, both of which future research may well prove to have important catalytic or synthetic applications.

■**Caution.** *Metal carbonyls are highly toxic and must be handled in a hood and with care.*

A. CARBONYL(η^5-CYCLOPENTADIENYL)(THIOCARBONYL) (TRIPHENYLPHOSPHINE)MANGANESE(I)

Procedure

Tetrahydrofuran (THF) (about 500 mL) is distilled[14] under nitrogen from sodium/benzophenone into a Pyrex ultraviolet irradiation vessel (capacity 575 mL) fitted with a water-cooled quartz finger containing a 100-W Hanovia high-pressure mercury lamp.* Dicarbonyl(η^5-cyclopentadienyl)(thiocarbonyl)manganese(I) (0.51 g, 2.32 mmole)[10] is dissolved in the freshly distilled tetrahydrofuran. The reaction vessel is wrapped in aluminum foil and placed in an ice-water bath, and the reaction mixture is then irradiated with ultraviolet light. (■**Caution.** *Exposure of the eyes to ultraviolet light must be avoided at all times.*) The progress of the reaction is monitored by following the changes in the ν_{CO} region (2150-1800 cm^{-1}) of the infrared spectrum of the reaction mixture using a pair of matched 1.0-mm NaCl cells. The samples are withdrawn from the reaction mixture through a side arm of the irradiation vessel by means of an air-tight syringe fitted with a 15 cm stainless steel hypodermic needle. The solution gradually turns dark red, and extensive decomposition occurs, as evidenced by the appearance of brown precipitate. After about 1½ hours, no further changes occur in the infrared spectrum. At this stage, the ν_{CO} bands at 2006 and 1954 cm^{-1} due to the starting material have virtually disappeared, and there is a new

*The design of the irradiation vessel used by the authors is similar to that described by Strohmeier[15] (see also Fig. 1. p. 194).

strong ν_{CO} band at 1912 cm^{-1}, due presumably to formation of the highly reactive $(\eta^5\text{-}C_5H_5)Mn(CO)(CS)(C_4H_8O)$ species. The irridation is stopped and the vessel is removed from the ice-water bath. Triphenylphosphine (1.10 g, 4.2 mmole) is now added to the dark-red solution, and the reaction mixture is heated slightly by passing warm tap water (35-40°) through the quartz finger. The reaction is allowed to continue until the infrared bands due to the product $(\eta^5\text{-}C_5H_5)Mn(CO)(CS)[(C_6H_5)_3P]$ at 1925 (ν_{CO}) and 1231 (ν_{CS}) cm^{-1} have ceased increasing in intensity (2-4 hr). The reaction mixture is filtered under nitrogen through a medium-porosity sintered-glass filter to remove the copious amount of light-brown decomposition product that forms. The tetrahydrofuran solvent is pumped off from the orange-brown filtrate at about 50° on a rotary evaporator to give a brown oil. Thin layer chromatography under nitrogen of a drop of this oil on a silica gel sheet (Eastman Kodak #13181) in a 4:1 hexane-acetone mixture shows the presence of $(C_6H_5)_3P$ ($R_f = 0.63$), $(\eta^5\text{-}C_5H_5)Mn(CO)_2(CS)$ ($R = 0.62$), and $(\eta^5\text{-}C_5H_5)Mn(CO)(CS)[(C_6H_5)_3P]$ ($R_f = 0.53$). The last complex is extracted from the brown oil by preparative thin layer chromatography under nitrogen on six activated silica gel plates (20 × 20 cm, coating thickness 0.75 mm*).[16] The eluant is 4:1 hexane-acetone. Since the product is air sensitive, particularly in solution, every effort must be made to minimize exposure of the compound to air during the following work-up procedure. All solvents should be freshly distilled under nitrogen prior to their use. The brown oil is taken up in a minimum quantity of dichloromethane and spotted onto the plates; the eluant is a 4:1 hexane-acetone mixture. The full extent of the orange $(\eta^5\text{-}C_5H_5)Mn(CO)(CS)[(C_6H_5)_3P]$ band on each plate is determined by exposing the plates to a low-intensity ultraviolet lamp.† The product bands are quickly scraped off with a spatula, the scrapings are combined, and the triphenylphosphine derivative is dissolved in dichloromethane (about 20 mL). The insoluble silica gel is filtered off under nitrogen through a medium-porosity, sintered-glass filter and the solvent is removed from the filtrate under reduced pressure at about 50° on a rotary evaporator to yield an oily orange solid. Dissolution of this oily solid in a minimum quantity of dichloromethane, followed by addition of a large excess of hexane, and subsequent evaporation of the solvents almost to dryness, affords orange crystals of the desired complex. These crystals are then dried *in vacuo* (25°/0.003 torr) for 9 hours. Yield 0.264 g [25% based on $(\eta^5\text{-}C_5H_5)Mn(CO)_2(CS)$]; dec 114-117° (uncorrected). *Anal.* Calcd. for C$_{25}$H$_{20}$OPSMn: C, 66.1; H, 4.4; S, 7.0. Found: C, 66.0; H, 4.5; S, 6.3.

---

*Machery, Nagel and Co. adsorbants used here may be purchased from Canlab, 8655 Delmeade Rd., Town of Mount Royal, Montreal Quebec, Canada H4T 1M3.
†Mineralight UVS.11, Ultra-Violet Products Inc., San Gabriel, CA 91778.

Properties

Carbonyl (η^5-cyclopentadienyl)(thiocarbonyl)(triphenylphosphine)manganese(I) is air sensitive, and the complex gradually turns green upon prolonged exposure to air. Although it is soluble in most common organic solvents, it is only slightly soluble in hydrocarbons. If the solvents are not deaerated, it decomposes rapidly. Although a detailed study of the decomposition reaction has not been carried out, it appears from changes in the ν_{CO} and ν_{CS} regions of the infrared spectrum that the major product is $(\eta^5\text{-}C_5H_5)Mn(CO)_2[(C_6H_5)_3P]$, with small amounts of unidentified thiocarbonyl-containing products also being produced. The principal infrared absorptions of $(\eta^5\text{-}C_5H_5)Mn(CO)(CS)[(C_6H_5)_3P]$ are: in CS_2 solution, ν_{CO} at 1925 (s); ν_{CS} at 2131 (s); in hexane, ν_{CO} at 1939 (s) and 1929 (s); ν_{CS} at 1236 (s); and in Nujol mull; ν_{CO} at 1913 (s) and ν_{CS} at 1213 (s) cm^{-1}. The appearance of *two* ν_{CO} bands for this monocarbonyl complex in hexane solution is attributed to the detection of two different conformational isomers resulting from restricted rotation of the phenyl groups within the $(C_6H_5)_3P$ ligand (see Reference 17). In the solid-state Raman spectrum, the ν_{CO} mode appears at 1913 cm^{-1} (Kr$^+$ laser, 647.1-nm excitation). The ^{1}H NMR spectrum in saturated CS_2 solution exhibits a complex phenyl resonance at τ 2.73 and an approximately 1:1 doublet centered at τ 5.75 due to coupling of the $\eta^5\text{-}C_5H_5$ protons with the ^{31}P atom of the triphenylphosphine ligand (J_{P-H}^3, 1.6 Hz).

B. [1,2-BIS(DIPHENYLPHOSPHINO)ETHANE](η^5-CYCLOPENTADIENYL)-(THIOCARBONYL)MANGANESE(I)

Procedure

As described in Section A, about 500 mL of tetrahydrofuran is distilled under nitrogen into the ultraviolet irradiation vessel. Dicarbonyl (η^5-cyclopentadienyl)-(thiocarbonyl)manganese(I) (0.51 g, 2.32 mmole) and 1,2-bis(diphenylphosphinoethane) [ethylenebis(diphenylphosphine), diphos] * (1.03 g, 2.55 mmole) are dissolved in the tetrahydrofuran. The irradiation process is carried out as above. After about 2½ hours, the infrared spectrum indicates that there is little or no $(\eta^5\text{-}C_5H_5)Mn(CO)_2(CS)$ remaining and the irradiation is stopped. The brown decomposition product that forms is filtered off under nitrogen through a medium-porosity, sintered-glass filter and then the solvent is pumped off from the yellow filtrate at about 50° on a rotary evaporator to give an orange oil. The

*Available from Strem Chem., P.O. Box 212, Danvers, MA 01923.

(η^5-C_5H_5)Mn(CS)(diphos) is extracted from this oil by chormatography on a silica gel column (2.5 × 35 cm) using a 3:2 benzene-hexane mixture as eluent. Thin layer chromatography on a silica gel sheet (Eastman Kodak #13181), using the same solvent mixture, reveals the presence of (η^5-C_5H_5)Mn(CO)$_2$(CS) (R_f = 0.92), diphos (R_f = 0.88), and (η^5-C_5H_5)Mn(CS)(diphos) (R_f = 0.72). Some difficulty may be experienced and it may be necessary to elute with dichloromethane-acetone mixtures, or even pure acetone, to remove the (η^5-C_5H_5)Mn(CS)(diphos) completely from the column. The presence of the strong ν_{CS} absorption at 1206 cm^{-1} and the absence of any ν_{CO} bands in the infrared spectra verify that the orange fractions collected contain (η^5-C_5H_5)Mn(CS)-(diphos). However, if the infrared spectra indicate the presence of a small amount of (η^5-C_5H_5)Mn(CO)$_2$(CS), it can be conveniently removed by high-vacuum sublimation (30°/0.001 torr) following the product work-up. Solvent removal from the combined orange fractions at about 50° on a rotary evaporator yields an oily orange solid, which, upon repeated recrystallization from dichloromethane-hexane mixtures following the same procedure described earlier for the isolation of (η^5-C_5H_5)Mn(CO)(CS)[(C_6H_5)$_3$P], affords the desired complex as an orange crystalline solid [0.532 g, 41% yield based on (η^5-C_5H_5)Mn(CO)$_2$(CS); dec. 197-199° (uncorrected)]. *Anal.* Calcd. for $C_{32}H_{29}P_2SMn$: C, 68.3; H, 5.2; P, 11.0 Found: C, 67.9; H, 5.6; P, 11.2.

Properties

[1,2-Bis(diphenylphosphino)ethane] (η^5-cyclopentadienyl)(thiocarbonyl)manganese(I) is air stable, both as a solid and in solution. It is very soluble in CS_2, acetone, and chlorinated organic solvents, moderately soluble in benzene, and only slightly soluble in alkanes. The prominent infrared bands in CS_2 solution are ν_{CS} at 1206 (s); δ_{MnCS} at 555 (m) and 543 (s) cm^{-1}. The ^{1}H NMR spectrum in saturated CS_2 solution exhibits a complex phenyl resonance with peaks at τ 2.30, 2.76, and 2.95, a broad CH_2 resonance with peaks at τ 7.4 and 7.6, and an approximately 1:2:1 triplet due to coupling of the η^5-C_5H_5 protons with the two equivalent ^{31}P atoms of the diphos ligand (J_{P-H}^3, 1.5 Hz) centered at τ 5.85. Preliminary studies on the reactions of this complex suggest that the sulfur atom of the thiocarbonyl ligand is strongly nucleophilic, and products most probably containing $-C{\equiv}S-$ bridges are formed, as in the case for (diphos)$_2$(OC)W$-$C${\equiv}$S$-$W(CO)$_5$, which is formed in the reaction of W(CO)(CS)-(diphos)$_2$ with W(CO)$_5$(acetone).[18]

References

1. M. C. Baird and G. Wilkinson, *Chem. Commun.,* **1966**, 267.
2. I. S. Butler and A. E. Fenster, *J. Organomet. Chem.,* **66**, 161 (1974).
3. I. S. Butler, *Acc. Chem. Res.,* **10**, 359 (1978).

4. J. L. de Boer, D. Rogers, A. C. Skapski, and P. G. H. Troughton, *Chem. Commun.*, **1966**, 756.
5. J. S. Field and P. J. Wheatley, *J. Chem. Soc., Dalton Trans.*, **1972**, 2269.
6. W. G. Richards, *Trans. Faraday Soc.*, **63**, 257 (1967).
7. N. Jonathan, A. Morris, M. O'Kada, D. J. Smith, and K. J. Ross., *Chem. Phys. Lett.*, **13**, 324 (1972).
8. D. L. Lichtenberger and R. F. Fenske, *Inorg. Chem.*, **15**, 2015 (1976).
9. I. S. Butler, N. J. Coville, and A. E. Fenster, *Inorg. Synth.*, **16**, 53 (1976).
10. I. S. Butler and A. E. Fenster, *J. Organomet. Chem.*, **51**, 307 (1973).
11. A. E. Fenster and I. S. Butler, *Inorg. Chem.*, **13**, 915 (1974).
12. N. J. Coville and I. S. Butler, *J. Organomet. Chem.*, **64**, 101 (1974).
13. I. S. Butler and N. J. Coville, *J. Organomet. Chem.*, **80**, 235 (1974).
14. For further details on the purification of tetrahydrofuran, see *Inorg. Synth.*, **12**, 317 (1970).
15. W. Strohmeier, *Angew. Chem. Int. Ed. Engl.*, **3**, 730 (1964).
16. O. Motl and L. Novotny, in *Laboratory Handbook of Chromatographic Methods*, O. Miker (ed.), D. Van Nostrand Co., Ltd., London, 1961, Chap. 4.
17. D. A. Brown, H. J. Lyons, and A. R. Manning, *Inorg. Chim. Acta*, **4**, 428 (1970).
18. B. D. Dombek and R. J. Angelici, *J. Am. Chem. Soc.*, **97**, 1261 (1975) and references cited therein.

44. DICARBONYL (η⁵-CYCLOPENTADIENYL)-(SELENOCARBONYL)MANGANESE(I)

$$(\eta^5\text{-}C_5H_5)Mn(CO)_3 + C_4H_8O \xrightarrow{h\nu} (\eta^5\text{-}C_5H_5)Mn(CO)_2(C_4H_8O) + CO$$

$$(\eta^5\text{-}C_5H_5)Mn(CO)_2(C_4H_8O) + CSe_2 + (C_6H_5)_3P \rightarrow$$

$$(\eta^5\text{-}C_5H_5)Mn(CO)_2(CSe) + (C_6H_5)_3PSe + C_4H_8O$$

Submitted by I. S. BUTLER,[†] D. COZAK,[†] S. R. STOBART,[‡]
and K. R. PLOWMAN[†]
Checked by P. V. YANEFF[§] and R. O. HARRIS[§]

Whereas the diatomic molecules carbon monoxide and carbon monosulfide can readily be synthesized, attempts to isolate the analogous carbon monoselenide

[†]Department of Chemistry, McGill University, P.O. Box 6070, Station A, Montreal, Quebec, Canada H3A 2K6.
[‡]Department of Chemistry, University of Victoria, Victoria, British Columbia, Canada V8W 2Y2.
[§]Department of Chemistry, Scarborough College, University of Toronto, Toronto, Ontario, Canada M1C 1A4.

species have been unsuccessful even at very low temperatures.[1] However, it has recently been established that the CSe molecule can be stabilized by coordination to transition metals and a few selenocarbonyl complexes have now been prepared, for example, $(\eta^6\text{-}C_6H_5CO_2CH_3)Cr(CO)_2(CSe)$,[2] $(\eta^5\text{-}C_5H_5)M(CO)_2$-(CSe) (M = Mn,[2] Re[3]), and $RuCl_2(CO)(CSe)[(C_6H_5)_3P]_2$.[4] X-Ray data for the ruthenium complex indicate that the bonding is similar to that in transition metal carbonyl and thiocarbonyl complexes, namely, the CSe ligand is attached to the ruthenium atom by way of the carbon atom and the Ru—C—Se moiety is essentially linear.[4] The synthesis presented here describes the preparation of a selenocarbonyl complex of manganese, $(\eta^5\text{-}C_5H_5)Mn(CO)_2(CSe)$. Similar methods may be used to prepare related chromium and rhenium complexes and future research in this area may well prove this method to be a general route to many other selenocarbonyl complexes of the transition metals.

■**Caution.** *Metal carbonyls are highly toxic and must be handled in a hood and with care. Since CSe_2 has an obnoxious smell and may have high toxicity it should also be handled carefully, in an efficient hood.*

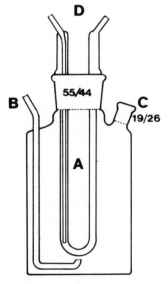

Fig. 1. Ultraviolet irradiation vessel (capacity 1800 mL): (*A*) water-cooled quartz lamp holder; (*B*) nitrogen inlet for purging and solution agitation; (*C*) nitrogen outlet and sampling port; (*D*) inlet and outlet for cooling water.

Procedure*

Dry tetrahydrofuran (1500 mL) is placed in an ultraviolet irradiation vessel (capacity 1800 mL) (see Fig. 1) similar to that described by Strohmeier.[5] Tricarbonyl(η^5-cyclopentadienyl)manganese(I)† (1.57 g, 7.6 mmole) is then dissolved in the solvent. After it is wrapped with aluminum foil, the irradiation vessel is placed in an ice-water bath and the reaction mixture is agitated with a stream of nitrogen. The ultraviolet source‡ located in the quartz finger of the irradiation vessel is then switched on. (■**Caution.** *Exposure of the eyes to the ultraviolet light must be avoided at all times.*) The solution gradually turns from pale yellow to deep wine red owing to the formation of the monosubstituted complex, $(\eta^5\text{-}C_5H_5)Mn(CO)_2(C_4H_8O)$. The progress of the reaction can be conveniently followed by monitoring changes in the ν_{CO} region of the infrared spectrum of the reaction mixture.** The irradiation is stopped when the infrared peaks due to the THF intermediate have ceased to increase in intensity (about 1 hr).†† Carbon diselenide (1.0 g, 5.8 mmole)‡‡ and excess triphenylphosphine (2.01 g, 7.7 mmole) are then added sequentially to the solution and the reaction mixture is allowed to stand at room temperature for 2 hours. The resulting dark-yellow solution is transferred to a 2-L, round-bottomed flask, and the solvent is removed under reduced pressure on a rotatory evaporator. The dark-brown oily residue is extracted with deaerated hexane (7 × 50 mL) and the combined extracts are evaporated to dryness to yield a yellow-brown solid.§ Extended high-vacuum sublimation of this crude reaction product onto an ice-water

---

*All the solvents should be distilled under nitrogen immediately prior to their use, and to minimize the possibility of decomposition, all transfers of materials and solutions should be performed routinely under a nitrogen atmosphere.

†Available from Strem Chem., P.O. Box 212, Danvers, MA 01923.

‡450-W Hanovia high-pressure mercury lamp.

**The highly reactive THF intermediate complex cannot be isolated from solution, but its presence is clearly indicated by the infrared spectrum of the reaction mixture in the ν_{CO} region: $(\eta^5\text{-}C_5H_5)Mn(CO)_3$, 2017 (s) and 1929 (vs); $(\eta^5\text{-}C_5H_5)Mn(CO)_2(THF)$, 1925 (s) and (1850 (s).

††The desired $(\eta^5\text{-}C_5H_5)Mn(CO)_2(CSe)$ product is extremely difficult to separate from the corresponding tricarbonyl complex and therefore it is essential to ensure conversion of as much as possible of the latter complex into the monosubstituted THF derivative. However, prolonged irradiation results in some disubstitution, and subsequent decomposition of this species leads to a decrease in the yield of the final product.

‡‡The CSe₂ is added directly from the shipping ampule (Strem Chemicals Inc., P.O. Box 212, Danvers, MA 01923) under a hood.

§A Dry Ice-acetone trap should be used to prevent escape of any CSe_2. Any unreacted CSe_2 appears in the last 50-100 mL of THF, from which it can be recovered by fractional distillation for proper disposal.

cooled finger inserted into the flask affords the desired selenocarbonyl complex as a golden-yellow crystalline solid, 0.54 g, 34% yield based on CSe_2. An analytical sample can be obtained by further sublimation (mp 65-66°, uncorrected). It is important that he crude reaction product be completely dry before the sublimation is begun, otherwise an oil forms on the cold finger. The product should not be exposed to prolonged evacuation because of its high volatility. Even after sublimation, the selenocarbonyl complex may contain a small amount of $(\eta^5\text{-}C_5H_5)Mn(CO)_3$ as an impurity; however, this may be removed by further fractional sublimations. To maximize the yield of the product, it is advisable to shield the compound from light as much as possible during the work-up procedure. *Anal.* Calcd. for $C_8H_5O_2SeMn$: C, 36.0; H, 1.9; Se, 29.6; Mn, 20.6; MW 268. Found: C, 36.1; H, 2.0; Se, 29.6; Mn, 20.3; Mw, 268 (mass spectroscopy, ^{80}Se). The residue remaining after the successive extractions with hexane contains triphenylphosphine selenide and various decomposition products. If desired, a pure sample of $(C_6H_5)_3PSe$ can be obtained by extraction of the residue with benzene and subsequent repeated recrystallizations from benzene-ethanol solutions (mp 188-189°, lit.[6] 187-189°). *Anal.* Calcd. for $C_{18}H_{15}PSe$: C, 63.3; H, 4.4; P. 9.1; Se, 23.2; MW, 342. Found: C, 63.4; H, 4.7; P, 8.8; Se, 23.3; MW, 342 (mass spectroscopy, ^{80}Se).

Properties

Dicarbonyl(η^5-cyclopentadienyl)(selenocarbonyl)manganese(I), $(\eta^5\text{-}C_5H_5)Mn(CO)_2(CSe)$, is air stable but slowly darkens when exposed to light. The complex is soluble in all common organic solvents but rapidly decomposes in solution if exposed to light. If the solutions are shielded from light, they are stable for weeks. A single sharp resonance at δ 4.8 ppm (from TMS) is observed in the ^1H NMR spectrum in CS_2 solution. the ^{13}C NMR spectrum in CS_2 solution at $-50°$ exhibits resonances at δ 85.8 (singlet, $\eta^5\text{-}C_5H_5$), 205.5 [singlet, $(CO)_2$], and 358.0 ppm (singlet, CSe). In its infrared spectrum, the absorptions due to CO and CSe stretching modes are: in CS_2 solution, ν_{CO} = 2010 (s) and 1960 (s), ν_{CSe} = 1107 (s); in hexane solution, ν_{CO} = 2015 (s) and 1965 (s), ν_{CSe} = 1113 (s); and in Nujol mull, ν_{CO} = 2012 (s) and 1962 (s), ν_{CSe} = 1115 (s) cm^{-1}. The major fragments in the mass spectrum at 70 eV are $(\eta^5\text{-}C_5H_5)Mn(CO)_2(CSe)^+$ (m/e 268), $(\eta^5\text{-}C_5H_5)Mn(CSe)^+$ (m/e 212) and $(\eta^5\text{-}C_5H_5)Mn^+$ (m/e 120); there is a conspicuous absence of the $(\eta^5\text{-}C_5H_5)Mn(CO)(CSe)^+$ (m/e 240) fragment. The complex readily undergoes photochemically induced CO substitution with various ligands such as $(C_6H_5)_3P$.[3]

References

1. R. Steudel, *Angew. Chem., Int. Ed. Engl.*, **6**, 635 (1967).
2. I. S. Butler, D. Cozak, and S. R. Stobart, *Chem. Commun.*, **1975**, 103.

3. D. Cozak, Ph.D. Thesis, McGill University, Montreal, Quebec, Canada, 1976.
4. G. R. Clark, K. R. Grundy, R. O. Harris and S. M. James, *J. Organomet. Chem.*, **90**, C37 (1975).
5. W. Strohmeier, *Angew. Chem. Int. Ed. Engl.*, **3**, 730 (1964).
6. R. R. Carlson and D. W. Meek, *Inorg. Chem.*, **13**, 1741 (1974).

45. ARENE THIOCARBONYL CHROMIUM(0) COMPLEXES

Submitted by GÉRARD JAOUEN* and GÉRARD SIMONNEAUX*
Checked by IAN S. BUTLER[†] and DANIEL COZAK[†]

In sharp contrast to metal carbonyl compounds, very few thiocarbonyl complexes have been prepared so far and these are known mainly for group VIII metals.[1] This is unexpected because the calculated strength of the M—CS bond is greater than that of the M—CO bond[2] (CS is favored over CO in terms of both σ and π bonding). Fenster and Butler suggested that synthetic difficulties may be the reason for this discrepancy, and they reported a method for the preparation of thiocarbonyl complexes of Mn(I).[3] Also, the formation of thiocarbonyls of zero-valent group VI metals was reported in the (arene)Cr(CO)$_3$[4] series (ben-, chrotrene) by way of an (arene)Cr(CO)$_2$(*cis*-cyclooctene) intermediate and in M(CO)$_6$ series (M = Cr, Mo, W) using Cl$_2$CS as a reagent.[5]

The preparations of several (arene)(thiocarbonyl)chromium(0) complexes bearing electron-donating and/or electron-withdrawing substituents on the ring are described here (**A-E**). This type of compound is useful in organometallic chemistry for problems related to stereochemistry around a chromium atom[6] and in organic chemistry in the activation of arene substituents with respect to alkylation.[7]

A	**B** [R = CO$_2$CH$_3$, R' = H]	**D** [R = CO$_2$CH$_3$, R' = CH$_3$]
	C [R = CO$_2$CH$_3$, R' = CH$_3$]	**E** [R = R' = CH$_3$]

*Laboratoire de Chimie des Organométalliques, E. R. A.–CNRS n° 477 Université de Rennes, 35031 Rennes Cedex–France.
[†] Department of Chemistry, McGill University, P. O. Box 6070, Station A, Montréal, Québec, Canada M 3C 3GI.

■**Caution.** *Because of the toxicity of the metal carbonyls and the need for venting CO, the reactions should be carried out in a fume chamber. Care must also be exercised with the CS_2, as it is extremely flammable.*

A. DICARBONYL-(η^6-*o*-DIMETHYLBENZENE)(THIOCARBONYL)-CHROMIUM(0)

$$\{\eta^6\text{-}o\text{-}(CH_3)_2(C_6H_4)\}Cr(CO)_3 + C_8H_{14} \xrightarrow{\text{UV}}$$

$$\{\eta^6\text{-}o\text{-}(CH_3)_2(C_6H_4)\}Cr(CO)_2(C_8H_{14}) + CO$$

$$\{\eta^6\text{-}o\text{-}(CH_3)_2(C_6H_4)\}Cr(CO)_2C_8H_{14} + P(C_6H_5)_3 + CS_2 \rightarrow$$

$$\{(\eta^6\text{-}o\text{-}(CH_3)_2(C_6H_4)\}Cr(CO)_2(CS) + C_8H_{14} + SP(C_6H_5)_3$$

Procedure

The photochemical reaction is carried out in a 200-mL Pyrex vessel fitted with a water-cooled quartz finger. The UV source (T.Q. 150 W Hanau high-pressure mercury lamp) is located inside the finger (Fig. 1). (■**Caution.** *Laboratory personnel must be shielded from the UV light.*) A 4-g sample (0.016 mole) of tricarbonyl(η^6-*o*-dimethylbenzene)chromium,[8] 30 mL of *cis*-cyclooctene and 150 mL of benzene are placed in the reaction and irradiated for 50 minutes at room temperature under a nitrogen stream. The yellow solution gradually becomes orange. The UV lamp is turned off and 3.9 g (0.016 mole) of triphenylphosphine dissolved in 100 mL of deoxygenated CS_2 (CS_2 is degassed by bubbling nitrogen through the solvent) is added to the flask. The solution is then allowed to stand overnight under nitrogen (all the operations dealing with the cyclooctene complex must be performed under nitrogen). After filtration and removal of the solvents under vacuum, the crude oil is purified by chromatography through a deaerated silica gel column (column: 4 × 60 cm; silica gel: Mallinckrodt A-R 100 mesh) using an 80:20 hexane-diethyl ether mixture as eluent. A 3.4-g sample of yellow dicarbonyl(η^6-*o*-dimethylbenzene)(thiocarbonyl)chromium is obtained (yield 84%; mp 100°) after removal of the solvent under vacuum, and 0.2 g of the starting material is recovered from the front-running yellow fraction. The complex may be recrystallized from diethyl ether-petroleum ether (30-70 mL) by leaving the solution in a dark cupboard. *Anal.* Calcd. for $C_{11}H_{10}CrO_2S$: C, 51.16; H, 3.87. Found: C, 51.36, H, 3.85.

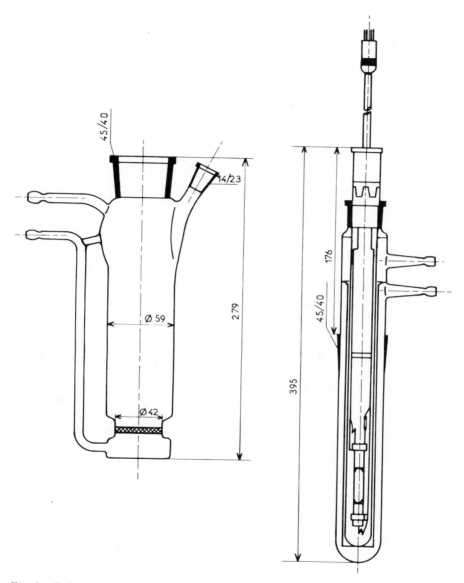

Fig. 1. Left: photochemical reactor. Right: water-cooled UV lamp for reactor.

Properties

The complex is air stable in the solid state but decomposes in solution upon exposure to sunlight. It is soluble in most common organic solvents. The infrared spectrum exhibits two metal carbonyl bands at 1908 and 1965 cm^{-1} and a thio-carbonyl absorption at 1207 cm^{-1} (CH_2Cl_2 solution). The NMR spectrum (δ_{CH_3} at 2.1 ppm, δ_{C-H} arene at 5.7; $CDCl_3$ solution) is consistent with the structure shown above (**A**). Progress of the reaction may be easily monitored by thin layer chromatography* (TLC plate: 20 ×20 × 1 mm, Kieselgel G. Typ. 60, Merck) using a 80:20 hexane-diethyl ether mixture as eluent.

B. DICARBONYL(METHYL η^6-BENZOATE)(THIOCARBONYL)-CHROMIUM(0)

$$(\eta^6\text{-}C_6H_5CO_2CH_3)Cr(CO)_3 + C_8H_{14} \xrightarrow{\text{UV}}$$
$$(\eta^6\text{-}C_6H_5CO_2CH_3)Cr(CO)_2(C_8H_{14}) + CO$$

$$(\eta^6\text{-}C_6H_5CO_2CH_3)Cr(CO)_2(C_8H_{14}) + P(C_6H_5)_3 + CS_2 \rightarrow$$
$$(\eta^6\text{-}C_6H_5CO_2CH_3)Cr(CO)_2(CS) + SP(C_6H_5)_3 + C_8H_{14}$$

Procedure

A solution of 5 g (0.018 mole) of tricarbonyl(methyl η^6-benzoate)chromium[8] and 30 mL of *cis*-cyclooctene in deoxygenated benzene (180 mL) is irradiated as described above for 2¼ hours at room temperature. The yellow-orange solution rapidly becomes deep red. After the irradiation is stopped, a solution of 4.5 g (0.018 mole) of triphenylphosphine in 100 mL of CS_2 free of oxygen (see above) is added to the flask under nitrogen. The mixture is allowed to stand 30 minutes at room temperature, filtered, concentrated under vacuum, and purified by chromatography with a silica gel column using a petroleum ether-diethyl ether, 75:25 eluent. The first eluted compound is the starting material (0.8 g recovered) and the major band is the red thiocarbonyl complex (3.2 g; yield 73%, mp 79°). The product is recrystallized from hexane-diethyl ether (50:50) to give prismatic crystals. *Anal.* Calcd. for $C_{11}H_8CrO_4S$: C, 45.83; H, 2.77. Found C, 45.98, H, 2.89.

*The checkers suggested that infrared spectroscopy might also be used to monitor the reactions. The appearance of the ν_{CO} band of the CO_2CH_3 functional group or the high-energy (A_r) ν_{CO} band of the carbonyl ligands may be conveniently used. Care should be taken, however, not to confuse the ν_{CO} bands of the metal carbonyl with the two combination bands due to the arene ring at 1953 and 1809 cm^{-1}.

Properties

The intermediate cyclooctene complex appears to be more reactive with respect to CS coordination and more sensitive to oxidation when the arene ring bears electron-withdrawing groups (e.g., CO_2CH_3). Dicarbonyl(methyl η^6-benzoate)-thiocarbonyl)chromium is air stable in the solid state and reasonably stable in solution.[9] The infrared spectrum exhibits metal carbonyl absorptions at 1980 and 1935 cm^{-1} and a metal thiocarbonyl stretch at 1215 cm^{-1} (Nujol) (these occur at 1978, 1932, and 1912 cm^{-1} in CH_2Cl_2 solution).[10] Irradiation of the compound in the presence of phosphite or phosphine leads to slow substitution of CO by these ligands, whereas the CS ligand remains inert to substitution. The crystal structure has been published.[11]

C. DICARBONYL(METHYL η^6-*m*-METHYLBENZOATE)(THIO-CARBONYL)CHROMIUM(0)

$$\{\eta^6\text{-}m\text{-}(CH_3)C_6H_4CO_2CH_3\}\,Cr(CO)_3 + C_8H_{14} \xrightarrow{UV}$$

$$\{\eta^6\text{-}m\text{-}(CH_3)C_6H_4CO_2CH_2\}\,Cr(CO)_2(C_8H_{14}) + CO$$

$$\{\eta^6\text{-}m\text{-}(CH_3)C_6H_4CO_2CH_3\}Cr(CO)_2C_8H_{14} + P(C_6H_5)_3 + CS_2 \rightarrow$$

$$\{\eta^6\text{-}m\text{-}(CH_3)C_6H_4CO_2CH_3\}Cr(CO)_2(CS) + C_8H_{14} + SP(C_6H_5)_3$$

Procedure

Under the conditions described above, racemic dicarbonyl(methyl η^6-*m*-methyl benzoate)(thiocarbonyl)chromium, mp 74°, may be obtained in 90% yield from the corresponding racemic tricarbonyl complex, mp 97°.[12] Thus 4.5 g of product are isolated starting with 4.8 g (0.017 mole) of tricarbonyl(methyl η^6-*m*-methylbenzoate)chromium irradiated for 4½ hours under nitrogen (0.1 g recovered from the chromatography; eluent: 80:20 petroleum ether-diethyl ether). If the optically active tricarbonyl complex $[\alpha]_D^{25} = -113°$, $c = 1.31$ g/L, benzene (S), mp 122°,[13,14] is used instead of the racemic complex, the thio-carbonyl complex, mp 100°, shows an optical rotation of $[\alpha]_D^{25} = -140°$, $c = 1.09$ g/L, CHCl$_3$, (S). The absolute configurations are shown in C. In this reaction, the planar chirality, due to the complexed disubstituted arene ring, remains untouched. So the absolute configuration of the thiocarbonyl complex can be acertained by correlation to that of the tricarbonyl complex.[14] The infrared spectrum exhibits carbonyl absorptions at 1976 and 1928 cm^{-1} and a thiocarbonyl stretch at 1208 cm^{-1} (CH_2Cl_2 solution). (NMR: δ_{CH_3} 2.33 ppm; $\delta_{CO_2CH_3}$ 4.07 ppm in CDCl$_3$ solution).

D. CARBONYL(METHYL η^6-*m*-METHYLBENZOATE)(THIOCARBONYL) (TRIPHENYL PHOSPHITE)CHROMIUM(0)

$$\{\eta^6\text{-}m\text{-}(CH_3)C_6H_4CO_2CH_3\}Cr(CO)_2(CS) + P(OC_6H_5)_3 \rightarrow$$

$$\{\eta^6\text{-}m\text{-}(CH_3)C_6H_4CO_2CH_3\}Cr(CO)(CS)P(OC_6H_5)_3 + CO$$

Procedure

A mixture of dicarbonyl(methyl η^6-*m*-methylbenzoate)(thiocarbonyl)chromium (2 g, 0.007 mole) and 16.5 g (0.058 mole) of triphenyl phosphite in 200 mL of deoxygenated benzene is irradiated under nitrogen for 3½ hours. The orange solution rapidly becomes red and some decomposition is observed. Filtration and removal of the solvent are followed by primary chromatography on 15 silica gel preparative TLC plates (TLC plates: 400 × 200 × 5 mm, Kiesel gel G, Typ. 60 Merck) using a 1:1 petroleum ether-diethylether mixture as the eluent. Three bands are developed, corresponding, respectively, to carbonyl (methyl η^6-*m*-methylbenzoate)(thiocarbonyl)bis(triphenylphosphite)chromium from the front (0.8 g), the mixture of the two diastereoisomers (1.2 g), and finally, the starting material (0.3 g).

The mixture of diastereoisomers is then dissolved in benzene (200 mL) and submitted to a second preparative TLC using a 2:6 benzene-hexane mixture as eluent (to avoid crystallization of the product on the plates, the quantity should not exceed 0.05 g/per plate). Diastereoisomer I (mp 140°, 0.5 g) is eluted first, followed by diastereoisomer II (mp 158°, 0.5 g). Yield 30% *Anal.* Calcd. for I $C_{29}H_{25}CrO_6PS$: C, 59.59; H, 4.31. Found: C, 59.47; H, 4.26.

Starting with the *S* optically active dicarbonyl thiocarbonyl complex, the optically active disubstituted complex, mp 129° $[\alpha]_D^{25} = -140°$, $c = 1.22$, CDCl$_3$, and diastereoisomers are isolated: I, mp 100°, $[\alpha]_D^{25} = +195°$, $c = 1.06$ g/L, CHCl$_3$ and II mp 119°, $[\alpha]_D^{25} = -276°$, $c = 1.12$ g/L, CHCl$_3$.

Properties

The relative amounts of the products depend on irradiation time, and the formation of the bis (triphenyl phosphite) chromium complex may be almost completely avoided by following the reaction with thin layer chromatography. The diastereoisomers are air stable and reasonably soluble in most organic solvents. Their infrared spectra exhibit a metal carbonyl band at 1925 cm^{-1} (ν_{CO} ester at 1729 cm^{-1}) and a metal thiocarbonyl band at 1925 cm^{-1} (CH$_2$Cl$_2$ solution). The NMR spectra. I: δ_{CH_3} at 1.87 ppm: $\delta_{CO_2CH_3}$ at 3.77 ppm; II: δ_{CH_3} at 1.77

ppm and $\delta_{CO_2CH_3}$ at 3.93 ppm (CDCl$_3$ solution), may be used for determination of purity with respect to separation of the diastereoisomers.

The planar chirality afforded by the substituted arene ring is removed by conversion of the ester function CO_2CH_3 into CH_3. The resulting η^6-*m*-dimethyl benzene complex **E** owes its optical activity to the chiral chromium center. Thus I, $[\alpha]_D^{25} = +195^\circ$ (0.02 g, 0.5×10^{-4} *M*), is dissolved in 50 mL of diethyl ether and reduced to carbonyl(η_6-*m*-dimethylbenzene)(thiocarbonyl)(triphenyl phosphite)-chromium, mp 161°, $[\alpha]_D^{25} = +223°$, c = 1.40 g/L, CHCl$_3$ solution, (0.018 g, yield 69%) at room temperature using equimolecular amount of LiAlH$_4$-AlCl$_3$ (0.380 g/1.33 g; 10^{-2} *M*) as reducing agent. The infrared spectrum contains ν_{CO} at 1915 cm^{-1} and ν_{CS} at 1205 cm^{-1} (CH$_2$Cl$_2$ solution) and the NMR spectrum exhibits the diastereotopic methyl groups at δ_{CH_3} at 2.13 and 1.86 ppm (CDCl$_3$ solution).

The (methyl η^6-*m*-methylbenzoate)(thiocarbonyl)bis(triphenyl phosphite)-chromium gives the following spectroscopic properties: ν_{CO} at 1723 cm^{-1} (ester), and ν_{CS} at 1158 cm^{-1} (CH$_2$Cl$_2$ solution); δ_{CH_3} at 1.53 ppm and $\delta_{CO_2CH_3}$ at 3.5 ppm (C$_6$D$_6$ solution).

The thiocarbonyl ligand is expected to be more polar and reactive than the analogous carbonyl ligand and the observed ease of nucleophilic attack at the carbon atom supports this idea.[5] However, few of its specific properties have been recognized, with the exception of strong π-acceptor character[2,3,9] and the possibility of bridging through C and S.[15,16]

References

1. I. S. Butler and A. E. Fenster, *J. Organomet. Chem.*, **66**, 161 (1974).
2. W. G. Richards, *Trans. Faraday Soc.*, **63**, 257 (1967).
3. A. E. Fenster and I. S. Butler, *Inorg. Chem.*, **13**, 915 (1974).
4. G. Jaouen, *Tetrahedron Lett.*, **1973**, 5159.
5. B. D. Dombek and R. J. Angelici, *J. Am. Chem. Soc.*, **95**, 7516 (1973).
6. (a) G. Simonneaux, A. Meyer, and G. Jaouen, *Chem. Comm.*, **1975**, 69; (b) G. Jaouen, A. Meyer, and G. Simonneaux, *Tetrahedron*, **31**, 1889, (1975).
7. G. Jaouen, A. Meyer, and G. Simonneaux, *Chem. Commun.*, **1975**, 813.
8. (a) M. D. Rausch, G. A. Moser, E. J. Zaiko, and A. L. Lipman, *J. Organomet. Chem.*, **23**, 85 (1970): (b) C. A. L. Mahaffy and P. L. Pauson, *Inorg. Synth.*, **19**, 154 (1979).
9. G. Jaouen and R. Dabard, *J. Organomet. Chem.*, **72**, 377, (1974).
10. P. Caillet and G. Jaouen, *J. Organomet. Chem.*, **91C**, 53 (1975).
11. J. Y. Saillard, G. Le Borgne, and D. Grandjean, *J. Organomet. Chem.*, **94**, 409 (1975).
12. A. Meyer, *Ann. Chim.*, **8**, 315 (1975) and **8**, 397, (1973).
13. R. Dabard, A. Meyer and G. Jaouen, *C. R. Acad. Sci., C*, **268**, 201 (1969).
14. (a) G. Jaouen and R. Dabard, *J. Organomet. Chem.*, **21**, P. 43 (1970); (b) M. A. Bush, T. A. Dullforce, and G. A. Simm, *Chem. Commun.*, **1969**, 1491.
15. B. D. Dombeck and R. J. Angelici, *J. Am. Chem. Soc.*, **96**, 7568 (1974).
16. A. Efraty, R. Arneri, and M. A. Huang, *J. Am. Chem. Soc.*, **98**, 639 (1976).

46. *trans*-[CHLORO(THIOCARBONYL)BIS-(TRIPHENYLPHOSPHINE)RHODIUM(I)]

$$Rh(PPh_3)_3Cl + CS_2 \xrightarrow{PPh_3} \textit{trans-}[RhCl(CS)(PPh_3)_2] + Ph_3PS$$

Submitted by M. KUBOTA* and C. O. M. HO*
Checked by Y. FUJII[†] and V. MAINZ[‡]

Baird and Wilkinson in 1966 reported the first syntheses of transition metal complexes containing the thiocarbonyl (CS) ligand.[1] The compounds *trans*-[RhX(CS)(PPh_3)_2], where X = Cl, Br, were obtained from the reaction of RhX(PPh_3)_3 and CS_2 in approximately 50% yield. A crystallographic study[2] revealed that the Rh—C(S) bond distance is 0.07 Å shorter than the Rh—C(O) bond distance in the analogous rhodium carbonyl complex. Other studies indicate that the CS ligand is a better σ-donor and π-electron pair acceptor than CO.[3] Triphenylphosphine and RhCl(PPh_3)_3 in a 1:1 mixture of methanol and carbon disulfide give an emerald-green solution from which a nearly quantitative yield of orange crystals of *trans*-[RhCl(CS)(PPh_3)_2] has been reported.[4] The synthesis presented here, which is an adaptation of the latter procedure, can be accomplished in 3 hours from the readily available starting compound RhCl(PPh_3)_3.[5] The compound *trans*-[RhCl(CS)(PPh_3)_2] is a useful starting material for the synthesis of a variety of thiocarbonyl complexes of rhodium(I) and rhodium(III).

Procedure

■**Caution.** *Carbon disulfide is highly flammable and toxic and should be vented out through an efficient fume hood.*

Chlorotris(triphenylphosphine)rhodium(I)[5] (1.42 g, 1.53 mmole), triphenylphosphine (4.0 g, 15.3 mmole), and a magnetic stirring bar are placed in a 50-mL, round-bottomed flask fitted with a gas-inlet tube. The flask is purged with nitrogen because the starting reagent RhCl(PPh_3)_3 decomposes in the presence of oxygen and triphenylphosphine in solution under room illumination reacts with oxygen.[6] (The product, *trans*-[RhCl(CS)(PPh_3)_2], does not react with oxygen.) Methanol (12 mL) that has been deaerated by bubbling nitrogen through it for 5 minutes is added. Freshly distilled carbon disulfide (10 mL) is

*Department of Chemistry, Harvey Mudd College, Claremont, CA 91711.
[†]Department of Chemistry, Univerisity of Illinois, Urbana, IL 61801.
[‡]Department of Chemistry, University of California, Berkeley, CA 94720.

then added in one portion. The flask is stoppered and stirred for 20 minutes, and then it is fitted with a condenser and heated at reflux for 30 minutes.

The solvent is evaporated at reduced pressure, using a rotary evaporator and water aspirator. The remaining solid is dried *in vacuo* for 1.5 hours. The orange product is washed with two 5-mL portions of methanol and four 5-mL portions of cold diethyl ether. The product is dried *in vacuo*. Yield 932 mg (86%). The product may be recrystallized to remove possible triphenylphosphine contamination by dissolving it in hot chloroform (50 mL), filtering if necessary, and then slowly adding 100 mL of ethanol (yield 50%). *Anal.* Calcd. for $RhC_{37}H_{30}ClP_2S$: C, 62.86; H, 4.28. Found: C, 62.66; H, 4.25. In addition to infrared absorption bands due to triphenylphosphine, *trans*-$[RhCl(CS)(PPh_3)_2]$ has a strong absorption band at 1300 cm^{-1} (Nujol) that is attributed to C—S stretching.

Properties

Orange crystals of *trans*-[chloro(thiocarbonyl)bis(triphenylphosphine)rhodium-(I)] melt with decomposition at 259-262° in air.* The compound is stable in air. It is soluble in dichloromethane, chloroform, toluene, and benzene; only slightly soluble in methanol, ethanol, and diethyl ether; and insoluble in hexane, cyclohexane, and other hydrocarbon solvents. The chloro ligand may be readily substituted by metathesis, for example, to form *trans*-$[Rh(NCS)(CS)(PPh_3)_2]$. The bromo derivative *trans*-$[RhBr(CS)(PPh_3)_2]$ has also been prepared from $RhBr(PPh_3)_3$. Like its carbonyl analogue, *trans*-$[RhCl(CS)(PPh_3)_2]$ oxidatively adds chlorine,[1] bromine,[1] and tetracyanoethylene (ethylenetetracarbonitrile).[7] The compound $[Rh(CS)(\eta^2\text{-}CS_2)(PPh_3)][BPh_4]$ has been isolated from solutions of *trans*-$[RhCl(CS)(PPh_3)_2]$ in methanol and carbon disulfide containing triphenylphosphine and sodium tetraphenylborate.[8]

References

1. M. C. Baird and G. Wilkinson, *Chem. Commun.,* **1966,** 267.
2. J. L. de Boer, D. Rogers, A. C. Skapski, and P. H. Troughton, *ibid.,* **1966,** 756.
3. I. S. Butler and A. E. Fenster, *J. Organomet. Chem.,* **66,** 161 (1974); also see M. A. Andrews, *Inorg. Chem.,* **16,** 496 (1977).
4. M. C. Baird, G. Hartwell, and G. Wilkinson, *J. Chem. Soc., A,* **1967,** 2037.
5. J. A. Osborn and G. Wilkinson, *Inorg. Synth.,* **10,** 67 (1967).
6. G. L. Geoffroy, D. A. Denton, and C. W. Eigenbrot, Jr., *Inorg. Chem.,* **15,** 2310 (1976).
7. W. H. Baddley, *J. Am. Chem. Soc.,* **88,** 4545 (1966); M. Haga, K. Kawakami, and T. Tanaka, *Inorg. Chem.,* **15,** 1946 (1976).
8. G. R. Clark, T. J. Collins, S. M. James, W. R. Roper, and K. G. Town, *Chem. Commun.,* **1976,** 475. The cationic complex may contain the bidentate sulfur-bound zwitterion $Ph_3P^+\text{--}CS_2^-$ ligand.

*The checkers observed a decomposition temperature of 250-252°, which agrees with the literature value.[4]

47. *trans*-[CHLORO(THIOCARBONYL)BIS- (TRIPHENYLPHOSPHINE)IRIDIUM(I)]

$$\textit{trans-}[Ir(PPh_3)_2(N_2)Cl] + CS_2 + Ph_3P \rightarrow$$

$$\textit{trans-}[Ir(PPh_3)_2(CS)Cl] + N_2 + Ph_3PS$$

Submitted by M. KUBOTA*
Checked by A. P. GINSBERG[†] and C. R. SPRINKLE[†]

Whereas physical and chemical evidence strongly indicate that CS is a better τ-donor and π-acceptor ligand than carbon monoxide,[1] methods for the synthesis of thiocarbonyl analogues of the well-known metal carbonyl complexes are still somewhat limited in scope. The thiocarbonyl complex *trans*-[Ir(PPh$_3$)$_2$(CS)Cl] , which is analogous to the well-known complex *trans*-[Ir(PPh$_3$)$_2$(CO)Cl] ,[2] was first prepared in low yield in 1968 by the reaction of carbon disulfide and Ir(PPh$_3$)$_3$Cl.[3] The synthesis of Ir(PPh$_3$)$_2$(CS)Cl from Ir(PPh$_3$)$_2$(N$_2$)Cl[4] can be accomplished[5-7] in 3 or 4 hours in yields up to 70%. The isolable intermediates in this reaction, Ir(PPh$_3$)$_2$Cl(C$_2$S$_5$) and Ir(PPh$_3$)$_2$Cl(CS)(CS$_3$), may also be converted in high yield to Ir(PPh$_3$)$_2$(CS)Cl. The thiocarbonyl complex Ir(PPh$_3$)$_2$(CS)Cl is a convenient starting material for a variety of thiocarbonyl iridium complexes.[3,6,7]

Procedure

■**Caution.** *Carbon disulfide is highly flammable and toxic and should be vented out through an efficient fume hood.*

The nitrogen complex *trans*-[Ir(PPh$_3$)$_2$(N$_2$)Cl] (680 mg, 0.87 mmole)[4] and a magnetic stirring bar are placed in a 100-mL, two-necked, round-bottomed flask. Nitrogen is slowly introduced from a tube fitted to one neck of the flask, and 4 mL of cold carbon disulfide (about 5°) is added in one portion through the other neck. The yellow nitrogen complex reacts rapidly with gas evolution to give an immediate dark-green solution which rapidly turns black. The reaction mixture is stirred to ensure that all of the nitrogen complex has reacted. The carbon disulfide is then evaporated with a stream of nitrogen, leaving a black product Ir(PPh$_3$)$_2$(C$_2$S$_5$)Cl.[5] If the starting nitrogen complex is not of high purity, the black Ir(PPh$_3$)$_2$(C$_2$S$_5$)Cl may be recrystallized by dissolving it in

*Department of Chemistry, Harvey Mudd College, Claremont, CA 91711.
[†] Bell Laboratories, Murray Hill, NJ 07974.

dichloromethane and precipitating it with excess methanol to give black micro-crystals.

Triphenylphosphine (2 g), 14 mL of chloroform, and 7 mL of methanol are added to the black solid $Ir(PPh_3)_2(C_2S_5)Cl$. The nitrogen flow is reduced to maintain only a slight positive pressure in the flask, and a water-cooled condenser is fitted to the other neck of the flask. The flask is heated with a heating mantle and the mixture is heated at reflux for 3 hours, during which time the reaction mixture turns from orange-brown to orange. After cooling, the orange crystals that form are collected on a medium-porosity filter. They are washed with four 5-mL portions of diethyl ether and dried *in vacuo*. Yield 485 mg (71%). The product may be recrystallized by dissolving in a minimum volume of chloroform and slowly adding twice that volume of methanol. *Anal.* Calcd. for $IrC_{37}H_{30}ClP_2S$: C, 55.8; H, 3.80. Found: C, 55.4; H, 3.66. In addition to absorption bands due to triphenylphosphine, *trans*-$[Ir(PPh_3)_2(CS)Cl]$ has infrared absorption bands at 1332 cm^{-1} (s) (Nujol) attributed to ν_{CS} and at 292 cm^{-1} (w) due to ν_{Ir-Cl}. Impure preparations of *trans*-$[Ir(PPh_3)_2(CS)Cl]$ have infrared bands at 1360, 1050, 868, and 838 cm^{-1}, which are due to intermediate products in the reaction.[5] These intermediate products may be converted to *trans*-$[Ir(PPh_3)_2(CS)Cl]$ by further heating at reflux with triphenylphosphine in chloroform-methanol.

Properties

trans-[Chloro(thiocarbonyl)bis(triphenylphosphine)iridium(I)]   forms   orange crystals that melt with decomposition at 260-270°* in air. The compound is stable in air, moderately soluble in chloroform, dichloromethane, and benzene, but only slightly soluble in methanol, diethyl ether, and hexane. Alkyl or aryl phosphine exchange gives[6] products such as *trans*-$[Ir[P(C_6H_{11})_3]_2(CS)Cl]$ and $[Ir(diphos)_2(CS)]^+$ [diphos = ethylenebis[diphenylphosphine] or 1,2-bis(diphenylphosphino)ethane. Carbon monoxide bubbled through a solution of *trans*-$[Ir(PPh_3)_2(CS)Cl]$ gives[6]   the   cationic   dicarbonyl   complex $[Ir(PPh_3)_2(CO)_2(CS)]^+$. With sodium tetrahydroborate(1-) and triphenylphosphine in ethanol, the complex gives $Ir(PPh_3)_3H(CS)$, and with carbon disulfide and sodium tetraphenylborate the complex gives the cationic complex,[3,8] $[Ir(PPh_3)_2(CS)(\eta^2\text{-}CS_2)]^+$. Like its carbonyl analogue, *trans*-$[Ir(PPh_3)_2(CS)Cl]$ gives adducts with CO, SO_2, BCl_3, HCl, tetracyanoethylene (ethylenetetracarbonitrile), fumaronitrile, and $[NO][BF_4]$,[3,7] but unlike its carbonyl analogue the less basic thiocarbonyl complex does not react with hydrogen, oxygen, or boron trifluoride.

---

*The checkers observed melting with decomposition in the 280-290° range.

References

1. I. S. Butler and A. E. Fenster, *J. Organomet. Chem.,* **51**, 101 (1974).
2. K. Vrieze, J. P. Collman, C. T. Sears, Jr., and M. Kubota, *Inorg. Synth.,* **11**, 101 (1968).
3. M. P. Yagupsky and G. Wilkinson, *J. Chem. Soc., A,* **1968**, 2813.
4. J. P. Collman, N. W. Hoffman, and J. W. Hosking, *Inorg. Synth.,* **12**, 8 (1970); R. J. Fitzgerald and H.-M. Lin, *ibid.,* **16**, 41 (1976).
5. M. Kubota and C. R. Carey, *J. Organomet. Chem.,* **24**, 491 (1970).
6. M. J. Mays and F. P. Stefanini, *J. Chem. Soc., A,* **1971**, 2747.
7. R. J. Fitzgerald, N. Y. Sakkab, R. S. Strange, and V. P. Narutis, *Inorg. Chem.,* **12**, 1081 (1973).
8. G. R. Clark, T. J. Collins, S. M. James, W. R. Roper, and K. G. Town, *Chem. Commun.,* **1976**, 475. The cationic complex may contain the $Ph_3P^+-CS_2^-$ ligand.

48. (η^5-CYCLOPENTADIENYL)NITROSYL CHROMIUM, MOLYBDENUM, AND TUNGSTEN COMPLEXES

Submitted by BRIAN W. S. KOLTHAMMER,* PETER LEGZDINS,*
and JOHN T. MALITO*
Checked by JOSEF TAKATS[†]

A. ALKYL AND ARYL(η^5-CYCLOPENTADIENYL)DINITROSYL CHROMIUM, MOLYBDENUM, AND TUNGSTEN COMPLEXES

Until recently, the only known alkyl and aryl(η^5-cyclopentadienyl)dinitrosyl complexes of the group VIB metals [i.e., (η^5-C$_5$H$_5$)M(NO)$_2$R, where M = Cr, Mo, W] were the methyl, ethyl, and phenyl derivatives of the chromium-containing compound. These derivatives can be prepared in low yields by the reaction of the appropriate Grignard reagent with (η^5-cyclopentadienyl)dinitrosylchromium bromide or iodide.[1] The recent utilization of alkyl- and arylaluminum compounds as alkylating or arylating agents now provides a high-yield synthetic route to a variety of such alkyl and aryl(η^5-cyclopentadienyl)dinitrosyl complexes [i.e., (η^5-C$_5$H$_5$)M(NO)$_2$R, M = Cr, Mo and R = CH$_3$, C$_2$H$_5$, CH$_2$CH(CH$_3$)$_2$, C$_6$H$_5$; M = W and R = CH$_3$, C$_6$H$_5$][2]. The syntheses described below involve reactions characterized by the following general equation:

*Department of Chemistry, The University of British Columbia, Vancouver, B.C., Canada, V6T 1W5.
†Department of Chemistry, The University of Alberta, Edmonton, Alberta, Canada T6G 2G2.

$$(\eta^5\text{-}C_5H_5)M(NO)_2Cl \ + \ R\text{-}Al{\displaystyle <} \quad \xrightarrow{\text{benzene}} \quad (\eta^5\text{-}C_5H_5)M(NO)_2R \ + \ [Cl\text{-}Al{\displaystyle <}]$$

Procedure

The complexes are all prepared similarly and the four procedures outlined below provide representative examples of the general synthetic method.

1. *(η⁵-Cyclopentadienyl)isobutyldinitrosylchromium.* A 200-mL, three-necked flask is fitted with a nitrogen inlet, an addition funnel, and a magnetic stirrer and it is thoroughly flushed with prepurified nitrogen. All of the manipulations in this and subsequent procedures are performed under prepurified nitrogen. The flask is charged with 1.0 g (4.7 mmole) of $(\eta^5\text{-}C_5H_5)Cr(NO)_2Cl^{3,5}$ and 25 mL of benzene (reagent grade dried by distillation from potassium metal and deaerated with prepurified nitrogen). A solution of 0.68 g (4.8 mmole) of hydridodiisobutylaluminum* in 30 mL of benzene is introduced by syringe into the addition funnel and is added dropwise over a period of 30 minutes to the stirred reaction mixture at room temperature. A red-brown oil deposits on the walls of the flask during this addition. Stirring is then continued until the infrared spectrum of the supernatant liquid no longer exhibits nitrosyl absorptions due to the initial reactant (about 2 hours). The reaction mixture is concentrated *in vacuo* to about 15 mL and the green supernatant solution is transferred by syringe onto a short (3 × 7 cm) column of alumina (Woelm neutral grade 1) made up in benzene. The column is eluted with benzene (about 150 mL) until the eluate is colorless and the bulk of the solvent is then removed from the eluate *in vacuo* (about 10^{-1} mm). Removal of the last traces of benzene under high vacuum (<0.005 mm for 2 hr) at room temperature affords 0.42 g (38% yield) of analytically pure $(\eta^5\text{-}C_5H_5)Cr(NO)_2(i\text{-}C_4H_9)$ as a green oil. *Anal.* Calcd. for $C_9H_{14}CrN_2O_2$: C, 46.15; H, 6.03; N, 11.96. Found: C, 45.87; H, 6.18; N, 11.78.

2. *(η⁵-Cyclopentadienyl)dinitrosylphenylmolybdenum.* A 200-mL, three-necked flask equipped with a nitrogen inlet, an addition funnel, and a magnetic stirrer is charged with 1.0 g (3.9 mmole) of $(\eta^5\text{-}C_5H_5)Mo(NO)_2Cl^{3,5}$ and 25 mL of benzene. A solution containing 0.35 g (1.4 mmole) of freshly prepared triphenylaluminum[4] in 75 mL of benzene is prepared[†] and is introduced by syringe into the addition funnel. This solution is added dropwise over a period of 30 minutes to the stirred reaction mixture at room temperature. After addition is complete, the mixture is stirred for 1 hour, during which time a small amount

---

*Available from Texas Alkyls Inc., Stauffer Chemical Co., Specialty Chemical Division, Westport, CT 06880. Solutions of various organoaluminum reagents in hydrocarbon solvents may also be purchased directly from Texas Alkyls Inc.

[†] Warming the mixture to 40-50° may be necessary to effect dissolution.

of a red-brown solid forms. The infrared spectrum of the final supernatant liquid should not exhibit any nitrosyl absorptions attributable to the starting material. The mixture is then concentrated *in vacuo* to about 15 mL and the supernatant green solution is transferred by syringe onto a 7 × 5 cm alumina column prepared in benzene. The solution is carefully chromatographed with benzene as solvent until the first yellow-green band is eluted (about 100 mL). The solvent is removed from the eluate under high vacuum (<0.005 mm) at room temperature for 4 hours. The dark-green oil that remains is an analytically pure sample of $(\eta^5\text{-}C_5H_5)Mo(NO)_2(C_6H_5)$ (0.60 g, 52% yield). *Anal.* Calcd. for $C_{11}H_{10}MoN_2O_2$: C, 44.30; H, 3.36; N, 9.40. Found: C, 44.48; H, 3.42; N, 9.22.

3. *(η⁵-Cyclopentadienyl)ethyldinitrosylmolybdenum.* A 100-mL three-necked flask fitted with a nitrogen inlet, an addition funnel, and a magnetic stirrer is charged with 1.28 g (5.00 mmole) of $(\eta^5\text{-}C_5H_5)Mo(NO)_2Cl^{3,5}$ and 20 mL of benzene and is cooled in an ice bath. A solution containing 0.57 g (5.0 mmole) of triethylaluminum in 15 mL of benzene is introduced by syringe into the addition funnel and is added dropwise to the stirred reaction mixture. The progress of the reaction is monitored by infrared spectroscopy and the triethylaluminum reagent is added until the nitrosyl absorptions due to the initial reactant disappear. (This reaction produces a dark-red oil as a by-product and if an excess of triethylaluminum is used, the yield of the desired product is markedly reduced.) The final reaction mixture is concentrated *in vacuo* to about 10 mL and the supernatant solution is transferred by syringe onto a 3 × 10 cm alumina column prepared in benzene. Elution of the column with benzene until the washings are colorless (about 150 mL) produces a bright-green eluate. The bulk of the solvent is then removed from the eluate *in vacuo* (about 10^{-1} mm). Removal of the last traces of benzene under high vacuum (<0.005 mm for 1.5 hr) at room temperature affords 0.72 g (57% yield) of $(\eta^5\text{-}C_5H_5)Mo(NO)_2(C_2H_5)$ as a dark-green oil. *Anal.* Calcd. for $C_7H_{10}MoN_2O_2$: C, 33.62; H, 4.03; N, 11.20. Found: C, 34.05; H, 4.26; N, 10.90.

4. *(η⁵-Cyclopentadienyl)methyldinitrosyltungsten.* A 200-mL, three-necked flask equipped with a nitrogen inlet and magnetic stirrer is charged with 1.72 g (5.00 mmole) of $(\eta^5\text{-}C_5H_5)W(NO)_2Cl^{3,5}$ and 80 mL of benzene. A solution of 0.72 g (10 mmole) of trimethylaluminum in 20 mL of benzene is introduced by syringe directly into the reaction flask. The mixture is stirred at room temperature for 24 hours, during which time a red oil deposits. The infrared spectrum of the final reaction mixture shows nitrosyl absorptions of approximately equal intensities attributable to both the initial reactant and the desired product. [Optimum yields of $(\eta^5\text{-}C_5H_5)W(NO)_2(CH_3)$ are obtained with this reaction time and further reaction leads only to extensive decomposition.] The final mixture is concentrated to about 10 mL *in vacuo* and the supernatant liquid is purified by chromatography on a 5 × 12 cm alumina column with benzene as eluent until the washings are colorless (about 150 mL). The eluate is evaporated

to dryness *in vacuo* and the resultant solid is sublimed at 35-40° (0.005 mm) onto a water-cooled probe to obtain 0.15 g (10% yield) of $(\eta^5\text{-}C_5H_5)W(NO)_2CH_3)$ as a green solid. Mp 96.0-96.5°. *Anal.* Calcd. for $C_6H_8N_2O_2W$: C, 22.24; H, 2.49; N, 8.65. Found: C, 22.48; H, 2.50; N, 8.51.

Properties

The properties of these compounds have been discussed previously.[2] The complexes are green oils or solids readily soluble in organic solvents. They are stable in air for short periods of time but are best stored under nitrogen at temperatures below 0°. The characteristic IR stretching frequencies for these compounds are listed below.

Compound	ν_{NO}(in hexane)(cm^{-1})
$(C_5H_5)Cr(NO)_2(i\text{-}C_4H_9)$	1775 (s); 1675 (s)
$(C_5H_5)Mo(NO)_2(C_6H_5)$	1745 (s); 1658 (s)
$(C_5H_5)Mo(NO)_2(C_2H_5)$	1735 (s); 1643 (s)
$(C_5H_5)W(NO)_2(CH_3)$	1705 (s); 1620 (s)

B. BIS[(η⁵-CYCLOPENTADIENYL)DINITROSYLCHROMIUM]

The compound $[(\eta^5\text{-}C_5H_5)Cr(NO)_2]_2$ was first prepared in low yields (<5%) by the reduction of $(\eta^5\text{-}C_5H_5)Cr(NO)_2Cl$ with sodium tetrahydroborate in a two-phase water-benzene system.[6] Recently, this complex was isolated in 75% yield from the zinc amalgam reduction of $(\eta^5\text{-}C_5H_5)Cr(NO)_2Cl$ in tetrahydrofuran over a period of 21 hours.[2] However, $[(\eta^5\text{-}C_5H_5)Cr(NO)_2]_2$ is synthesized most conveniently by the reduction of the above-mentioned chloro complex with sodium amalgam in benzene as outlined below.

$$2(\eta^5\text{-}C_5H_5)Cr(NO)_2Cl + 2Na/Hg \xrightarrow{\text{benzene}} [(\eta^5\text{-}C_5H_5)Cr(NO)_2]_2 + 2NaCl$$

Procedure

A 200-mL, three-necked flask fitted with a gas inlet and an egg-shaped magnetic stirring bar is thoroughly flushed with prepurified nitrogen. (All subsequent manipulations are performed under nitrogen in a well-ventilated fume hood.) The flask is charged with 4 mL of mercury, and an approximately 1% sodium amalgam is formed by the addition of 0.5 g (20 mmole) of sodium in about 0.1-g portions. The freshly prepared amalgam is then covered with 60 mL of benzene (reagent grade dried by distillation from potassium metal and deaerated with prepurified nitrogen). A solution of 1.06 g (5.00 mmole) of $(\eta^5\text{-}C_5H_5)Cr\text{-}(NO)_2Cl^{3,5}$ in 60 mL of benzene is transferred by syringe onto the *vigorously* stirred amalgam at room temperature. The original yellow-green solution becomes deep red and a gray solid precipitates as the reaction progresses. The reaction mixture is stirred until the infrared spectrum of the supernatant liquid indicates the absence of the initial reactant (about 1.5 hr). (Reaction beyond this point leads to decomposition of the desired product.) The final mixture is allowed to settle, and the supernatant liquid is removed by syringe from the amalgam and is filtered through 3 × 3 cm of Celite supported on a medium-porosity Schlenk-type filter frit. The filtrate is concentrated *in vacuo* to about 20 mL and is transferred to a short (2 × 5 cm) column of alumina (Woelm neutral grade 1) in benzene. Elution of the column with benzene until the washings are colorless (about 120 mL) produces a deep-purple eluate. The eluate is taken to dryness *in vacuo* and the remaining purple solid is dried thoroughly under high vacuum (<0.005 mm for 2 hr) at room temperature to obtain 0.50 g (56% yield) of analytically pure $[(\eta^5\text{-}C_5H_5)Cr(NO)_2]_2$. *Anal.* Calcd. for $C_{10}H_{10}Cr_2N_4O_4$: C, 33.91; H, 2.85; N, 15.82. Found: C, 33.78; H, 2.75; N, 15.82.

Properties

The properties of this compound have been discussed previously.[6] The complex is a purple-black solid readily soluble in organic solvents. The solid is stable in air for short periods of time but indefinitely under nitrogen. It melts at 146.5-147.0° with some decomposition and its IR spectrum in CH_2Cl_2 exhibits ν_{NO} at 1667 (s) and 1512 (m) cm^{-1}.

References

1. T. S. Piper and G. Wilkinson, *J. Inorg. Nucl. Chem.,* **3**, 104 (1956).
2. J. K. Hoyano, P. Legzdins and J. T. Malito, *J. Chem. Soc., Dalton Trans.,* **1975**, 1022.
3. P. Legzdins and J. T. Malito, *Inorg. Chem.,* **14**, 1875 (1975).
4. W. Strohmeier and K. Hümpfner, *Chem. Ber.,* **90**, 2339 (1957).
5. J. K. Hoyano, P. Legzdins, and J. T. Malito, *Inorg. Synth.,* **18**, 129 (1978).
6. R. B. King and M. B. Bisnette, *Inorg. Chem.,* **3**, 791 (1964).

49. OLEFIN COMPLEXES OF PLATINUM

Submitted by J. L. SPENCER*
Checked by S. D. ITTEL[†] and M. A. CUSHING, JR.[†]

Significant advances in organonickel chemistry followed the discovery of *trans,trans,trans*-(1,5,9-cyclododecatriene)nickel, Ni(cdt), and bis(1,5-cycloocta-diene)nickel Ni(cod)$_2$ by Wilke *et. al.*[1] In these and related compounds, in which only olefinic ligands are bonded to the nickel, the metal is especially reactive both in the synthesis of other compounds and in catalytic behavior. Extension of this chemistry to palladium and to platinum has hitherto been inhibited by the lack of convenient synthetic routes to zero-valent complexes of these metals in which mono- or diolefins are the only ligands. Here we described the synthesis of bis(1,5-cyclooctadiene)platinum, tris(ethylene)-platinum, and bis(ethylene)(tricyclohexylphosphine)platinum. The compound Pt(cod)$_2$ (cod = 1,5-cyclooctadiene) was first reported by Müller and Göser,[2] who prepared it by the following reaction sequence:

$$PtCl_2(cod) \xrightarrow[\text{(b) MeOH, } -50°]{\text{(a) } i\text{-Pr MgBr, } -40°} Pt(i\text{-Pr})_2(cod) \xrightarrow[\text{cod}]{uv} Pt(cod)_2$$

The method described below provides a more reliable synthetic route to this compound.

A. BIS(1,5-CYCLOOCTADIENE)PLATINUM(0)

$$2Li + C_8H_8 \rightarrow Li_2C_8H_8$$

$$Li_2C_8H_8 + Pt(C_8H_{12})Cl_2 + C_8H_{12} \rightarrow Pt(C_8H_{12})_2 + 2LiCl + C_8H_8$$

Procedure

Dichloro(1,5-cyclooctadiene)platinum(II) may be prepared from hexachloro-platinic acid,[3] by heating bis(benzonitrile)dichloroplatinum(II)[4] in 1,5-cyclo-

---

*Department of Inorganic Chemistry, The University of Bristol, Bristol BS8 1TS, England.
[†]Central Research and Development Department, E.I. du Pont de Nemours and Co., Inc., Experimental Station, Wilmington, DE 19898.

octadiene at $145°C$ for 5 minutes or from potassium tetrachloroplatinate.[5] The complex $PtCl_2(cod)$ has a very low solubility in the reaction mixture and must be finely ground to ensure complete reaction. Commercial 1,5-cyclooctadiene is often wet and contaminated with other impurities and should be distilled from sodium immediately prior to use. The 1,3,5,7-cyclooctatetraene* should also be distilled. All solvents are dried and distilled under nitrogen. In particular, peroxide-free diethyl ether is first dried over sodium wire and then distilled under nitrogen from lithium tetrahydroaluminate. (■**Caution.** *This distillation should be carried out behind a shield, and the still pot must not be allowed to go to dryness.*)

1. *Preparation of (1,3,5,7-cyclooctatetraene)dilithium* $(Li_2C_8H_8)$. Lithium foil (0.7 g, 100 mmole) is suspended under nitrogen in dry diethyl ether (80 mL) in a magnetically stirred 100-mL, two-necked, round-bottomed flask at $0°$. A 2.5-g sample (24 mmole) of 1,3,5,7-cyclooctatetraene is added and the mixture is stirred for 16 hours. The small quantity of white precipitate is allowed to settle, an aliquot of the orange solution is removed with a syringe, and the molarity is checked by hydrolysis and titration against standard acid. A saturated solution of (1,3,5,7-cyclooctatetraene)dilithium is about 0.24 M. (■**Caution.** *The solution is no more flammable than diethyl ether but solid $Li_2C_8H_8$ is pyrophoric in air.*)

2. *Preparation of bis(1,5-cyclooctadiene)platinum(0).* A 250-mL, three-necked, round-bottomed flask is charged with 3.7 g (10 mmole) of dichloro(1,5-cyclooctadiene)platinum(II) and equipped with a Teflon magnetic stirring bar, a pressure-equalizing dropping funnel, and a stopcock adapter connected by a T-piece to a supply of dry nitrogen and a mineral oil bubbler. The apparatus is flushed with nitrogen, and 15 mL of deoxygenated 1,5-cyclooctadiene is added through the third neck, which is then closed. The flask and its contents are cooled in a Dry Ice-isopropyl alcohol bath at $-40°$. The diethyl ether solution of (cyclooctatetraene)dilithium (10 mmole) is transferred to the dropping funnel with a hypodermic syringe and then added dropwise to the rapidly stirred slurry over a period of 45 minutes, while the temperature is maintained at -50 to $-30°$. When the addition is complete, the cream-colored mixture is allowed to warm to $0°$ (30 min). The dropping funnel is replaced by a stopcock adapter connected to a vacuum line by a large solvent trap, cooled by liquid nitrogen, and the ether is distilled off at reduced pressure. Excess 1,5-cyclooctadiene and cyclooctatetraene are then removed at $10°$ (0.01 torr) until the residue is quite dry. Nitrogen is readmitted to the flask, and the pale-tan solid is extraced with five 50 mL portions of toluene at $20°$. The extract is filtered under nitrogen through

---

*Available from BASF Aktiengesellschaft, 6700 Ludwigschafen, Germany or Aldrich Chemical Co., 940 W. St. Paul Ave. Milwaukee, WI 53233.

a short column (8 × 2.5 cm) of alumina* into a 500-mL, round-bottomed flask. After the column is washed with a further 50 mL of toluene, the combined filtrate and washing are evaporated at reduced pressure (20°) to a volume of approximately 15 mL. The mother liquor is then removed with a syringe from the off-white crystals, which are washed with diethyl ether (four 5-mL portions) and dried, first in a stream of nitrogen, then under vacuum (20°, 0.01 torr, 1 hr) to give 1.6-2.4 g (40-60% yield) of $Pt(C_8H_{12})_2$. *Anal.* Calcd. for $C_{16}H_{24}Pt$: C, 47.4; H, 5.9. Found: C, 47.7; H, 5.7. The material so prepared is generally pure enough for most purposes. Pure white crystals may be obtained by dissolution in a large volume of petroleum ether (approximately 80 mL/mmole), filtration through alumina, and recrystallization at $-78°$.

Properties

Bis(1,5-cyclooctadiene)platinum is appreciably more oxidatively and thermally stable than the nickel analogue, and the dry solid may be handled safely in air.[6] The 1,5-cyclooctadiene groups are readily displaced by a range of other ligands, including phosphines, ethylene, strained olefins, and isocyanides. In several instances it has been found that the order of addition is important in these displacement reactions, the best results being obtained when the $Pt(C_8H_{12})_2$ is added slowly to a solution of the ligand.

The ^{1}H NMR spectrum (C_6D_6) shows resonances at τ 5.80 [m, 8H, CH, J_{PtH} at 55 Hz] and 7.81 [m, 16H, CH_2]. The ^{13}C NMR spectrum (C_6D_6, proton decoupled) shows resonances at δ 73.3 ppm [C=C, J_{PtC} at 143 Hz] and δ 33.2 ppm [CH_2, J_{PtC} at 15 Hz].

B. TRIS(ETHYLENE)PLATINUM(0)

$$Pt(C_8H_{12})_2 + 3C_2H_4 \rightarrow Pt(C_2H_4)_3 + 2C_8H_{12}$$

Procedure

Ethylene (C.P. grade) may be used directly from the cylinder without further pruification.

■**Caution.** *The entire preparation is performed under an atmosphere of ethylene. Also, the physiological properties of the product, which is quite volatile, have not been investigated. Therefore all operations should be carried out in a well-ventilated hood.*

An 80-mL Schlenk tube,[7] with magnetic stirring bar and containing 20-mL of petroleum ether (bp 40-60°), is flushed continuously with ethylene and cooled in ice for 10 minutes. Bis(1,5-cyclooctadiene)platinum is added in 0.1-g

*BDH, Brockman activity II.

portions, with rapid stirring, until no more dissolves readily (about 1.1 g). (If too much bis(1,5-cyclooctadiene)platinum is added, or if it is added too quickly, an insoluble white precipitate forms.) The tube is then temporarily sealed (to prevent large volumes of ethylene from dissolving in the solution) and cooled with a toluene slush bath (−96°) for 40 minutes, during which time a mass of white crystals of product forms. When crystallization is complete, ethylene is readmitted, another Schlenk tube is attached to the first by means of a curved glass transfer tube, and the supernatant liquid is decanted into the second tube. The tubes are separated, and the crude solid is redissolved in 9 mL of petroleum ether at room temperature, and the solution is filtered through a sintered-glass frit (porosity 3, fine) into a clean Schlenk tube. The product is again crystallized at −96°, and the supernatant liquid is decanted into the tube containing the first mother liquor. One further recrystallization from pentane or light petroleum ether (5 mL of bp 30-40°) gives, after drying at −40° (0.01 torr for 1 hr), a 0.42-g (56%) yield of white crystals of tris(ethylene)platinum(0). *Anal.* Calcd. for $C_6H_{12}Pt$: C, 25.8; H, 4.3. Found: C, 25.2; H, 4.5.

A considerable quantity of the $Pt(cod)_2$ (20%) may be recovered by passing the combined mother liquors through a small pad of alumina and evaporating the filtrate to dryness at reduced pressure.

Properties

Tris(ethylene)platinum(0) is stable for several hours at 20° under 1 atm of ethylene and keeps for many weeks at −20°.[6] In the absence of ethylene, decomposition to metallic platinum occurs in minutes at room temperature. The complex is quite volatile, with a vile smell, and sublimes slowly at 20° in an atmosphere of ethylene onto a cold finger at 0°.

Despite the obvious lability of the complex, the ^1H NMR (C_6H_6, C_2H_4 1 atm) shows a well-defined singlet, τ6.94, with ^{195}Pt satellites (J_{PtH} at 57 Hz), as well as the signal for free ethylene at τ4.72, although at 40° both signals begin to broaden.

Other tris(olefin)platinum(0) complexes (where olefin represents a strained olefin such as bicyclo[2.2.1]heptene, dicyclopentadiene, or *trans*-cyclooctene) may be similarly obtained by direct displacement of 1,5-cyclooctadiene, often in quantitative yield.[6]

C. BIS(ETHYLENE)(TRICYCLOHEXYLPHOSPHINE)PLATINUM(0)

$$Pt(C_8H_{12})_2 + 3C_2H_4 \rightarrow Pt(C_2H_4)_3 + 2C_8H_{12}$$

$$Pt(C_2H_4)_3 + (C_6H_{11})_3P \rightarrow Pt(C_2H_4)_2P(C_6H_{11})_3 + C_2H_4$$

Procedure

This preparation should be carried out in a well-ventilated hood.

To a 250-mL two-necked round-bottomed flask, fitted with a magnetic stirring bar, is added 90 mL of petroleum ether (bp 30-40°), cooled to 0°. The apparatus is flushed with ethylene until the solvent is saturated. Bis(1,5-cyclooctadiene)-platinum (3.8 g, 9.2 mmole) is added in small (0.2 g) portions, each portion being allowed to dissolve before the next is added. A petroleum ether solution of tricyclohexylphosphine* (2.6 g, 9.2 mmole in 15 mL) is added slowly. After a brief evolution of gas, crystallization of the product begins. The ethylene source is removed, and the flask is flushed with a slow stream of nitrogen for 1 hour. The supernatant liquid is decanted into a clean flask, and the crystals are washed well with cold petroleum ether (four 10-mL portions), the washings being added to the mother liquor. The white crystals may be dried at 20° (0.05 torr) for 1 hour. *Anal.* Calcd. for $C_{22}H_{41}PPt$: C, 49.7; H, 7.8. Found: C, 49.5; H, 8.2.

A second crop of crystals may be obtained as follows. The combined mother liquor and washings are evaporated to complete dryness at reduced pressure, and the residue is dissolved in petroleum ether (50-100 mL) under an atmosphere of ethylene. This solution is filtered through a short column of alumina (2 × 5 cm), under ethylene, into a round-bottomed flask. Evaporating the solvent to a small volume gives the second crop of $Pt(C_2H_4)_2P(C_6H_{11})_3$; combined yield 4.2 g (85%).

Properties

Bis(ethylene)(tricyclohexylphosphine)platinum(0) is a white crystalline solid that may be handled safely in air, but should be stored in an inert atmosphere. It is slightly soluble in petroleum ether and readily dissolves in aromatic solvents to give reasonably stable solutions, even in the absence of dissolved ethylene. The ^1H NMR spectrum (C_6H_6) shows a sharp singlet with ^{195}Pt satellites at τ 7.25 $(C_2H_4, J_{PtH} = 57$ Hz) and a broad featureless resonance centered at τ 8.5 (C_6H_{11}). The complex undergoes a variety of reactions to afford organoplatinum compounds. With triorganosilanes and germanes, it gives binuclear complexes $[Pt(H)(MR_3)(P(C_6H_{11})_3)]_2$ (M = Si, Ge; R = Me, Et, Cl, OEt), and with allyltrimethylstannane it gives the η^3-allyl complex $Pt(SnMe_3)(C_3H_5)(P(C_6H_{11})_3)$.[8]

References

1. P. W. Jolly and G. Wilke, *The Organic Chemistry of Nickel,* Vol. 1 and 2, Academic Press Inc., New York, 1974, 1975.
2. J. Müller and P. Göser, *Angew. Chem. Int. Ed.,* **6**, 364 (1967).
3. D. Drew and J. R. Doyle, *Inorg. Synth.,* **13**, 48 (1972).

*Available from Strem Chemicals Inc., P.O. Box 212, Danvers, MA 01923.

4. F. R. Hartley, *Organomet. Chem. Rev., A,* **6,** 119 (1970).
5. J. X. McDermott, J. F. White, and G. M. Whitesides, *J. Am. Chem. Soc.,* **98,** 6521 (1976).
6. M. Green, J. A. K. Howard, J. L. Spencer and F. G. A. Stone, *J. Chem. Soc. Dalton Trans.,* 1977, 271; *Chem. Commun.,* 1975, 3, 449.
7. D. F. Shriver, *The Manipulation of Air-Sensitive Compounds,* McGraw-Hill Book Co., New York, 1969.
8. M. Green, J. A. K. Howard, J. Proud, J. L. Spencer, C. A. Tsipis, and F. G. A. Stone, *Chem. Commun.,* 1976, 671.

50. DI-μ-CHLORO-BIS(η^4-1,5-CYCLOOCTADIENE)-DIRHODIUM(I)

Submitted by G. GIORDANO* and R. H. CRABTREE[†]
Checked by R. M. HEINTZ,[‡] D. FORSTER,[‡] and D. E. MORRIS[‡]

Di-μ-chloro-bis(η^4-1,5-cyclooctadiene)dirhodium(I), [RhCl(1,5-C_8H_{12})]$_2$, has been prepared in 60% yield by reducing rhodium trichloride hydrate in the presence of excess olefin in aqueous ethanol.[1] In the present preparation the yield has been greatly increased (to 94%). Two related complexes, [RhCl(1,5-C_6H_{10})]$_2$[2] and [RhCl(C_6H_{12})$_2$]$_2$, are similarly prepared in high yield from 1,5-hexadiene and 2,3-dimethyl-2-butene, respectively.

Such diene complexes can be used to prepare homogeneous hydrogenation catalysts *in situ*, especially where a variable tertiary phosphine/rhodium ratio is required[3] or where an asymmetric tertiary phosphine is employed for asymmetric synthesis.[4] The cyclooctadiene complex is also the starting point for the preparation a number of complexes of the type [Rh(1,5-C_8H_{12})L$_2$]$^+$ (L represents a variety of P— and N— donor ligands) of interest in homogeneous catalysis.[5]

Procedure

$$2RhCl_3 + 2C_8H_{12} + 2CH_3CH_2OH + 2Na_2CO_3 \rightarrow$$

$$[RhCl(C_8H_{12})]_2 + 2CH_3CHO + 4NaCl + 2CO_2 + 2H_2O$$

A 100-mL, two-necked, round-bottomed flask is fitted with a reflux condenser connected to a nitrogen bubbler. The flask is charged with 2.0 g (7.6 mmole) of

*Present address: Istituto di Chimica Generale, Università, Via Venezian 21, 20133 Milano, Italy.
[†]Sterling Chemistry Laboratory, Yale University, New Haven, CT 06520.
[‡]Monsanto Chemical Co, St Louis, MO 63166.

rhodium trichloride trihydrate[§] (a generous loan of the Compagnie des Métaux Précieux) and 2.2 g (7.7 mmole) of sodium carbonate decahydrate.* Under nitrogen, 20 mL of deoxygenated ethanol-water (5:1) and 3 mL of 1,5-cyclo-octadiene[†] are added and the mixture is then heated at reflux with stirring for 18 hours, during which time the product precipitates as a yellow-orange solid. The mixture is cooled and immediately filtered and the product is washed with pentane and then with methanol-water (1:5) until the washings no longer contain chloride ion. The product is dried *in vacuo.* Yield 1.67 g (94%). *Anal.* Calcd. for $C_{16}H_{24}Cl_2Rh_2$: C, 38.97; H, 4.91; Cl, 14.38; Rh, 41.74. Found: C, 39.01; H, 4.80; Cl, 14.08.

Properties

Di-μ-chloro-bis(η⁴-1,5-cyclooctadiene)dirhodium(I) is a yellow-orange, air-stable solid. It can be used directly as obtained for preparative purposes[5] or as a precursor for homogeneous catalysts.[3,4] It can be recrystallized from dichloromethane-diethyl ether to give orange prisms. The compound is soluble in dichloromethane somewhat less soluble in acetone and insoluble in pentane and diethyl ether. Characteristic strong bands occur in the infrared spectrum at 819, 964, and 998 cm^{-1} (Nujol mull). The cyclooctadiene vinylic protons resonate in the ¹H NMR spectrum at τ 5.7 and the allylic protons at τ 7.4-8.3 (deuteriochloroform solution). Other physical properties are given by Chatt.[1]

Analogous Complexes

The 1,5-hexadiene complex, $[RhCl(C_6H_{10})]_2$, may be prepared by this method with a reaction time of 24 hours. The temperature should not exceed 40° to avoid the deposition of metallic rhodium. Under these conditions the yield is 85% of analytically pure product.

The new 2,3-dimethyl-2-butene complex, $[RhCl(C_6H_{12})_2]_2$, is prepared by a similar method, in which the solution is left at 20° for 30 days instead of being refluxed. A trace of metallic rhodium is deposited. This is removed by recrystallization from dichloromethane-diethyl ether to give an analytically pure product; yield 75%.

References

1. J. Chatt and L. M. Venanzi, *J. Chem. Soc.,* 1957, 4735.

[§] Available from Alfa Products, Ventron Corp., P.O. Box 299, Danvers, MA 01923.
*The checkers obtained an off-color (olive-green) product when using sodium carbonate. However, in the absence of sodium carbonate they repeatedly obtained the expected yellow-orange product in good yields (90-94%) and time periods (18 hr).
[†] Available from Aldrich Chemical Co., 940 W. St. Paul Ave., Milwaukee, WI 53233.

2. G. Winkhaus and H. Singer, *Chem. Ber.,* **99**, 3602 (1966).
3. J. A. Osborn and G. Wilkinson, *J. Chem. Soc., A,* **1968**, 1054.
4. L. Horner, H. Buethe, and H. Siegel, *Tetrahedron Lett.,* 4023 (1968); H. B. Kagan, and T-P. Dang, *J. Am. Chem. Soc.,* **94**, 6429 (1972); W. S. Knowles, J. J. Sabacky, and B. D. Vineyard, *Chem. Commun.,* **1972**, 10.
5. R. R. Schrock and J. A. Osborn, *J. Am. Chem. Soc.,* **93**, 2397 (1971).

51. (η^3-ALLYL)PALLADIUM(II) COMPLEXES

Submitted by YOSHITAKA TATSUNO,* TOSHIKATSU YOSHIDA,* and SEIOTSUKA*
Checked by NAJEEB AL-SALEM[†] and BERNARD L. SHAW[†]

(η^3-Allyl)(η^5-cyclopentadienyl)palladium(II), first prepared by B. L. Shaw,[1] is a labile organopalladium compound useful for preparations of various Pd(0) complexes. The present preparation from bis(η^3-allyl)di-μ-chloro-dipalladium-(II) follows the method of Shaw.

A. BIS(η^3-ALLYL)DI-μ-CHLORO-DIPALLADIUM(II)[2]

$$2Na_2PdCl_4 + 2CH_2=CHCH_2Cl + 2CO + 2H_2O \rightarrow$$

$$(\eta^3\text{-}C_3H_5)_2Pd_2Cl_2 + 4NaCl + 2CO_2 + 4HCl$$

Procedure

■**Caution.** *The preparation should be performed in a well ventilated hood.*

A 200-mL, two-necked, round-bottomed flask equipped with a magnetic stirring bar, a gas inlet tube, and a condenser topped with a bubbler is charged with an aqueous solution of palladium(II) chloride (4.44 g, 25 mmole) and sodium chloride (2.95 g, 50 mmole) in 10 mL of H_2O, followed by methanol (60 mL) and allyl chloride (3-chloro-1-propene) (6.0 g, 67 mmole). Carbon monoxide is passed slowly (2-2.5 L/hr) under stirring through the reddish-brown solution by way of a gas-inlet tube for 1 hour. The bright yellow suspension thus obtained is poured into water (300 mL) and extracted with chloroform (2 ×

*Department of Chemistry, Faculty of Engineering Science, Osaka University, Toyonaka, Osaka, Japan 560.
[†]School of Chemistry, University of Leeds, Leeds LS2 9JT, England.

100 mL). The extract is washed with water (2 X 150-mL portions) and dried over calcium chloride. Evaporation under reduced pressure (20 torr) gives yellow crystals. Yield 4.3 g (93%). The crude product can be used without further purification. The analytically pure sample can be obtained by recrystallization from a mixture of dichloromethane/hexane. The compound decomposes at 155-156°. *Anal.* Calcd. for $C_6H_{10}Cl_2Pd_2$: C, 19.49; H, 2.73. Found: C, 19.60; H, 2.75.

Properties

The air-stable, yellow, crystalline compound is soluble in benzene, chloroform, acetone, and methanol. The ^1H NMR spectrum ($CDCl_3$) shows two doublets at δ 3.03 (anti CH_2, J = 12.0 Hz) and 4.10 ppm (syn CH_2, J = 7.1 Hz), and a triplet at δ 5.48 ppm (CH) in a relative ratio of 2:2:1. A variety of reactions of this compound are summarized in several reviews.[3]

B. (η³-ALLYL)(η⁵-CYCLOPENTADIENYL)PALLADIUM(II)

$$(\eta^3\text{-}C_3H_5)_2Pd_2Cl_2 + 2NaC_5H_5 \rightarrow 2Pd(\eta^3\text{-}C_3H_5)(\eta^5\text{-}C_5H_5) + 2NaCl$$

Procedure

■**Caution.** *(η³-Allyl)(η⁵-cyclopentadienyl)palladium is volatile and has an unpleasant odor. As the toxity is unknown, all manipulations should be carried out in an efficient hood. All solvents are dried over sodium metal and distilled under nitrogen.*

A tetrahydrofuran (THF) solution of sodium cyclopentadienide[4] is prepared by adding freshly distilled cyclopentadiene[5] to a sodium suspension in THF. The concentration of the resulting pale-pink solution can be determined by titration with acid.

In a 300-mL, three-necked flask equipped with a three-way stopcock, a pressure-equalizing dropping funnel, and a Teflon-coated magnetic stirring bar, is placed bis(η³-allyl)di-μ-chloro-dipalladium(II) (9.9 g, 27 mmole). The flask is evacuated and filled with nitrogen three times. THF (100 mL) and benzene (100 mL) are added through the three-way stopcock under nitrogen, with a syringe, to give clear yellow solution. The flask is then cooled with an ice-sodium chloride mixture to -20°. A THF solution of sodium cyclopentadienide (54 mmole in 28 mL of THF) is transferred by syringe to a nitrogen-flushed dropping funnel and is then added dropwise to the cooled solution with stirring at -20°. The solution changes slowly from yellow to a dark red. After 1 hour the ice bath is removed and the temperature of the reaction mixture is allowed to reach room tempera-

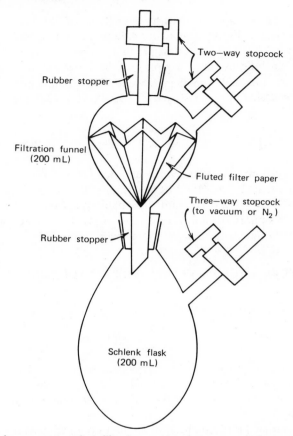

Fig. 1. Simple apparatus for filtering under inert atomsphere. The flow of nitrogen or arsgon is controlled by a finger over the base of the needle.

ture with stirring. The stirring is continued for an additional 30 minutes. The solvents are removed by distillation *in vacuo* (30-60 torr) to give a dark-red solid. If the pressure is lower than 30 torr, a considerable amount of the palladium complex sublimes at 25°. The solid residue is extracted with hexane (80 mL) and the extract is filtered through a dried filter paper under a nitrogen at- mosphere in a filtration funnel as shown in Fig. 1.* The red filtrate is evaporated *in vacuo* (30-60 torr), affording red needles of $(\eta^3$-allyl)$(\eta^5$-cyclopentadienyl)- palladium(II). The yield is about 9.2 g (80%).[†]

---

*As a regular glass filter will become clogged, it is not suitable for this filtration.
[†]The checker used a mechanical stirrer and obtained the a 98% yield.

The product can be used for most preparations of palladium(0) complexes. This compound can be further purified by sublimination at 40° under 30 torr, with slightly decreased yield. *Anal.* Calcd. for $C_8H_{10}Pd$: C, 45.20; H, 4.74. Found: C, 45.20; H, 4.77.

Properties

(η^3-Allyl)(η^5-cyclopentadienyl)palladium(II) is an easily sublimed compound with an unpleasant odor. It forms red, needlelike crystals that decompose at 61°. In the solid state it is fairly stable, although it decomposes gradually at room temperature to give a black solid that is insoluble in hexane. It is therefore recommended that the complex be stored below -20° under nitrogen. The ^1H NMR spectrum (C_6D_6) shows signals at δ 2.14 (2H, doublet, J = 11 Hz), 3.11 (2H, doublet, J = 6 Hz), and 4.63 ppm (1H, complex) for the η^3-allyl protons, and a signal at δ 8.1 ppm (5H, singlet) for the cyclopentadienyl ring protons.

This compound reacts readily with alkyl isocyanides to give a cluster "Pd(CNR)₂",[6] and with bulky alkyl phosphines to give two coordinated palladium(0) complexes.[7]

References

1. B. L. Shaw, *Proc. Chem. Soc.,* **1960**, 247.
2. W. T. Dent, R. Long, and A. J. Wilkinson, *J. Chem. Soc.,* **1964**, 1585.
3. P. M. Maitlis, *The Organic Chemistry Palladium,* Vol. I, Academic Press Inc., New York and London, 1971; F. A. Hartley, *The Chemistry of Platinum and Palladium,* Applied Science Publishers Ltd., London, 1973.
4. F. G. A. Stone, R. West ed., *Advances in Organometallic Chemistry,* Vol. 2, Academic Press Inc., New York and London, 1964; p. 365; R. B. King and F. G. A. Stone, *Inorg. Synth.,* 7, 99 (1963).
5. R. B. Moffett, *Org. Synth.,* Coll. Vol. 4, 238 (1963).
6. E. O. Fischer and H. Werner, *Chem. Ber.,* 95, 703 (1962); S. Otsuka, A. Nakamura, and Y. Tatsuno, *J. Am. Chem. Soc.,* 91, 6994 (1969).
7. S. Otsuka, T. Yoshida, M. Matsumoto, and K. Nakatsu, *J. Am. Chem. Soc.,* 98, 5850 (1976).

52. (η⁵-CYCLOPENTADIENYL)HYDRIDOZIRCONIUM COMPLEXES

Submitted by P. C. WAILES* and H. WEIGOLD*
Checked by JEFFREY SCHWARTZ† and CHU JUNG†

*Division of Applied Organic Chemistry, CSIRO, Melbourne, 3001, Australia.
†Department of Chemistry, Princeton University, Princeton, NJ, 08540.

The dihydridozirconium complex, $ZrH_2(\eta^5\text{-}C_5H_5)_2$, was prepared originally[1] by the action of trialkylamines on the bis(tetrahydroborate), $Zr(BH_4)_2(\eta^5\text{-}C_5H_5)_2$. Below are described more convenient methods for the preparation of the dihydride and of the chlorohydrido derivative, $ZrHCl(\eta^5\text{-}C_5H_5)_2$, from readily available starting materials. These compounds are useful intermediates in the preparation of alkyl- and alkenylzirconium compounds,[2,3] and under hydrogen pressure they are active catalysts for hydrogenation of olefins and acetylenes.[2] The "hydrozirconation" reactions require $ZrHCl(\eta^5\text{-}C_5H_5)_2$ as starting material.[4]

A. BIS(η^5-CYCLOPENTADIENYL)DIHYDRIDOZIRCONIUM

$$2ZrCl_2(\eta^5\text{-}C_5H_5)_2 + H_2O \rightarrow [ZrCl(\eta^5\text{-}C_5H_5)_2]_2O + 2HCl$$

$$[ZrCl(\eta^5\text{-}C_5H_5)_2]_2O + Li[AlH_4] \rightarrow 2ZrH_2(\eta^5\text{-}C_5H_5)_2 + LiCl + \tfrac{1}{2}[AlOCl]_2$$

1. μ-Oxo-bis[chlorobis(η^5-cyclopentadienyl)zirconium]

The method of preparation described here is an adaption of the original,[5,6] which gives consistently high yields.

Procedure

To a solution of $ZrCl_2(\eta^5\text{-}C_5H_5)_2$* (29 g, 0.1 mole) in dichloromethane (250 mL) in a conical flask is added aniline (10 mL) and water (1.3 mL) with shaking. A white precipitate of aniline hydrochloride forms immediately. After it is chilled for several hours in a refrigerator the suspension is filtered. Occasionally large crystals of the product are present at this stage and these are dissolved by addition of more CH_2Cl_2. The filtrate is evaporated to small volume, and light petroleum ether (100 mL, bp 30-40°) is added to precipitate the product. After filtration and washing with petroleum ether, white crystals of the oxo-bridged compound are obtained (25.5 g, 97% yield); these slowly turn pink on storage. The melting point varies with the rate of heating, but if the sample in an evacuated capillary is placed in a melting point apparatus preheated to 260°, a melting point around 305° should be obtained.

*Available from Arapahoe Chemicals Division, Syntex Inc., 2855 Walnut St., Boulder, CO 80302.

2. Bis(η⁵-cyclopentadienyl)dihydridozirconium

In this reaction purified tetrahydrofuran (THF) is required, as well as a standardized solution of Li[AlH$_4$] in tetrahydrofuran. The THF from freshly opened bottles is distilled from Li[AlH$_4$]. (**■Caution.** *Li[AlH$_4$] is a hazardous material and must be handled in dry conditions and in small quantitites. Serious explosions can occur when impure THF is purified if it contains peroxides (see Reference 7).*

Standardized solutions of Li[AlH$_4$] are prepared by stirring Li[AlH$_4$] in purified THF for several hours under nitrogen or argon and filtering through Celite (previously baked out at 140° and degassed under vacuum) using the apparatus shown in Fig. 1, followed by hydrolysis of an aliquot of this solution with dilute acid and accurate measurement of the hydrogen evolved. This is best done on a vacuum line using a Töpler pump, but simpler methods should also suffice.

Procedure

To a solution of [ZrCl(η⁵-C$_5$H$_5$)$_2$]$_2$O (17.7 g, 33.4 mmole) in purified THF (200 mL) in a 500-mL flask of the type shown in Fig. 1, a clear solution of Li[AlH$_4$] in THF (20 mL of 1.7 M, 34 mmole) is added dropwise from a hypodermic syringe with stirring. An atmosphere of nitrogen or argon is maintained at all times. A white precipitate slowly appears, but precipitation is not complete for several hours. The mixture is set aside overnight and then filtered anaerobically as in Fig. 1, giving the dihydrido complex as an almost colorless

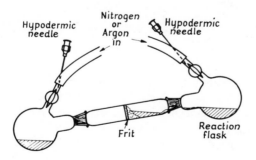

Simple apparatus for filtering under inert atmosphere. The flow of N$_2$ or Ar is controlled by a finger over the base of the needle.

Fig. 1. Filtration apparatus.

microcrystalline solid (yield 10.1 g, 66%). *Anal.* Calcd. for $C_{10}H_{12}Zr$: ash (ZrO_2), 55.14%; hydrolyzable H, 2.00 g-atom/mole. Found: ash, 55.5%; hydrolyzable H, 1.89 g-atom/mole. The yield of dihydride never exceeds 66%; the remainder of the zirconium is believed to be a complex with an aluminum compound (see Properties).

B. CHLOROBIS(η^5-CYCLOPENTADIENYL)HYDRIDOZIRCONIUM

$$ZrCl_2(\eta^5\text{-}C_5H_5)_2 + LiAlH(t\text{-}BuO)_3 \rightarrow ZrHCl(\eta^5\text{-}C_5H_5)_2 + LiCl + Al(t\text{-}BuO)_3$$

Procedure*

The apparatus and procedures are similar to those in the preparation above and a 1L flask is used. A solution of lithium tri-*tert*-butoxyhydridoaluminate[†9] (28.6 g, 113 mmole) in purified THF (100 mL) is added slowly to a solution of $ZrCl_2(\eta^5\text{-}C_5H_5)_2$ (32.9 g, 113 mmole) in THF (500 mL) with stirring. After complete addition, stirring is continued for 1 hour, after which the mono-hydrido complex is collected by anearobic filtration (Fig. 1) and washed with THF (yield 26.3 g, 90%). *Anal.* Calcd. for $C_{10}H_{11}ClZr$: ash (ZrO_2), 47.77%; hydrolyzable H, 1.00 g-atom/mole; Cl, 13.75. Found: ash, 47.0%; hydrolyzable H, 1.02 g-atom/mole; Cl, 13.4. One-quarter mole of Li[AlH_4] may be used instead of the tri-*tert*-butoxy hydrido complex, but the essential control of stoichiometry is more difficult (see Properties).

Other hydridozirconium derivatives that have been prepared from aluminum hydrides are $ZrH(CH_3)(\eta^5\text{-}C_5H_5)_2$, $ZrH(AlH_4)(\eta^5\text{-}C_5H_5)_2$, the complex $ZrH_2(\eta^5\text{-}C_5H_5)_2 \cdot [ZrH(\eta^5\text{-}C_5H_5)_2]_2O$, and deuterido derivatives corresponding to all of these hydrido complexes.[8]

Properties

The bis(η^5-cyclopentadienyl)zirconium hydrides are colorless solids that hydrolyze in water. Accurate measurement of the hydrogen thus evolved is a sensitive method of analysis. Alternatively, reaction with CH_2Cl_2 in a stoppered NMR tube and quantitative estimation of the CH_3Cl so formed can be used. The compounds are associated through bridging hydrido ligands, which explains their insolubility and the low infrared frequencies of the metal-hydrogen bands [1520 and 1300 cm^{-1} for $ZrH_2(\eta^5\text{-}C_5H_5)_2$ and 1390 cm^{-1} for $ZrHCl(\eta^5\text{-}C_5H_5)_2$]. All

---

*An interesting alternative procedure is to stir $ZrCl_2(\eta^5\text{-}C_5H_5)_2$ in tetrahydrofuran with 0.5 g-atom of magnesium turnings. A red color develops and after 3-5 days a 30% yield of $ZrHCl(\eta^5\text{-}C_5H_5)_2$ can be recovered by filtration.

†Available from Alfa Products, Ventron Corp., P.O. Box 299, Danvers, MA 01923.

dissolve readily in excess Li[AlH$_4$] solution, probably forming zirconium-aluminum complexes with bridging hydrido ligands. All but the dihydrido complex slowly develop a pink color when exposed to light and therefore appear to be photosensitive.

References

1. R. K. Nanda and M. G. H. Wallbridge, *Inorg. Chem.,* 3, 1798 (1964).
2. P. C. Wailes, H. Weigold and A. P. Bell, *J. Organomet. Chem.,* 43, C32 (1972).
3. P. C. Wailes, H. Weigold and A. P. Bell, *J. Organomet. Chem.,* 27, 373 (1971).
4. J. Schwartz and J. A. Labinger, *Angew. Chem. Int. Ed. Engl.,* 15, 333 (1976).
5. E. Samuel and R. Setton, *C. R.,* 256, 443 (1963).
6. A. F. Reid, J. S. Shannon, J. M. Swan, and P. C. Wailes, *Aust. J. Chem.,* 18, 173 (1965).
7. *Inorg. Synth.,* 12, 317 (1970).
8. P. C. Wailes and H. Weigold, *J. Organomet. Chem.,* 24, 405 (1970).
9. H. C. Brown and R. F. McFarlin, *J. Am. Chem. Soc.,* 80, 5372 (1958).

53. TETRACARBONYL[OCTAHYDROTRIBORATO(1−)]-MANGANESE, (CO)$_4$Mn(B$_3$H$_8$)

$$[(CH_3)_4N] [B_3H_8] + (CO)_5MnBr \rightarrow (CO)_4Mn(B_3H_8) + CO + [(CH_3)_4N] Br$$

Submitted by STEVEN J. HILDEBRANDT* and DONALD F. GAINES*
Checked by T. F. MOORE† and J. D. ODOM†

In the past 10 years there has been increasing interest and activity[1] in the syntheses of transition metal complexes containing borane or anionic hydroborate ligands. Compounds of this type are often referred to as metalloboranes and are considered to be boron hydride polyhedra or polyhedral fragments in which one or more boron atom positions are occupied by metal atoms. Since bonding in the borane ligands is formally electron deficient, the bonding of the borane ligand to the metal is of considerable interest. The chemistry of this class of complexes has not been studied extensively, but preliminary work suggests notable differences from organotransition metal complexes. The octahydrotriborate(1−) ligand, [B$_3$H$_8$]$^−$, forms complexes with a variety of transition metals.[2] Presented here is the high-yield synthesis of (CO)$_4$Mn(B$_3$H$_8$),[3] a stable, volatile, and reactive metalloborane, whose chemistry has been studied more extensively than that of other transition metal octahydrotriborate(1−) com-

*Department of Chemistry, University of Wisconsin, Madison, WI 53706.
†Department of Chemistry, University of South Carolina, Columbia, SC 29208.

plexes. Analogous rhenium and iron species, $(CO)_4Re(B_3H_8)$ and $(\pi-C_5H_5)$ $(CO)Fe(B_3H_8)$, can be prepared when the procedure outlined below is slightly modified.[3]

Procedure

■**Caution.** *All reagents and products are toxic. All operations should be carried out in a vacuum line and/or in a good fume hood.*

Standard high-vacuum and/or inert-atmosphere techniques are employed throughout this synthesis.[4] The vacuum system should be capable of attaining a vacuum of about 10^{-5} torr. The fractionation train should consist of at least five U-traps with the bulb of each trap having an approximate liquid capacity below the U of 50 mL (in practice, bulbs of about 25 mL in volume are acceptable, but the smaller size necessitates dividing the product solution into two fractions before purification—see below). It is convenient to run small-scale preparations of $(CO)_4Mn(B_3H_8)$ in evacuated reaction flasks. A suitable reaction flask can be constructed by sealing a 300-mL, round-bottomed Pyrex flask to a high vacuum Teflon valve to which is attached a ᔦ joint adapter for transferring the product into the vacuum line after the reaction is complete. (Kontes #K-826500 and Ace Glass #8195 are excellent valves for the reaction flask.) O-Ring seats are preferable to a Teflon seat. Right angle valves with a large bore (8-12 mm) should be used to facilitate addition of solid reagents. (■**Caution.** *Because of the evolution of carbon monoxide, the volume of the reaction flask should be sufficient to allow for pressures not to exceed 1.2 atm.*)

The reaction flask is charged with 1.50 g (0.013 mole) of $[(CH_3)_4N] [B_3H_8]$[5] and 2.75 g (0.010 mole) of $(CO)_5MnBr$,[6] along with a Teflon-coated magnetic stirring bar and about twenty-five 4-mm borosilicate glass beads. Because of limited solubility of $[(CH_3)_4N] [B_3H_8]$ in CH_2Cl_2 it is advantageous to grind the octahydrotriborate(1-) salt in a mortar prior to introducing it into the reaction flask. The flask is then attached to the vacuum line and evacuated. The solid reagents are thoroughly dried by evacuation at room temperature for at least 15 minutes [longer pumping times should be avoided since the $(CO)_5MnBr$ slowly sublimes into the vacuum line, thereby decreasing the yield]. About 40 mL of CH_2Cl_2 (reagent grade, dried over 3-Å molecular sieves) is distilled *in vacuo* into the reaction flask, the Teflon valve is closed, and the solution is stirred at room temperature. The reaction is complete in about 72 hours, as indicated by the cessation of carbon monoxide evolution. The progress of the reaction is easily monitored by freezing the reaction flask in liquid nitrogen and observing the carbon monoxide pressure on a mercury manometer (vapor pressure of CO = 400 torr at $-196°$). Since carbon monoxide evolution is slow, the flask should be opened to as small a volume as possible between the flask and the manometer to maximize the pressure reading. The carbon monoxide produced is pumped away, the flask valve is closed, and the reaction is continued at room temperature. After the reaction is complete, the product mixture is allowed to evaporate and is condensed in a U-trap at $-196°$ on the vacuum line. If the bulbs on the U-traps are not large enough to accommodate the total solvent volume the product mixture can be split in two portions. The CH_2Cl_2 can be separated from the product by distillation through a $-45°$ (chlorobenzene slush) U-trap and condensation in a U-trap cooled to $-196°$. The $(CO)_4Mn(B_3H_8)$ is then purified by distillation through a $0°$ (ice-water) U-trap and condensation in a $-36°$ (1,2-dichloroethane slush) U-trap. The product is a yellow-orange liquid (mp $+4°$) having a vapor pressure of about 1 torr at $20°$. The yield of the purified product is approximately 70% [based on $(CO)_5MnBr$].

■**Caution.** *The residues left behind in the reaction flask should be cautiously destroyed under nitrogen in the hood, behind a safety shield, by the slow addition of 20% 1-pentanol-hexane solution until no further reaction is visible.*

Other reactants can successfully be substituted without changing the procedure outlined here. For instance, $Cs[B_3H_8]$ and $(CO)_5MnBr$ react in diethyl ether solvent to give a 53% yield of the desired product in 70 hours.

Larger scale preparations (up to 0.100 mole) have been carried out on the bench top (*in the hood!*) under inert atmosphere conditions (a slow nitrogen purge through a oil bubbler), using the appropriate scale-up in reactant and solvent quantities. The reaction is complete in about 72 hours. The bulk of the solvent can be removed by evaporation at or below room temperature under reduced pressure on the bench top [heating should be avoided because it leads to significant product decomposition and the formation of $HMn(CO)_5$ and $Mn_2(CO)_{10}$ impurities]. The final purification of the product should be accomplished on the vacuum line.

Properties

Characterization of the product is most easily achieved by infrared spectroscopy. The low-resolution gas-phase infrared spectrum (10-cm cell, $P \approx 1$ torr) of $(CO)_4Mn(B_3H_8)$ exhibits the following prominent bands: 2560 (m), 2550 (m, sh), and 2500 (mw) ($B-H_{terminal}$); 2200 (w), 2150 (m), 2070 (s), 2010 (s), and 1980 (w, sh) (carbonyl); and 1825 (w) cm^{-1} ($B-H-B$). The ^1H and ^{11}B NMR spectra of $(CO)_4Mn(B_3H_8)$ show that the octahydrotriborate(1-) ligand is bound to the manganese atom in a bidentate fashion through hydrogen bridges (see Fig. 1). The ^{11}B NMR spectrum [C_6D_6 solvent, $BF_3\cdot O(C_2H_5)_2$ external reference] shows a low-field triplet at -0.4 ppm (B_2) and a high-field multiplet at $+42.2$ ppm ($B_{1,3}$). The ^1H NMR spectrum (C_6D_6 solvent, TMS reference) consists of five resonances [δ 3.25 (H_2 or H_4), 2.83 (H_4 or H_2), 1.17 ($H_{1,3}$),

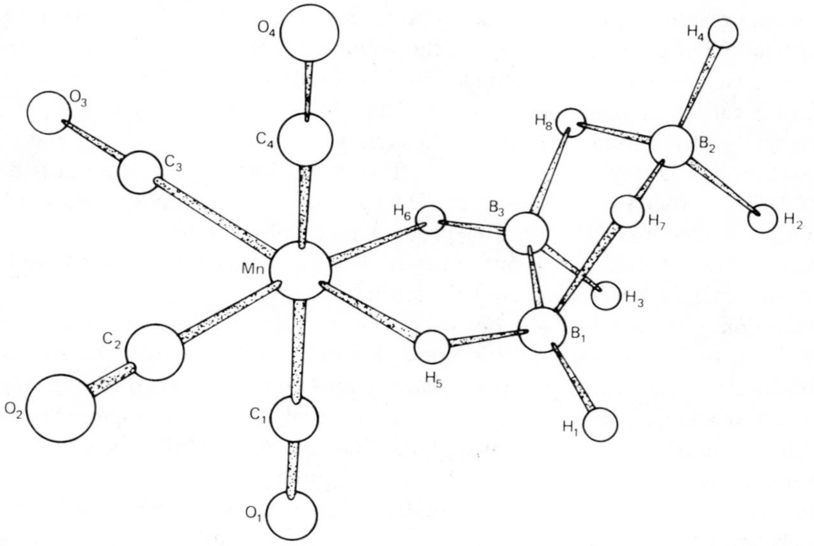

Fig. 1.

-0.68 ($H_{7,8}$) and -12.01 ($H_{5,6}$) ppm]. The high-field resonance is very characteristic of those hydrogen atoms bridging between the metal and the adjacent boron atoms. Representative NMR spectra are reproduced in Reference 3. The 70 EV mass spectrum of $(CO)_4Mn(B_3H_8)$ exhibits a parent ion in addition to the mass envelopes corresponding to the loss of each of the four carbon monoxide ligands. As a further check on the identity of the product the exact mass of the parent ion, $[(^{12}C^{16}O)_4{}^{55}Mn^{11}B_3{}^1H_8]^+$, was found to be 208.0086 (calcd. 208.0083).

Tetracarbonyl[octahydrotriborato(1-)] manganese is an air-sensitive liquid that decomposes slowly (about 5% in 4 days) at room temperature as a neat liquid in a vacuum. It is soluble in benzene, toluene, dichloromethane, diethyl ether, and tetrahydrofuran (THF), but decomposes upon heating at reflux in these solvents (especially in THF). Gas-phase thermal decarbonylation or solution photodecarbonylation of $(CO)_4Mn(B_3H_8)$ yields the novel and reactive compound $(CO)_3Mn(B_3H_8)$, in which the octahydrotriborate(1-) ligand is tridentate.[3]

References

1. P. A. Wegner, in *Boron Hydride Chemistry,* E. L. Muetterties (ed.), Academic Press Inc., New York, 1975 p. 431.
2. (a) F. Klanberg and L. J. Guggenberger, *Chem. Commun.,* **1967,** 1293; (b) F. Klanberg, E. L. Muetterties, and L. J. Guggenberger, *Inorg. Chem.,* 7, 2272 (1968); (c) L. J. Guggenberger, *ibid.,* 9, 367 (1970); (d) S. J. Lippard and D. A. Ucko, *ibid.,* 7, 1051 (1968); (e) S. J. Lippard and K. M. Melmed, *ibid.,* 8, 2755 (1969); (f) E. L. Muetterties, W. G. Peet, P. A. Wegner, and C. W. Alegranti, *ibid.,* 9, 2447 (1970); (g) H. Beall, C. H. Bushweller, W. J. Dewkett, and M. Grace, *J. Am. Chem. Soc.,* 92, 3484 (1970); (h) C. H. Bushweller, H. Beall, M. Grace, W. Dewkett, and H. Bilofsky, *ibid.,* 93, 2145 (1971); (i) A. R. Kane and E. L. Muetterties, op. cit., 93, 1041 (1971); (j) L. J. Guggenberger, A. R. Kane, and E. L. Muetterties, *ibid.,* 94, 5665 (1972).
3. D. F. Gaines and S. J. Hildebrandt, *J. Am. Chem. Soc.,* 96, 5574 (1974).
4. D. F. Shriver, *The Manipulation of Air-Sensitive Compounds,* McGraw-Hill Book Co., New York, 1969.
5. W. J. Dewkett, M. Grace, and H. Beall, *Inorg. Synth.,* 15, 115 (1974).
6. E. W. Abel and G. Wilkinson, *J. Chem. Soc.,* **1959,** 1501.

Chapter Five

MAIN GROUP COMPOUNDS

54. DIHYDRO(ISOCYANO)(TRIMETHYLAMINE)BORON (TRIMETHYLAMINE-ISOCYANOBORANE ADDUCT)

$$2\,[(CH_3)_3N]\,BH_3 \; + \; I_2 \; \rightarrow \; 2\,[(CH_3)_3N]\,BH_2I \; + \; H_2$$

$$[(CH_3)_3N]\,BH_2I \xrightarrow[\text{H}_2\text{S}]{\text{AyCN}} [(CH_3)_3N]\,BH_2(NC)$$

Submitted by JOSÉ L. VIDAL*[†] and G. E. RYSCHKEWITSCH[†]
Checked by VIC B. MARRIOTT[‡] and H. C. KELLY[‡]

Reports on cyanoboron species describing their formation and hydrolysis,[1,2] their ability to act as ligands toward metal centers,[3,4] and their function as a source of new boron compounds[5] have been published recently. Amine-cyanoborane adducts also have been studied in similar fashion.[5-8] These compounds have been prepared by reaction of alkali cyanotrihydroborate with hydrogen chloride[7] or by the exchange reaction of trimethylamine-iodoborane adduct with alkali metal cyanide.[8] The product resulting in all these cases is the boron-carbon bonded isomer.

The boron-nitrogen bonded isomer, trimethylamine-isocyanoborane adduct,[9-13]

*Union Carbide Corporation, P. O. Box 8361, South Charleston, WV 25303, current address.
[†]Department of Chemistry, University of Florida, Gainsville, FL 32603.
[‡]Department of Chemistry, Texas Christian University, Fort Worth, TX, 76129.

has been shown to coordinate with post-transition metal ions[10] and with other boron species to give adducts and tetrahedral boron cations,[11] to displace carbon monoxide from transition metals complexes,[13] and to resemble chemically the carbon isocyanides.[11,12] Isocyanoborane adducts are thus expected to have a chemistry as broad as carbon isocyanides. However, the lack of availability of pure alkali trihydroisocyanoborate salts precludes a preparation similar to that reported for amine-cyanoborane adducts.[5-7]

The present method is the only documented synthesis of trimethylamine-isocyanoborane adduct, the first boryl isocyanide reported. The procedure may be general and thus useful for the preparation of other amine-isocyano and amine-pseudohaloborane adducts. Readily available materials are used, and a high yield of crude or purified product is obtained in an estimated 10-hour working period. Schlenk techniques are advocated for this work, but if Schlenk-ware is not available solutions may be transferred quickly in air and subsequently handled under a nitrogen atmosphere.

Procedure

■**Caution.** *This synthesis should be carried out in a well-ventilated hood because of the use of highly poisonous hydrogen sulfide. Usual safety measures recommended for handling cyanide salts should be carefully observed during the synthesis and the disposal of the residues. Exposure of any of those products to acids could result in liberation of highly poisonous hydrogen cyanide.*

A 500-mL, three-necked, round-bottomed flask equipped with a mechanical stirrer, a 200-mL pressure-equalizing funnel, and an adapter containing a 2-mm vacuum stopcock is charged with 124.0 g (0.926 mole) of silver cyanide, and the system is purged with nitrogen gas. Dry dichloromethane, 180 mL (previously distilled from P_4O_{10} under nitrogen) is added, and the mixture is stirred for 30 minutes while being cooled in an ice-water-salt bath (Fig. 1).

Trimethylamine-iodoborane adduct (42.80 g, 0.215 mole)* is prepared by previously published procedures[14] or by dissolving sublimed trimethylamine-borane adduct[15] (15.70 g, 0.215 mole) in 100 mL of dry dichloromethane and adding solid iodine (27.30 g, 0.1075 mole) in portions. The resulting light-yellow solution is allowed to stand for 30 minutes while being purged with nitrogen. It is then checked by proton NMR to ensure the total conversion of the initial borane. Trimethylamine-iodoborane is the only detected aminoborane species as indicated by the resonance band of this species at 2.63 ppm downfield of tetra-methylsilane. If conversion is not complete, additional iodine is added in small portions until the disappearance of the band of $[(CH_3)_3N]BH_3$ at 2.54 ppm.

---

*The checkers advocate the use of a solution prepared from freshly sublimed trimethyl-amine-iodoborane adduct.

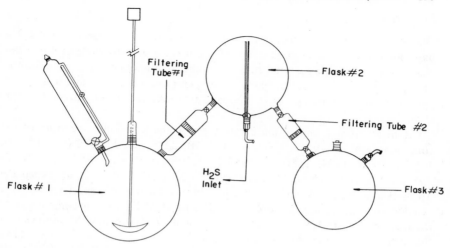

Fig. 1. Schlenk system used in the synthesis of trimethylamine-iaocyano-borane.

The solution is then added dropwise to the stirred mixture over a period of 45 minutes, and stirring is continued for an additional 1 hour. The constant-rate addition funnel is substituted by a 200-mL, medium-porosity sintered-glass frit, and the mixture is filtered under nitrogen pressure. The filtrate is collected in a previously evacuated nitrogen-purged 500-mL three-necked, round-bottomed flask equipped with a gas dispersion tube and two adapters, each attached to a medium-porosity, sintered-glass frit and a 2-mm vacuum stopcock. The receiving flask is cooled with a convenient refrigerant such as an ice-salt mixture and partially evacuated during the filtration.

The filtrate is then warmed to room temperature, and hydrogen sulfide is bubbled into the solution until no more silver sulfide precipitates. This solid is separated by filtration through a medium-porosity, sintered-glass frit also connected to a previously evacuated, nitrogen-purged, and ice-salt-cooled 500-mL, two-necked, round-bottomed flask bearing an adapter fitted with a 2-mm vacuum stopcock. The filtrate is warmed to room temperature, and the stirred solution is evaporated under vacuum to give a solid product, which is further dried for another 1½ hours. The yellow-brown crystalline solid, mp 43-45°, is crude dihydro(isocyano)(trimethylamine)boron. Yield 16.0 g (75.8% based on trimethylamine-borane).

Further purification is achieved by vacuum sublimation of the crude material at an oil-bath temperature of 40°. A white crystalline product is obtained, mp 46°, with a yield of about 31%. *Anal.* Calcd. for $C_4H_{11}BN_2$: C, 49.04; H, 11.32; N, 28.60; B, 11.03. Found: C, 48.10; H, 11.30; N, 28.59; B, 11.07.

Properties

Trimethylamine-isocyanoborane adduct is soluble in many organic solvents. The pure compound can be stored under nitrogen for 6 months without noticeable change, but the crude material turns hard and yellow under similar conditions. Thermal isomerization occurs easily at temperatures above $50°$, and it is accompanied by decomposition, as in the case of attempted acid-induced isomerization.

The main infrared absorption bands noted in the KBr pellet spectrum of the compound are: 3010, 2970, 2430 (s), 2350, 2285, 2130, 1475 (s), 1460 (s), 1450 (w, sh), 1404, 1250 (s), 1185 (s), 1105 (s), 975 (s), 855 (s), 799 (s), 705 cm^{-1} (all values $\pm$ 5 cm^{-1}). The bands at 2430 and 1185 cm^{-1} are assignable to B–H stretching and bending vibrations in coordinated BH_2 species, whereas a cyanide stretching mode is probably indicated at 2130 cm^{-1}. A comparison of the spectra of $[(CH_3)_3N] BH_2(CN)$[5,7,10,12,15] and $[(CH_3)_3N] BH_2(NC)$ shows that the assignable cyanide absorption is located at lower energies (2210 vs. 2130 cm^{-1}) and is several times more intense in the latter species. Similar differences are observed between alkyl cyanides and isocyanides.[16] A comparison of data for $[(CH_3)_3N] BH_2(CN)$ and $[(CH_3)_3N] BH_2(NC)$ with those for $[H_3BCN]$ and $[H_3BNC]^-$ show that similar relative results are observed for the cyanide infrared stretching vibrations.

Further differences between the two isomers are found in the ^1H and ^{11}B NMR spectra. Proton resonances are found, respectively, for $[(CH_3)_3N] BH_2(CN)$ and $[(CH_3)_3N] BH_2(NC)$ at 2.70 and 2.63 ppm downfield of tetramethylsilane, added as internal standard to the dichloromethane solutions.[12] Boron-11 resonances for $[(CH_3)_3N] BH_2(CN)$ and $[(CH_3)_3N] BH_2(NC)$, respectively, are found at 33.3 and 25.8 ppm upfield of $B(OCH_3)_3$, used as external standard. Both signals are 1:2:1 triplets, with a $J_{^{11}B-H}$ for $[(CH_3)_3N] BH_2(CN)$ and $[(CH_3)_3N] BH_2(NC)$ of 108 and 118 Hz, respectively.

CN stretching frequencies, boron-11 chemical shifts, and boron-hydrogen coupling constants are for $[H_3B(CN)]^-$: ν_{CN^-} at 2180 cm^{-1}; ^{11}B signal, 62.5 ppm; $J_{^{11}B-H}$, = 90 Hz; and for $[H_3B(NC)]^-$: ν_{CN^-} at 2070 cm^{-1}, ^{11}B, 45.5 ppm; $J_{^{11}B-H}$ = 94 Hz.

References

1. J. R. Berschied, Jr. and K. F. Purcell, *Inorg. Chem.,* 9, 624 (1970).
2. M. M. Kreevoy and J. E. C. Hutchins, *J. Am. Chem. Soc.,* 91, 4329 (1969).
3. S. J. Lippard and P. S. Welcker, *Inorg. Chem.,* 11, 6 (1972).
4. D. G. Holah, A. N. Hughes, B. C. Hui, and K. Wright, *Inorg. Nucl. Chem. Lett.,* 9, 835 (1973).
5. S. S. Uppal and H. C. Kelly, *Chem. Commun.,* 1970, 1619.

6. D. L. Reznicek and N. E. Miller, *Inorg. Chem.,* **11**, 858 (1972).
7. C. Weidig, S. S. Uppal, and H. C. Kelly, *Inorg. Chem.,* **13**, 1763 (1974).
8. P. J. Bratt, M. P. Brown, and K. R. Seddon, *J. Chem. Soc., Dalton Trans.,* **1974**, 2161.
9. José L. Vidal, Ph. D. Dissertation, University of Florida, 1974.
10. José L. Vidal and G. E. Ryschkewitsch, *Chem. Commun.,* **1976**, 192.
11. José L. Vidal and G. E. Ryschkewitsch, *Inorg. Chem.,* **16**, 1673 (1977).
12. José L. Vidal and G. E. Ryschkewitsch, *Inorg. Chem.,* **16**, 1898 (1977).
13. José L. Vidal and G. E. Ryschkewitsch, *J. Inorg. Nucl. Chem.,* **39**, 1097 (1977).
14. G. E. Ryschkewitsch and J. W. Wiggins, *Inorg. Synth.,* **12**, 120 (1970).
15. J. Bonham and R. S. Drago, *ibid.,* **9**, 8 (1967).
16. L. Malatesta and F. Bonati, *Isocyanide Complexes of Metals,* Wiley-Interscience, New York, 1969.

55. 1-METHYLDIBORANE(6)

$$3B_2H_6 + 2CH_3Li \rightarrow 2CH_3B_2H_5 + 2LiBH_4$$

Submitted by ROGER K. BUNTING,* FRANCIS M. JUNGFLEISCH,[†] CHARLES L. HALL[‡] and SHELDON G. SHORE[§]
Checked by CHARLES B. UNGERMANN[#] and THOMAS ONAK[#]

1-Methyldiborane(6) has previously been prepared by the thermal equilibration of mixtures of diborane(6) and trimethylborane,[1] and by a pressure reaction of trimethylborane with diborane(6), where the latter is generated *in situ* from a tetrahydroborate salt and acid.[2] These procedures are complex requiring long reaction times for full equilibration of the reaction mixture, and the 1-methyldiborane(6) is produced as only a minor component among all possible substitution products. The method presented here is a single-step synthesis, and the yield of 1-methyldiborane(6) (20%) is significantly greater than from other methods (8%).

■**Caution.** *Diborane(6) and 1-methyldiborane(6) inflame explosively on contact with the atmosphere, and methyllithium may inflame on exposure to moist air.*

*Department of Chemistry, Illinois State University, Normal, IL 61761.
†DuPont Corporation, Wilmington, DE 19898.
‡Eastman Kodak Co., Rochester, NY 14650.
§Department of Chemistry, Ohio State University, Columbus, OH 43210.
#Department of Chemistry, California State University, Los Angeles, CA 90032.

Procedure

A diethyl ether solution containing 20 mmoles of methyllithium* is placed in a 50-mL reaction vessel equipped with a stopcock and standard taper joint. The vessel is attached to a vacuum line stopcock equipped with a mercury blow-out manometer,[3] and the contents are frozen at $-196°$ (liquid nitrogen) prior to evacuation of the vessel.

Thirty millimoles of diborane(6)[†4] is measured out on the vacuum line by expanding it into a calibrated vessel of at least 1-L capacity and then is recondensed at $-196°$. Under static vacuum the diborane(6) is then distilled and condensed onto the methyllithium at $-196°$.

With the reaction flask closed off from the vacuum line, the bath at $-196°$ is replaced by a bath at $-111°$ (CS_2 slush bath), which results in a gradual increase in pressure to approximately 50 torr as observed on the blow-out manometer. As the reaction proceeds, the pressure slowly diminishes and stabilizes after about 2 hours. When no further pressure change is observed, the reaction mixture is frozen again at $-196°$. The reaction vessel is then warmed by removing the cold bath, while the contents are subjected to pumping of the volatile products through successive cold traps at -126 (methylcyclohexane slush bath), -140, (Skelly F) and $-196°$.

Seven to eight millimoles of unreacted diborane(6) may be recovered from the trap at $-196°$, whereas pure 1-methyldiborane(6) is contained in the trap at $-140°$. Transfer of this product from its trap should be carried out by distillation from a bath at $-111°$. The yield of 1-methyldiborane(6) is 3 mmole, 20% of the theoretical value based on consumed diborane(6).

This preparative procedure may be successfully scaled up by at least a factor of 3, with due caution being taken to monitor the pressure in the reaction vessel.

Properties

The density of liquid 1-methyldiborane(6) at $-126°$ is 0.546 g/mL, or 13.1 mmole/mL.[5] Its vapor pressure is 55 torr at $-78.5°$[1]; the compound is thermally unstable at this temperature and slowly disproportionates to diborane(6) and 1,1-dimethyldiborane(6).[1,6] Although the infrared spectrum can be used to identify this compound, similarities to the spectra of the disproportionation products present difficulties in assaying purity.[7] The ^{11}B NMR spectrum is useful for identifying and assaying 1-$CH_3B_2H_5$. At $-110°$, the ^{11}B NMR

---

*Available from Alfa Products, Ventron Corp., P.O. Box 299, Danvers, MA 01923. The yield of 1-$CH_3B_2H_5$ is very much dependent on the purity of the methyllithium. Only freshly prepared methyllithium or material that has been stored in a refrigerator should be used.

† Available from the Callery Chemical Co., Callery, PA 16024.

spectrum of a neat sample consists of a doublet of triplets [(δ -11.3 ppm, with respect to $BF_3O(C_2H_5)_2$], arising from a terminal hydrogen atom spin coupling with boron atom 1(J = 130 Hz), and two bridge hydrogen atoms spin coupling with boron atom 1 (J = 41 Hz); and a triplet of triplets (δ -29.1 ppm), arising from spin coupling of two terminal hydrogen atoms with boron atom 2 (J = 134 Hz) and two bridge hydrogen atoms spin coupling with boron atom 2 (J = 44 Hz).

References

1. H. I. Schlesinger and A. O. Walker, *J. Am. Chem. Soc.*, **57**, 621 (1935).
2. L. H. Long and M. G. H. Wallbridge, *Chem. Ind. (Lond.)*, **1959**, 295.
3. D. F. Shriver, *The Manipulation of Air-Sensitive Compounds*, McGraw-Hill Book Co., New York, 1969.
4. (a) I. Shapiro, H. G. Weiss, M. Schmich, S. Skolnik, and G. B. L. Smith, *J. Am. Chem. Soc.*, **74**, 901 (1952). (b) G. F. Freeguard, and L. H. Long, *Chem. Ind. (Lond.)*, **1965**, 471. (c) A. D. Norman and W. L. Jolly, *Inorg. Synth.*, **11**, 15 (1968). (d) R. Köster and P. Binger, *ibid.*, **15**, 141 (1974).
5. F. M. Jungfleisch, PhD Thesis, Ohio State University, 1973.
6. L. Van Alten, G. R. Seely, J. Oliver, and D. M. Ritter, *Adv. Chem. Ser.*, **32**, 107 (1961).
7. (a) W. J. Lehmann, C. O. Wilson, J. F. Ditter, and I. Shapiro, *Adv. Chem. Ser.*, **32**, 139 (1961). (b) W. J. Lehmann, C. O. Wilson, Jr., and I. Shapiro, *J. Chem. Phys.*, **32**, 1088 (1960). (c) *ibid.*, **34**, 12 (1961).

56. 1,6-DIBORACYCLODECANE, $B_2H_2(C_4H_8)_2$
[1,2:1,2-BIS(TETRAMETHYLENE)DIBORANE(6)]

$$n\text{-}CH_2=CH-CH=CH_2 + n\text{-}H_3B{:}S(CH_3)_2 \xrightarrow[0°]{THF} (C_4H_8BH)_n + n\text{-}S(CH_3)_2$$

I

II

Submitted by SURENDRA U. KULKARNI* and HERBERT C. BROWN*
Checked by STEPHEN J. BACKLUND† and GEORGE ZWEIFEL†

*Richard B. Wetherill Laboratory of Purdue University, West Lafayette, IN 47907.
†Department of Chemistry, University of California, Davis CA 95616.

An exceptionally stable transannular boron-hydrogen-boron bridge compound, 1,6-diboracyclodecane (**II**), results from the hydroboration of 1,3-butadiene followed by thermal depolymerization and isomerization of the largely polymeric product (**I**).[1] This cyclic diboron compound is apparently the first molecule containing a transannular $>BH_2B<$ bridge and exhibits a number of unusual characteristics. It serves as a starting material for the preparation of other organoboron derivatives proceeding through symmetrical and unsymmetrical cleavage of the $>BH_2B<$ bridge.[2,3]

The first synthesis of 1,6-diboracyclodecane (**II**) was achieved by Köster,[4] who assigned the "bisborolane" structure. Brown and coworkers first suggested[5] and later proved[1] the presently accepted structure, also supported by Young and Shore.[2] Available methods for the preparation of **II** consist of the hydroboration of 1,3-butadiene with diborane(6),[2,4] with borane-tetrahydrofuran,[1,5] and with borane-trialkylamine,[4] followed by a thermal treatment to transform the initially formed polymeric materials. Considering the simplicity of the procedure, the borane-tetrahydrofuran method was originally the method of choice for a laboratory-scale preparation. In recent years, however, borane-methyl sulfide (BMS)* has become commercially available. Its use[6,7] further simplifies the preparation. We describe here the preparation of **II**, utilizing a modification of the earlier method,[1] with BMS as the hydroborating agent.

■**Caution.** *The reaction should be carried out in a well-ventilated hood and away from flames, since BMS, 1,3-butadiene, organoborane by-products, and methyl sulfide are flammable, odoriferous, and possibly toxic substances. BMS and the initial hydroboration product are sensitive to air and moisture. The inert atmosphere technique[8] is therefore recommended. Peroxide-free tetrahydrofuran (THF) should be distilled from a small quantity of lithium tetrahydroaluminate according to the procedures in Reference 8. BMS is a very concentrat-*

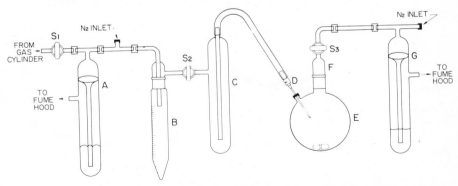

Fig. 1. Apparatus for the hydroboration of 1,3-butadiene.

*Available from Aldrich Chemical Co., 940 W. St. Paul Ave., Milwaukee, WI 53233.

ed reagent (approximately 10 M in BH_3), and the use of Neoprene gloves while handling this reagent is recommended.

Procedure

A 25-mL graduated tube (B), a 50-mL dry safety trap (C), and a 250-mL reaction flask (E) fitted with a septum inlet are assembled and connected to the bubblers (A and G) as shown in Fig. 1. Mercury is placed in bubbler A and paraffin oil in bubbler G. Stopcock S_2 is opened and 14.8 mL (200 mmole, d^{-78} = 0.73 g/mL, bp -4.4°) of 1,3-butadiene is condensed through S_1 into the tube B maintained at -78° using a Dry Ice-acetone bath. With the use of hypodermic syringes the reaction flask (E) is charged with 19.8 mL (200 mmole) of BMS* and 80 mL of THF and cooled in an ice bath. Stopcock S_1 is closed, S_2 and S_3 are kept open, and needle D is placed below the level of solution in E. (Care should be taken that no back up of the solution occurs through D.) Butadiene is allowed to distill into the reaction flask by lowering the Dry Ice-acetone bath.† When all of the butadiene has been condensed into the reaction flask (E) (about 2 hr), B and C are flushed with nitrogen to ensure the transfer of residual butadiene. The gas inlet needle (D) is removed and the mixture is stirred for 1 hour at room temperature to complete the reaction.

The connecting tube (F) is quickly replaced by a simple distillation assembly, while a flow of nitrogen is maintained through the side arm of E. (■**Caution.** *Delay in replacing the distillation assembly may result in the escape of methyl sulfide and other volatile materials into the atmosphere as well as the introduction of some atmospheric oxygen into the flask.*) The THF is distilled off at atmospheric pressure while the receiving flask is cooled. Then the borane is distilled, using a water aspirator (oil-bath temperature, 140°), into a 250-mL, round-bottomed flask fitted with septum inlet. The distillate [yield 11.6 g, 86%, bp 92-94°/18 torr, n_D^{20} 1.4795; IR: 2520 (m), 1610 (s), 1560 (ms) cm^{-1}] contains some active hydride species as an impurity. Some of the material remains as nonvolatile, viscous pot residue even after heating to 200°.

The distillate may be purified by selective oxidation of the active hydride impurity. While nitrogen is passed through the septum inlet, the flask containing the distillate is disconnected from the assembly and a magnetic stirring bar is introduced and fitted with a connecting tube leading to a fume hood. The flask is cooled in an ice bath and 50 mL of water is added dropwise with stirring. (■**Caution.** *Approximately 1.5 L of hydrogen is evolved. The reaction may be relatively vigorous in the initial stage.*) Ten milliliters of 0.5 M sodium hydroxide

---

*Commercial BMS estimated to be 10.1 M in BH_3 by hydrolysis[8] is used directly.
†The checkers recommend that the Dry Ice-acetone bath be removed completely. This reduces the time of evaporation to about 1 hour; however, it may promote the escape of butadiene through bubbler G.

is added, followed by the dropwise addition of 5 mL of 30% hydrogen peroxide. When the evolution of gas ceases (about 0.5 hr), stirring is continued for 1 hour at room temperature. The contents of the flask are transferred to a 250-mL separatory funnel and extracted with pentane (three 75-mL portions). This combined organic layer is washed with water (three 50-mL portions) and dried over anhydrous sodium sulfate. The solvent is distilled off and 8.8 g (65% yield) of 1.6-diboracyclodecane is obtained by distillation under reduced pressure (88-89°/16 torr) using a water aspirator.

Properties

1,6-Diboracyclodecane is a colorless liquid, bp 88-89°/16 torr, n_D^{20} 1.4895 (lit.[1] bp 59-60°/5 torr, n_D^{20} 1.4886), miscible with the common organic solvents but insoluble in water. It is exceptionally inert for a tetrasubstituted diborane. Thus it is stable to air, water, methanol, and acetic acid at room temperature, fails to hydroborate simple olefins without heating, and resists oxidation by cold alkaline hydrogen peroxide. However, the transannular BH_2B bridge exhibits interesting cleavage reactions. Ammonia and amines form unsymmetrical and symmetrical cleavage products, respectively,[2] whereas potassium hydride forms an organoborane anion containing a single transannular hydrogen bridge.[3]

The infrared spectrum shows a very strong band at 1610 cm^{-1} attributed to the

transannular $\begin{array}{c} H \\ >B \diagdown \diagup B< \\ H \end{array}$ bridge[1] in contrast to the band at 1560 cm^{-1}

exhibited by, 1,1,2,2-tetraethyldiborane(6) and similar derivatives. The ^{1}H NMR spectrum in CCl_4 consists of an unresolved signal at δ 0.80 ppm due to the $-CH_2-$ group adjacent to boron and a slightly resolved triplet at δ 1.57 ppm due to the other methylene group, both integrating for equal numbers of protons. The ^{11}B NMR spectrum in THF shows a broad unresolved resonance at δ -28.5 ppm from $BF_3 \cdot OEt_2$ [lit.[2] δ -28.5 ppm].

References

1. E. Breuer and H. C. Brown, *J. Am. Chem. Soc.,* **91**, 4164 (1969).
2. D. E. Young and S. G. Shore, *J. Am. Chem. Soc.,* **91**, 3497 (1969).
3. W. R. Clayton, D. J. Saturnino, P. W. R. Corfield, and S. G. Shore, *Chem. Commun.,* **1973**, 377.
4. R. Köster, *Angew. Chem.,* **72**, 626 (1960).
5. G. Zweifel, K. Nagase, and H. C. Brown, *J. Am. Chem. Soc.,* **84**, 183 (1962).
6. L. M. Braun, R. A. Braun, H. R. Crissman, M. Opperman, and R. M. Adams, *J. Org. Chem.,* **36**, 2388 (1971).
7. C. F. Lane, *J. Org. Chem.,* **39**, 1437 (1974).
8. H. C. Brown, *Organic Syntheses via Boranes,* Wiley & Sons, Inc., New York, 1975, Chapt. 9

57. TETRABUTYLAMMONIUM μ-HYDRO-DIHYDRO-BIS(μ-TETRAMETHYLENE)DIBORATE(1−)
$[(C_4H_9)_4N][B_2H_3(C_4H_8)_2]$

Submitted by M. YAMAUCHI,* D. J. SATURNINO† and S. G. SHORE‡
Checked by L. A. PEACOCK§ and R. A. GEANANGEL§

Single boron-hydrogen-boron bridge species have been proposed as transient intermediates in Lewis base cleavage reactions of diborane(6) and organodiboranes.[1] Anionic species containing such a bridge system have been isolated and characterized.[2] However, the tetrabutylammonium salt of μ-hydro-dihydro-bis-μ-(tetramethylene)diborate(1−), $[B_2H_3(C_4H_8)_2]^-$, is the only species whose structure has been determined by X-ray crystallography.[3] The preparation of this crystalline salt is described here.

The procedure described for reaction 1 is suitable for preparing a wide variety of complex hydroborate anions,[3b,4a,4b] whereas the procedure for reaction 2 is suitable for preparing stable crystalline salts of large, reactive anions.[2,3b,5]

Procedure

In a glove box that has a recirculating system to remove water and oxygen,[6] a

*East Michigan University, Ypsilanti, MI 48197.
†Mott Community College, Flint, MI 48597.
‡Ohio State University, Columbus, OH 43210.
§University of Houston, Houston, TX 77004.

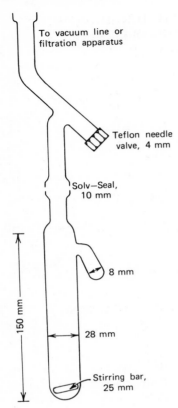

To vacuum line or
filtration apparatus

Teflon needle
valve, 4 mm

Solv—Seal,
10 mm

8 mm

150 mm

28 mm

Stirring bar,
25 mm

Fig. 1.   Reaction vessel *A*.

10-mmole sample of oil-free KH* is transferred to the main section of reaction vessel *A* (Figure. 1) and 6.90 mmole (1.10 mL) of 1,2:1,2-bis(tetramethylene) diborane(6),[1,7,8] $B_2H_2(C_4H_8)_2$, is added through the side arm. Vacuum-line techniques[6] are used to condense 10 mL of anhydrous tetrahydrofuran (THF) into the KH. While the KH suspension is being stirred at 0°, the organoborane is poured slowly into the KH, and the last traces are washed into the reaction mixture with THF. The mixture is stirred at 0° for 15 minutes and then at room temperature until dissolution of KH ceases. About 10 minutes at room temperature are required for formation of the anion when highly reactive KH, at least 95% pure, is used. The use of less than 95% pure KH should be avoided because reaction times tend to be excessive (several hours). The progress of the reaction may be followed by monitoring the disappearance of the very

*Research Organic/Inorganic Chemical Corp., 11686 Sheldon Street, Sun Valley, CA 91352.

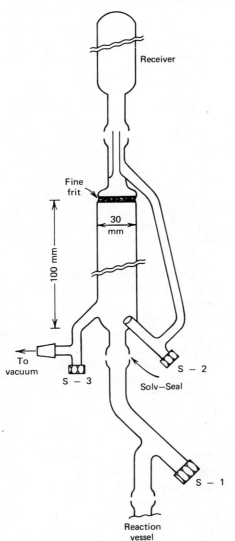

Fig. 2. Vacuum-line filtration apparatus.

intense, infrared active, bridge stretching frequency of $B_2H_2(C_4H_8)_2$ at 1610 cm^{-1}.

To prepare $[(C_4H_9)_4N] [B_2H_3(C_4H_5)_2]$, THF is withdrawn from the reaction mixture by means of a vacuum line until the volume remaining is 5 mL. The capped vessel is then removed from the vacuum line, and a 6.40-mmole (2.357 g) sample of $[(C_4H_9)_4N]I$ is added to the open end of the valve. The vessel is attached to a fine-frit, vacuum-line filtration apparatus (Fig. 2), and the system is evacuated slowly. About 3 mL of anhydrous CH_2Cl_2 is distilled into the $[(C_4H_9)_4N]$ I by cooling the solids with glass wool dipped in liquid nitrogen. (■**Caution.** *The Teflon stem of the valve and the Teflon bushing in the Solv-Seal joint* should not be cooled for these may contract enough to cause leakage of air into the system.*) The iodide salt is dissolved completely in the CH_2Cl_2 by allowing the mixture to warm to room temperature. This solution is slowly drained into the potassium salt solution while the latter is stirred rapidly at $0°$. An additional 12 mL of CH_2Cl_2 is used to transfer all of the $[(n-C_4H_9)_4N]$ I into the reaction vessel. The milky suspension that forms as soom as the two solutions are mixed is stirred at room temperature for an hour. A small bar magnet is placed outside the vessel to keep the stirring bar from moving and the solids are allowed to settle. The colorless solution is carefully decanted onto the frit by rotating the filtration assembly about the joint attached to the vacuum line. Care in this operation is essential, for the finely divided KI may pass through and clog the frit. The small amount of KI that is inevitably carried over is allowed to settle on the frit. To accelerate filtration, *S*-2 is closed, and the receiver is cooled to $0°$. The crude product is isolated by opening *S*-2 and *S*-3 and slowly removing the solvent by vacuum from the filtrate at room temperature. At this stage the product may be an oil, for some unreacted $K[B_2H_3(C_4H_8)_2]$ remains and it is a solvated liquid.

To recrystallize the product, the vessel and its contents are transferred under a stream of dry nitrogen to a medium-frit vacuum-line filtration apparatus. The crude product is dissolved in 5 mL of CH_2Cl_2, the solution is cooled to $-78°$, and while it is stirred rapidly, 15 mL of anhydrous $(C_2H_5)_2O$ is slowly added to precipitate the salt. A small bar magnet is placed outside the vessel to hold the stirring bar, about 0.75 atm of dry nitrogen is added, and *S*-3 is closed. The nitrogen atmosphere reduces splattering when the cold suspension comes in contact with warm parts of the apparatus in subsequent operations. Closing *S*-3 keeps the apparatus from being blown off the vacuum line should there be an accidental pressure buildup. To minimize dissolution of the product due to warming during filtration, an aluminum foil cup is made to hold Dry Ice around the frit. Then the filtration apparatus is cooled by swabbing it with glass wool dipped in liquid nitrogen, and the suspension is poured rapidly onto the frit by rotating the apparatus about the joint attached to the vacuum line. Dry Ice is packed into the aluminum foil cup to keep the mixture cold. The first few drops of filtrate are allowed to fall into the receiver and evaporate to drive out some of the nitrogen. Then valve *S*-2 is closed and the receiver is cooled to $-78°$ to accelerate filtration. Fresh $(C_2H_5)_2O$ is used to wash the product and solvent is removed at room temperature under high vacuum. Yield 1.8 g (75%). Elemental analysis: $C_{24}H_{55}B_2N$. Found: $C_{24.0}H_{54.2}B_{2.12}N_{1.11}$.

---

*Available from Fischer and Porter Co., Lab-Crest Div., Warminister, PA 18974.

Properties

Tetrabutylammonium μ-hydro-dihydro-bis-μ-(tetramethylene)-diborate(1-) is a white, crystalline solid that is stable at room temperature in an inert atmosphere. It is soluble in CH_2Cl_2, $CHCl_3$, $C_2H_2Cl_4$, primary amines, and secondary amines. Its solubility in $(C_2H_5)_2O$, THF, and $NH_3(l)$ is very low. The ^{11}B NMR spectrum is a doublet at δ = +4.9 ppm ($\delta_{BF_3O(C_2H_5)_2}$ = 0 ppm), J_{BH_t} = 70 Hz. The infrared spectrum of the anion has two bands in the B—H stretching region, one at 2230 cm^{-1} and the other at 2045 cm^{-1}.

References

1. D. E. Young and S. G. Shore, *J. Am. Chem. Soc.,* **91**, 3497 (1969).
2. R. K. Hertz, H. D. Johnson, II, and S. G. Shore, *Inorg. Chem.,* **12**, 1875 (1973) and references cited therein.
3. (a) D. J. Saturnino, W. R. Clayton, R. Nelson, M. Yamauchi, and S. G. Shore, *J. Am. Chem. Soc.,* **97**, 6063 (1975); (b) W. R. Clayton, D. J. Saturnino, P. W. R. Corfield and S. G. Shore, *Chem. Commun.,* **1973**, 377.
4. (a) C. A. Brown, *J. Am. Chem. Soc.,* **95**, 982 (1973); (b) *ibid.,* 4100 (1973).
5. V. T. Brice, H. D. Johnson, II, D. G. Denton, and S. G. Shore, *Inorg. Chem.,* **11**, 1135 (1972).
6. D. F. Shriver, *The Manipulation of Air-Sensitive Compounds,* McGraw-Hill Book Co., New York, 1969.
7. G. Zweifel, K. Nagase, and H. C. Brown, *J. Am. Chem. Soc.,* **84**, 183 (1962).
8. S. U. Kulkarni and H. C. Brown, *Inorg. Synth.,* **19**, 239 (1979).

58. HEXABORANE(10)

$$B_5H_9 + Br_2 \xrightarrow{\text{room temp.}} \text{1-BrB}_5H_8 + HBr \tag{1}$$

$$\text{1-BrB}_5H_8 + KH \xrightarrow[-78°]{(CH_3)_2O} K^+[\text{1-BrB}_5H_7]^- + H_2 \tag{2}$$

$$K^+[\text{1-BrB}_5H_7]^- + \tfrac{1}{2}B_2H_6 \xrightarrow[-78°]{(CH_3)_2O} B_6H_{10} + KBr \tag{3}$$

Submitted by R. J. REMMEL,* H. D. JOHNSON, II,[†] V. T. BRICE,[‡] and S. G. SHORE [§]
Checked by D. F. GAINES[#]

*Department of Chemistry, University of Alabama in Birmingham, Birmingham, AL 35294.
[†]Department of Chemistry, Northern Illinois University, DeKalb, IL 60115.
[‡]Department of Chemistry, University of Montana, Bozeman, MT 59715.
[§]Department of Chemistry, The Ohio State University, Columbus, OH 43210.
[#]Department of Chemistry, University of Wisconsin, Madison, WI 53106.

Hexaborane(10) was among the first boron hydrides isolated by Alfred Stock and his coworkers.[1] Until relatively recently, however, B_6H_{10} remained unstudied because the lack of adequate preparative methods. Early syntheses suffered either from low yields[1-4] or from requirements for starting materials that were difficult to obtain.[5-8]

A preparative procedure for B_6H_{10} that gave quantities (25% yields) large enough for detailed investigations was reported in 1971.[9] An improved synthesis, presented in 1973,[10] that involves reactions 1-3 above is the basis for the procedure given below. This procedure gives yields greater than 50% of theory based upon B_5H_9. Pentaborane(9) is commercially available and B_2H_6 may be either purchased or prepared in good yield.[11,12] The B_6H_{10} prepared from the reaction sequence above is easily purified. Furthermore, by starting with 2-$CH_3B_5H_8$, 2-$CH_3B_6H_9$ can be prepared, using a sequence of reactions analogous to reactions 1-3.[10]

■**Caution.** *Because the compounds are spontaneously flammable in air, great care should be exercised in handling B_5H_9, B_2H_6, B_6H_{10}, and 1-BrB_5H_8. Although it is possible to improve the yield of 1-BrB_5H_8 through the use of a Friedel-Crafts catalyst such as $AlBr_3$, the probability of forming $Br_2B_5H_7$ also is increased. Our limited experience with this compound suggests that it can detonate violently above room temperature. As a further word of caution, 1-BrB_5H_8 reacts explosively with hexamethylenetetraamine above $90°$.[13] The preparation presented here should be undertaken by chemists familiar with vacuum-line techniques. Bromine should be introduced by syringe into the reaction vessel in an adequate fume hood; heavy rubber gloves should be worn during this addition. Small quantitites of boron hydride wastes may accumulate in the cold traps that protect the pump station of the vacuum line. The contents of these traps should be handled with extreme caution. They should be distilled into a removable trap containing an amine (triethylamine, trimethylamine, or pyridine). The trap is warmed to $-78°$ and aollowed to stand for an hour or so. Dry nitrogen is then introduced from the vacuum line into the trap. The trap is then removed from the vacuum line, placed in the hood, and allowed to come to room temperature. Similar treatment should be given to used reaction vessels before they are exposed to air.*

Procedure

The vacuum line used in the following preparations is similar to that described by Shriver.[14] It consists of a pump station, a main reaction manifold with six reaction stations, a fractionation manifold with four U-traps and a reaction station at each end, a McLeod Guage, and a Töpler pump. The pump station employs a two-stage mechanical forepump and a two-stage mercury diffusion pump. Operating vacuum is 1.0×10^{-5} torr. Teflon valves are employed throughout.

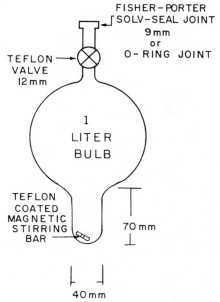

FISHER-PORTER
SOLV-SEAL JOINT
9 mm
or
O-RING JOINT

TEFLON
VALVE
12 mm

1
LITER
BULB

TEFLON
COATED
MAGNETIC
STIRRING
BAR

70 mm

40 mm

Fig. 1. Reaction bulb.

Bromine (80 mmole 12.8 g) is introduced by syringe into a 1-L reaction bulb (Fig. 1) containing a Teflon-coated magnetic stirring bar. The bulb is then placed on the vacuum line, cooled to $-196°$, and evacuated. Pentaborane(9)* (78 mmole, 7.8 mL liquid volume at $0°$) is distilled from a calibrated buret on the vacuum line and condensed into the reaction vessel, which is then allowed to warm to room temperature while the contents are vigorously stirred. A $-196°$ bath is kept close at hand in the event the reaction proceeds too vigorously. After several hours, only a slight color remains.

The volatile substances are then passed through a removable U-trap (Fig. 2) maintained at $-30°$ and a second U-trap maintained at $-196°$. The 1-BrB_5H_8 condenses in the removable U-trap, which is disconnected from the vacuum system and weighed. The yield of 1-BrB_5H_8[15] (54 mmole) is determined from the tared weight of this trap. The trap is then reconnected to the vacuum line and the 1-BrB_5H_8 is passed into a second removable U-trap. This trap is equipped with a rotatable side arm (Fig. 3). The side arm contains 54 mmole of KH†, added to the apparatus in an inert-gas-filled glove box. The KH is a freely flowing powder from which the mineral oil has been removed by repeated washing with

*Callery Chemical Company, Callery, PA 16024.

†Available from Research Organic/Inorganic Chemical Corp., 507-519 Main St., Belleville, NJ 07109.

9 mm FISHER / PORTER
SOLV−SEAL JOINT
or
O− RING JOINT

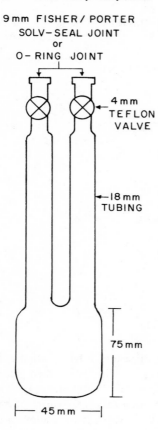

4 mm
TEFLON
VALVE

←18 mm
TUBING

75 mm

⊢— 45 mm —⊣

Fig. 2.   Removable U-trap.

dry pentane either in an inert-gas-filled box or by employing an extractor on the vacuum line. Dimethyl ether (42 mL),[*] previously dried over LiAlH$_4$ at −78°, is condensed into the vessel, which is then allowed to warm to −78°. The KH is tipped into the solution a few millimoles at a time. The hydrogen evolved is pumped away periodically. (Care should be taken not to allow the pressure in the system to exceed 1 atm.) When all the KH, has been added, the solution is allowed to stir at −78° for 1 hour. Diborane(6),[†] 27 mmole, is allowed to expand into the vessel at −78° and the vessel is then warmed to −35° and maintained there for 1 hour. During this period a white solid, KBr, precipitates. The dimethyl ether is then pumped into a U-trap maintained at −196° while the reaction vessel remains at −78°.

[*]Available from Matheson Gas Products, P.O. Box 85, 932 Paterson Plank Road, East Rutherford, MJ 07073.
[†]Available from Callery Chemical Company, Callery, PA 16024.

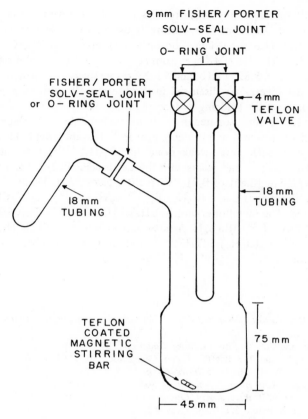

Fig. 3. Removable U-trap with rotatable side arm.

The reaction vessel is then warmed to $-35°$ and opened to U-traps maintained at room temperature, -78, and $-196°$. The bulk of the B_6H_{10} is collected in the $-78°$ trap at this time. The $-35°$ bath is then transferred from the reaction vessel to the trap at room temperature and the vessel is allowed to warm to room temperature while pumping through the -35, -78, and $-196°$ U-traps. Typical yields are in excess of 70% based on $1\text{-}BrB_5H_8$. The product should be stored in a sealed vessel at $-78°$ to retard decomposition.

Properties

The vapor pressure of the product is 7.5 torr at $0°$. No impurities are detectable by ^{11}B NMR spectroscopy. Hexaborane(10) is a colorless liquid that is air and moisture sensitive. It is stable in a vacuum at room temperature for several days.

The ^{11}B NMR spectrum of B_6H_{10} presents a simple positive means of identifying this compound. At room temperature it exhibits doublets at -14.1 (basal boron atoms) and 51.8 ppm (apical boron atom) relative to $BF_3 \cdot O(C_2H_5)_2$ in the area ratio 5:1. The ^1H NMR spectrum contains basal terminal, bridging, and apical resonances at τ 5.82, 11.10, and 11.22, respectively. The low-temperature ^{11}B and ^1H NMR spectra are discussed in the literature[16] with emphasis on the fluxional character of the bridging hydrogen atoms.

Hexaborane(10) is deprotonated by various nucleophiles.[17-20] Deprotonation occurs through the removal of a bridging proton.[18] Hexaborane(10) is a stronger Brönsted acid than B_5H_9, but a weaker one than $B_{10}H_{14}$.[17,18] It is protonated to form the $B_6H_{11}^+$ ion[19,21] and reacts with the Lewis acids BCl_3[19] and $i\text{-}B_9H_{15}$[22,23] to form adducts. Treatment of B_6H_{10} with BBr_3 gives 2-BrB_6H_9 in 59% yield.[24] Adducts with the Lewis bases $(C_6H_5)_3P$,[20,25] $(CH_3)_3N$,[19,20] and $(CH_3)_3P$[21,26] also are reported. The metalloboranes $[(C_6H_5)_3P]\,CuB_6H_9$,[27] $\mu\text{-}Fe(CO)_4B_6H_{10}$,[28] $trans\text{-}[Pt(B_6H_{10})_2Cl_2]$,[29] $Rh(B_6H_{10})_2(\text{acetylacetonate})$,[28] $[Rh(B_6H_{10})_2Cl]\,_2$,[28] $[Ir(B_6H_{10})_2Cl]\,_2$,[28] $Mg(B_6H_9)_2 \cdot 2THF$,[29,30] and $(\eta^5\text{-}C_5H_5)_2TiB_6H_9$[29] are prepared from hexaborane(10).

References

1. A. Stock, *Hydrides of Boron and Silicon,* Cornell University Press, Ithaca, New York, 1933.
2. P. L. Timms and C. S. G. Phillips, *Inorg. Chem.,* **3**, 297 (1964).
3. W. V. Kotlensky and R. Schaeffer, *J. Am. Chem. Soc.,* **80**, 4157 (1958).
4. H. A. Beall and W. N. Lipscomb, *Inorg. Chem.,* **3**, 1783 (1964).
5. J. L. Boone and A. B. Burg, *J. Am. Chem. Soc.,* **80**, 1519 (1958).
6. J. L. Boone and A. B. Burg, *J. Am. Chem. Soc.,* **81**, 1766 (1959).
7. J. L. Boone, U.S. Patent 3,110,565 (Nov. 12, 1963).
8. J. Dobson and R. Schaeffer, *Inorg. Chem.,* **7**, 402 (1968).
9. R. A. Geanangel, H. D. Johnson II, and S. G. Shore, *Inorg. Chem.,* **10**, 2563 (1971).
10. H. D. Johnson, II, V. T. Brice, and S. G. Shore, *Inorg. Chem.,* **12**, 689 (1973).
11. I. Shapiro, H. Weiss, M. Schmich, S. Skolnik, and G. Smith, *J. Am. Chem. Soc.,* **74**, 901 (1952).
12. K. C. Naiman and G. E. Ryschkewitsch, *Inorg. Nucl. Chem. Lett.,* **6**, 765 (1970).
13. P. M. Tucker, T. Onak, J. B. Leach, *Inorg. Chem.,* **9**, 1430 (1970).
14. D. F. Shriver, *The Manipulation of Air-Sensitive Compounds,* McGraw Hill Book Co., New York, 1969, Chap. 1.
15. L. H. Hall, V. V. Subanna, and W. S. Koski, *J. Am. Chem. Soc.,* **86**, 3969 (1964).
16. V. T. Brice, H. D. Johnson, II, and S. G. Shore, *J. Am. Chem. Soc.,* **95**, 6629 (1973).
17. H. D. Johnson, II, Ph.D. Dissertation, The Ohio State University, 1971.
18. H. D. Johnson, II, S. G. Shore, N. L. Mock, and J. C. Carter, *J. Am. Chem. Soc.,* **91**, 2131 (1969).
19. G. L. Brubaker, Ph.D. Dissertation, The Ohio State University, 1971.
20. G. L. Brubaker, M. L. Denniston, S. G. Shore, J. C. Carter, and F. Swicker, *J. Am. Chem. Soc.,* **92**, 7216 (1970).
21. H. D. Johnson, II, V. T. Brice, G. L. Brubaker, and S. G. Shore, *J. Am. Chem. Soc.,* **94**, 6711 (1972).

22. J. Rathke and R. Schaeffer, *J. Am. Chem. Soc.,* **95**, 3402 (1973).
23. J. Rathke and R. Schaeffer, *Inorg. Chem.,* **13**, 3008 (1974).
24. R. J. Remmel, Ph.D. Dissertation, The Ohio State University, 1975.
25. R. E. Williams and F. J. Gerhert, *J. Am. Chem. Soc.,* **87**, 3513 (1965).
26. M. L. Denniston, Ph.D. Dissertation, The Ohio State University, 1970.
27. V. T. Brice nad S. G. Shore, *Chem. Commun.,* **1970**, 1312.
28. A. Davidson, D. D. Traficante, and S. S. Wreford, *J. Am. Chem. Soc.,* **96**, 2802 (1974).
29. D. L. Denton, Ph.D. Dissertation, The Ohio State University, 1973.
30. D. L. Denton, W. R. Clayton, M. Mangion, S. G. Shore, and E. A. Meyers, *Inorg. Chem.,* **15**, 541 (1976).

59. DIMETHYLZINC

$$2(CH_3)_3Al + Zn(OCOCH_3)_2 \rightarrow (CH_3)_2Zn + 2(CH_3)_2AlOCOCH_3$$

Submitted by A. LEE GALYER* and GEOFFREY WILKINSON*
Checked by E. C. ASHBY[†] and MONSIEF BELL ASSOUED[†]

Previously published methods for the synthesis of dimethylzinc, a useful alkylating agent, include the reaction of dimethylmercury with metallic zinc,[1] the reaction of a zinc-copper couple with methyl iodide,[2] and the Grignard method.[3] The reaction of trimethylaluminum with zinc(II) halides or alkoxides can be used,[4] but it is more convenient to use zinc(II) acetate, which is very readily obtained by dehydrating the commercial dihydrate with boiling acetic anhydride or by the reaction[5]:

$$Zn(NO_3)_2 \cdot 4H_2O + \text{excess } (CH_3CO)_2O \rightarrow$$

$$Zn(OCOCH_3)_2 + 4CH_3COOH + \text{oxides of nitrogen}$$

The procedure reported below is adapted from a patent.[6]

Procedure

■**Caution.** *Trimethylaluminum and dimethylzinc burn spontaneously in air and react very violently with water. This reaction should be carried out in a good hood, and personnel should wear adequate protective clothing. A container of vermiculite or other suitable dry powder extinguisher should be immediately*

*Royal College of Science, London, SW7 2AY, England.
[†]Georgia Institute of Technology, Atlanta, GA 30332.

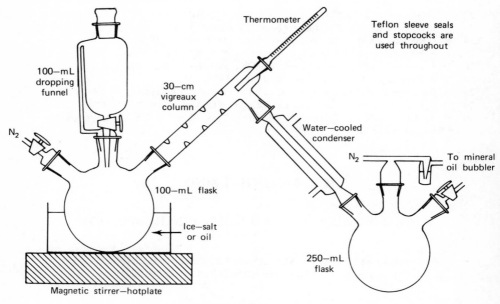

Fig. 1.   Reaction flask and receiver.

available. Safe procedures for the handling of pyrophoric liquid metal alkyls have been published.[7]

The apparatus, consisting of a 1-L three-necked flask fitted with nitrogen inlet, a 100-mL graduated, pressure-equalizing dropping funnel with a Teflon stopcock, and a 30-cm Vigreaux column leading by way of a water-cooled condenser to the 250-mL receiving flask (Fig. 1), is assembled from carefully dried components and flushed thoroughly with nitrogen. Seventy grams (0.382 mole) of anhydrous zinc(II) acetate is placed, together with a magnetic stirring bar, in the reaction flask and covered with 300 mL of dry, oxygen-free decalin. The flask and contents are cooled to about −10° with ice-salt, and 75 mL (0.782 mole) of trimethylaluminum* is placed in the dropping funnel and slowly added to the stirred mixture. After the addition is complete (about 1 hr), the yellowish mixture and the bath are at about 10°. Stirring is continued at this temperature until the solution is practically clear (about 2 hr), at which time the salt-water bath is replaced by an oil bath, the receiving flask is surrounded by ice-water, and the dropping funnel is replaced by a stopper while a positive purge of nitrogen is maintained through the center neck of the reaction flask. The reaction mixture is heated to 115-120° with stirring, whereupon the dimethylzinc distills

*Ethyl Corporation, Baton Rouge, LA.

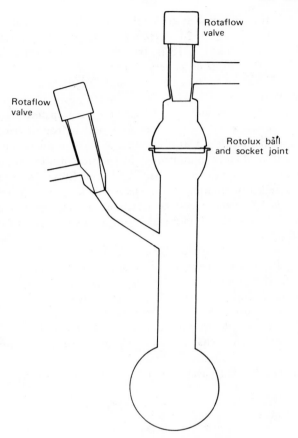

Rotaflow
valve

Rotaflow
valve

Rotolux ball
and socket joint

Fig. 2.   Storage container.

as a colorless, fuming mobile liquid at a still head temperature of 44°. Yield 32 g
(88%).

The receiver can be disconnected from the apparatus under a positive pressure
of nitrogen from both nitrogen inlets and the neck is closed with a stopcock.
The fuming reactive residues in the reaction flask may be destroyed by dilution
with petroleum ether (bp 60-80°), followed by slow addition of pentanol or
hexanol while cooling in ice-salt. After all reaction has ceased this mixture may
be poured into a large volume of ice for disposal.

Properties

Dimethylzinc is a clear colorless liquid; mp $-29.2°$, bp 44°. The ^{1}H NMR consists of a single peak at τ 9.49 (in benzene),[8] the ^{13}C NMR has a single resonance at δ -4.2 (in CS_2) with $^1J_{C-H}$ = 121. 6 Hz,[9] and the IR exhibits peaks at 614 (vs), 710 (vs), 1190 (s, doublet), 1315 (s, two shoulders), 2840 (sh) and 2935 (vs, doublet) cm^{-1}.[10] Dimethylzinc is very reactive toward oxygen, water, and common stopcock greases, but it may be stored indefinitely in ampules or under nitrogen in Schlenk-type tubes[11] with Teflon sleeve seals and valves. Small amounts of trimethylaluminum and dimethylzinc may be transferred by syringe, in which case the operator must be prepared to expect a drop or two of liquid to fall flaming from the needle tip. It is essential to work in a good hood with a dry nonflammable floor. Larger amounts may be transferred using Subaseals and stainless steel tubing under nitrogen pressure.[7,11] The Ethyl Corporation can supply copies of detailed instructions for the use of their newer type of 225 g lb lecture bottle together with the necessary attachments. The same company has published details of the handling and reactivity of aluminum alkyls from laboratory to industrial scale.*

A convenient laboratory-scale storage device consists of a Schlenk tube (about 100 mL) with a Teflon valve in the side arm and a socket joint on top of the body of the tube. The stopper is the corresponding Teflon-seated ball joint attached to a small Teflon valve (Fig. 2), which allows the air in the space above the socket to be purged with nitrogen prior to closing the tube. The ball and socket joint is preferable to the common cone and socket arrangement because the metal oxide crust that invariably forms in the joint is more easily broken.

Dimethylzinc is commercially available from Alfa Inorganics and K. and K. Laboratories.

References

1. E. Frankland and B. F. Duppa, *Annalen,* **130**, 118 (1864); *J. Chem. Soc.,* **17**, 30 (1864).
2. H. Sarroos and M. Morgana, *J. Am. Chem. Soc.,* **66**, 693 (1944); N. K. Hota and C. J. Willis, *J. Organomet. Chem.,* **9**, 169 (1967). See also R. J. H. Clark and M. A. Coles, *Inorg. Synth.,* **16**, 122 (1970).
3. D. T. Hurd, *J. Org. Chem.,* **13**, 711 (1948).
4. K. Ziegler, British Patent 836,734; *Chem. Abstr.,* **55**, 3435e (1961); A. P. Gray, A. B. Callear, and F. H. Edgecomb. *Can. J. Chem.,* **41**, 1502 (1963); T. Mole, *Organomet. React.,* **1**, 46 (1970).
5. G. Brauer, *Handbook of Preparative Inorganic Chemistry,* Vol. 2, 2nd ed., Academic Press Inc., New York, 1965, p. 1087.
6. S. M. Blitzer and T. H. Pearson, U.S. Patent 2,969,381; *Chem. Abstr.,* **55**, 9282 (1961).
7. D. W. Harvey, A. Meistress, and T. Mole, *Aust. J. Chem.,* **27**, 1639 (1974).
8. M. L. Maddox, S. L. Stafford, and H. D. Kaesz, *Adv. Organomet. Chem.,* **3**, 1 (1965).
9. F. J. Weigert, M. Winokur, and J. D. Roberts, *J. Am. Chem. Soc.,* **90**, 1566 (1968).

*Ethyl Corporation, PB 3500/I, NEW 568.

10. H. S. Gutowsky, *J. Chem. Phys.,* **17**, 128 (1949).
11. D. F. Shriver, *The Manipulation of Air-Sensitive Compounds,* McGraw-Hill Book Co., New York, 1969.

60. ELECTROCHEMICAL SYNTHESIS OF INDIUM(III) COMPLEXES

Submitted by JACOB J. HABEEB* and DENNIS G. TUCK*
Checked by BRUNO JASELSKIS[†]

The conventional preparative routes to anionic, neutral, or cationic complexes of indium start with the metal, which is dissolved in a suitable mineral acid to give a solution from which hydrated salts can be obtained by evaporation. These hydrates react with a variety of neutral or anionic ligands in nonaqueous solvents, and a wide range of indium(III) complexes have been prepared in this manner.[1] Alternatively, the direct high-temperature oxidation of the metal by halogens yields the anhydrous trihalides, which are again convenient starting materials in synthetic work. In the former case, the initial oxidation of the metal is followed by isolation, solution reaction, precipitation, and recrystallization.

The syntheses described below allow such a cycle of operations to be circumvented. As reported previously,[2] the electrochemical oxidation of indium metal at room temperature in an organic solvent system yields neutral, anionic, or cationic complexes of indium(III) within 2-3 hours, usually in good crystalline form. These syntheses are significantly quicker and more convenient than the procedures used in this laboratory in previous work.

General Procedure

The reactions are carried out in a 200-mL tall-form beaker, with a tightly fitting rubber stopper through which the platinum electrode leads are inserted; gas inlet and outlet tubes can be inserted as required. The cathode is a platinum wire carrying a 2 × 2 cm platinum sheet. The anode is a platinum wire onto which a shot of indium is beaten to form a 1 × 1 cm plate. The electrodes are placed 1-2 cm apart in the liquid phase, which is a mixture of organic solvents.

The applied voltage in these experiments lies between 15 and 50 V. The original work involved a Coutant LQ 50/50 power supply, delivering up to 50 V and 500 mA; any dc unit satisfying these conditions could presumably be used.

*Department of Chemistry, University of Windsor, Windsor, Ontario N9B 3P4, Canada.
[†]Department of Chemistry, Loyola University of Chicago, Chicago, IL 60626.

The syntheses described are typical of many others that can be carried out in this way.

■**Caution.** *Care is needed in the operation of cells at such voltages, and warning notices should be posted.*

A. INDIUM(III) CHLORIDE

$$In_{(anode)} + 3/2\ Cl_{2(g)} \rightarrow InCl_{3(s)}$$

■**Caution.** *This synthesis and that in Section B involve chlorine gas, and all operations should be performed in a well-vented hood.*

Procedure

Indium metal (0.5 g) is maintained at +50 V in a solution phase of 100 mL of benzene-methanol (3:1, v/v) containing 10 mg of tetraethylammonium perchlorate. Benzene is dried with sodium-naphthaquinone and is distilled before use (all under nitrogen); methanol is analytical grade reagent. The cell must be cooled in an ice bath throughout the experiment, since appreciable amounts of heat are evolved.

The cell is flushed with dry nitrogen, and a mixture of chlorine and nitrogen is then passed through the solution. The rate of flow of chlorine is approximately one bubble per second from a 2-mm id tube; the nitrogen flows through a separate inlet at approximately twice this rate. Under these conditions, a current of 100 mA is recorded.

After 2½ hours the gas flow is terminated and the power supply is disconnected. The layer of brown oil formed in the bottom of the vessel contains no indium and can be rejected. The supernatant is carefully removed to a 100-mL round-bottomed flask, and the solvent is removed by pumping *in vacuo* at room temperature until the volume is reduced to about 15 mL. The resultant white solid is collected and washed quickly with several small volumes of diethyl ether; although indium(III) chloride is soluble in diethyl ether, the rate of dissolution is fairly slow, and the losses in this washing are tolerable. The collection and washing operations are carried out under dry nitrogen. Yield 0.23 g (26%, based on indium consumed). *Anal.* Calcd. for $InCl_3$: In, 51.9; Cl 48.1. Found: In (atomic absorption), 51.9; Cl (silver nitrate-potassium thiocyanate titration) 48.8.

Properties

Indium trichloride is a colorless crystalline solid, mp 586°, that is highly deliquescent and is soluble in donor organic solvents. The crystal structure[3] and vibrational spectrum[4,5] have been reported.

B. TRIS(DIMETHYL SULFOXIDE)INDIUM(III) CHLORIDE [TRICHLORO-TRIS(DIMETHYL SULFOXIDE)INDIUM(III)]

$$In_{(anode)} + 3/2\ Cl_{2(g)} + 3(CH_3)_2SO_{(l)} \rightarrow InCl_3 \cdot 3[(CH_3)_2SO]_{(s)}$$

Procedure

Indium metal (0.85 g) is maintained at +15 V in a solution phase of 100 mL of 50:50 benzene-dimethyl sulfoxide (dmso). Benzene is purified as in Section A above; dimethyl sulfoxide is dried over 4A molecular sieves before use. The cell is cooled in an ice bath throughout the experiment. Chlorine gas is bubbled slowly through the solution phase (about one bubble per second from a 2-mm tube) for 2 hours. At the end of this period, the solution is brown, and most of the indium has dissolved; approximately 0.1 g of corroded material remains.

The colorless crystalline product that accumulates during the electrolysis is collected and can be recrystallized from hot methanol or ethanol (25 mL). The crystals are collected, washed with diethyl ether (2 × 25 mL), and dried *in vacuo* for 1 hour. Yield 2.8 g (92%). *Anal.* Calcd. for $C_6H_{18}O_3S_3InCl_3$: In, 25.3; C. 23.4; Found: In, 25.3; Cl, 23.1.

Properties

Tris(dimethyl sulfoxide)indium(III) chloride is a white crystalline nonhygroscopic compound, soluble in alcohols, ethyl acetate, and nitromethane. Decomposition occurs at 130°. The infrared spectrum and the results of thermal stability studies have been reported.[6] The presence of dmso can be verified from the infrared spectrum,[6] which shows C—H vibrations, and $\nu_{S=O}$ at 945, 960, and 995 cm^{-1}.

C. INDIUM(III) BROMIDE

$$In_{(anode)} + 3/2\ Br_{2(sol)} \rightarrow InBr_{3(s)}$$

Procedure

■**Caution.** *This synthesis (and that in Section D) involve liquid bromine, and all operations should be carried out in a well-vented hood. Liquid bromine should be handled with extreme care.*

The preparation is essentially identical to that described for $InCl_3$ (Sec. A), except that 2 g of bromine liquid and 10 mg of $[Et_4N]$ Br are added to the solution phase instead of a stream of chlorine gas and $[Et_4N]ClO_4$. The collection and subsequent treatment of the product are as for $InCl_3$. Yield 1.1 g (79%). *Anal.* Calcd. for $InBr_3$: In, 32.4; Br, 67.6. Found: In, 32.1; Br, 66.9.

Properties

Indium tribromide is a white hygroscopic crystalline solid, mp 436°; the vibrational spectrum suggests that each indium is surrounded octahedrally by six bromine atoms.[4,5] The solubility in organic solvents is similar to that for indium(III) chloride (see above).

D. TRIS(DIMETHYL SULFOXIDE)INDIUM(III) BROMIDE [TRIBROMO-TRIS(DIMETHYLSULFOXIDE)INDIUM(III)]

$$In_{(anode)} + 3/2\ Br_{2(sol)} + 3(CH_3)_2SO_{(l)} \rightarrow InBr_3 \cdot 3[(CH_3)_2SO]_{(s)}$$

Procedure

The preparation is essentially identical to that described for $InCl_3 \cdot 3dmso$, except that 2 g of bromine is added to the solution phase in place of a stream of chlorine gas. Yield 3.6 g (77%). *Anal.* Calcd. for $C_6H_{18}O_3S_3InBr_3$: In, 19.5; Br, 40.7. Found: In, 19.8; Br, 40.6.

Properties

The compound is a white crystalline solid whose thermal, spectroscopic, and solubility properties are similar to those of the chloride analogue.[6]

E. BIS(TETRAETHYLAMMONIUM) PENTACHLOROINDATE(III)

$$In_{(anode)} + 3/2\ Cl_{2(g)} + 2[Et_4N]Cl_{(sol)} \rightarrow [Et_4N]_2[InCl_5]_{(sol)}$$

Procedure

Indium metal (0.3 g) is maintained at +50 V in a solution phase of 100 mL benzene-methanol (3:1, v/v) to which 2 g of tetraethylammonium chloride is added. The solvents are purified as described in Section A above. Chlorine is bubbled through the solution as in Section B. Under these conditions, a current of 30 mA is recorded.

After 2 hours, the gas flow and the power supply are disconnected. No solid is visible at this point, but on transferring the solution phase to a suitable vacuum line and reducing the volume of solution by half, fine pale-yellow crystals of $[Et_4N]_2[InCl_5]$ appear. These are collected, washed with diethyl ether (two 25-mL portions), and dried *in vacuo* for 1 hour. A further crop of crystals can be obtained by adding an equal volume of diethyl ether to the filtrate. Yield 1.3 g (90%). *Anal.* Calcd. for $C_{16}H_{40}N_2InCl_5$: In, 20.8; Cl, 32.1. Found: In, 20.8; Cl, 32.1.

Properties

Tetraethylammonium pentachloroindate(III) is a white crystalline solid, mp 285° (dec.), slightly soluble in ethanol at 20° (more so at the boiling point) and also soluble in acetone and dichloromethane. The IR[7,8] spectrum shows absorptions at 294 (s), 282 (s), 268 (s), 152 (sh), and 142 cm^{-1}; Raman emissions have been reported[9,10] at 294 (s), 287 (sh), 194 (w), 167 (m), 123 (m), and 106 (m) cm^{-1}. The crystal structure determination[11,12] shows that the anion is essentially square-based pyramidal, an unusual stereochemistry for main group elements MX_5 species.

F. (TRIS(2,4-PENTANEDIONATO)INDIUM(III) [TRIS(ACETYLACETONATO)INDIUM(III)]

$$In_{(anode)} + 3C_5H_8O_{2(l)} \rightarrow In(C_5H_7O_2)_{3(sol)} + 3/2\ H_{2(g)}$$

Procedure

Indium metal (0.8 g) is maintained at +50 V in a solution phase of 80 mL of 2,4-pentanedione and 5 mL of methanol containing 50 mg tetraethylammonium perchlorate. Under these conditions, a current of 100 mA is recorded. Electrolysis is continued for 2 hours at room temperature; the metal is almost completely consumed in this period. The final solution is pale yellow, suggesting some decomposition of the diketone.

The product $[In(C_5H_7O_2)_3]$ may be obtained from the final solution in either one of two ways: (*a*) by reducing the volume of the solution by 50% *in vacuo*, or (*b*) by adding diethyl ether to the solution. In either case, the white crystals obtained are washed (diethyl ether, 4 × 25 mL) and dried *in vacuo*. Yield 2.6 g (92%). *Anal.* Calcd. for $C_{15}H_{21}O_6In$: In, 27.9. Found: In, 27.8.

Properties

Tris(acetylacetonato)indium(III) is a white crystalline compound, mp 186°, insoluble in water but soluble in most organic solvents.[13] The NMR spectrum[14] in $CDCl_3$ (10% w/v solution) has resonances at τ 4.59 (γ CH) and 7.98 (CH_3); in dimethyl sulfoxide-d_6 these values change to 4.51 and 8.06. Vibrations at 433 and 444 cm^{-1} have been assigned to ν_{M-O} modes.[15]

References

1. A. J. Carty and D. G. Tuck, *Prog. Inorg. Chem.*, **19**, 245 (1975).
2. J. J. Habeeb and D. G. Tuck, *Chem. Commun.*, **1975**, 808.
3. D. H. Templeton and G. F. Carter, *J. Phys. Chem.*, **58**, 940 (1954).
4. N. N. Greenwood, D. J. Prince, and B. P. Straughan, *J. Chem. Soc., A*, **1968**, 1694.
5. I. R. Beattie and J. R. Horder, *J. Chem. Soc., A*, **1969**, 2655.
6. A. J. Carty and D. G. Tuck, *J. Chem. Soc., A*, **1966**, 1081.
7. J. Gislason, M. H. Lloyd, and D. G. Tuck, *Inorg. Chem.*, **10**, 1907 (1971).
8. D. F. Shriver and I. Wharf, *Inorg. Chem.*, **8**, 2167 (1969).
9. S. R. Leone, B. Swanson, and D. F. Shriver, *Inorg. Chem.*, **9**, 2189 (1970).
10. D. M. Adams and R. R. Smardzewski, *J. Chem. Soc., A*, **1971**, 714.
11. D. S. Brown, F. W. B. Einstein, and D. G. Tuck, *Inorg. Chem.*, **8**, 14 (1969).
12. G. Joy, A. P. Gaughan, I. Wharf, D. F. Shriver, and J. P. Dougherty, *Inorg. Chem.*, **14**, 1795 (1975).
13. C. J. Morgan and H. D. K. Drew, *J. Chem. Soc.*, **1921**, 1058.
14. A. J. Carty, D. G. Tuck, and E. Bullock, *Can. J. Chem.*, **43**, 2559 (1965).
15. R. E. Hester and R. A. Plane, *Inorg. Chem.*, **3**, 513 (1964).

61. BIS[(TRIMETHYLSILYL)METHYL] MAGNESIUM

$$Me_3Si(CH_2Cl) + Mg \xrightarrow{Et_2O} [(Me_3Si)CH_2] MgCl$$

$$[(Me_3Si)CH_2] MgCl + 1,4\text{-dioxane} \xrightarrow{Et_2O}$$

$$\tfrac{1}{2}[(Me_3Si)CH_2]_2Mg + \tfrac{1}{2}MgCl_2 \cdot 2(1,4\text{-dioxane})$$

Submitted by RICHARD A. ANDERSEN* and GEOFFREY WILKINSON*
Checked by M. F. LAPPERT[†] and R. PEARCE[‡]

*Royal School of Science, London SW7 2AY, England.
[†]School of Molecular Sciences, University of Sussex, Brighton BN1 9QJ, England.
[‡]I.C.I. Limited, Corporate Laboratory–Runcorn, P.O. Box 11, The Heath, Cheshire, WA7 4QE, England.

Dialkylmagnesium compounds have been prepared by three methods on the laboratory scale: (1) exchange between dialkylmercury and magnesium metal,[1,2] (2) reaction of alkyllithium compounds with magnesium chloride,[3] and (3) disproportionation of a Grignard reagent with 1,4-dioxane.[4] The first method suffers from the general unavailability of alkylmercury compounds, through it is the best method for small-scale preparations of halide-free R_2Mg. The second method requires specially activated $MgCl_2$ and the preparation of alkyllithium compounds. The third method is the best available procedure for moderate-scale preparations, although the final product contains trace quantities of chloride.

The synthesis described below is a combination of those reported for t-Bu_2Mg[5] and Me_2Mg.[6] The procedure has been used for syntheses of $(t$-$BuCH_2)_2Mg$, $[(PhMe_2C)CH_2]_2Mg$, Me_2Mg, and Ph_2Mg and is doubtless of general applicability. The dialkylmagnesium compounds are superior alkylating agents towards transition metal halides relative to Grignard reagents in that higher yields and cleaner reactions are generally observed.[7]

Procedure

All reactions are performed in an atmosphere of nitrogen. An oven-dried, 1-L, three-necked flask, equipped with dropping funnel (500 mL), mechanical stirrer (with water-cooled bearing), and reflux condenser, is charged with Grignard grade magnesium turnings (14 g, 0.58 g-atom) and 150 mL of diethyl ether (distilled from sodium-benzophenone, or lithium tetrahydridoaluminate). (Chloromethyl)-trimethylsilane* (61 g, 0.50 mole) dissolved in diethyl ether (350 mL) is added to the dropping funnel. A few drops of the solution are added to start the reaction (if no reaction occurs, a few drops of 1,2-dibromoethane will initiate the exothermic reaction), and the remaining solution is added at a rate that will maintain a slow but steady reflux. After addition is complete, the solution, which varies from colorless to dark brown, is stirred for 3 hours. Hydrolysis of a 1-mL aliquot and titration with standard HCl (phenolphthalein indicator) gives a yield of 90% (0.45 mole).[8]

■**Caution.** *1,4-Dioxane is weakly carcinogenic and the operations should be carried out in a hood.*

1,4-Dioxane (45 mL, 0.52 mole), distilled from sodium under nitrogen, is added dropwise with rapid stirring over about 1-2 hours. After addition is complete the reflux condenser is replaced by a stopper and the dropping funnel is replaced

*Purchased from Peninsular Chemical Research Inc., P.O. Box 1466, Gainsville, FL 32602, or Fluorochem Ltd., Dinting Vale Trading Estate, Glossop, Derbyshire, England.

by a nitrogen inlet. The white suspension is stirred rapidly for 12 hours. The suspension is transferred to centrifuge tubes by way of 2.5 cm stainless steel tubing and the tubes are centrifuged at about 2500 rpm for 30 minutes. It It should be noted that simple filtration, even when using a filter aid (e.g., Celite), is generally difficult. The colorless supernatant liquid is transferred with stainless steel tubing to a 500-mL, two-necked flask, and the diethyl ether is removed in vacuum, with magnetic stirring. The white solid in the centrifuge tubes is washed with diethyl ether (100 mL), centrifuged, and transferred to the two-necked flask. After removal of all the diethyl ether the white solid so produced is heated in vacuum at $100\text{-}140°/10^{-2}$ torr for 24 hours to remove 1,4-dioxane. The white solid is dissolved in 200 mL of diethyl ether, and a 1-mL aliquot is titrated, as above, giving a yield of 70-80%. The ^{1}H NMR spectrum of the diethyl ether solution contains resonances at +1.30 ppm (s) due to Me_3SiCH_2Mg and +2.95 ppm (s) due to Me_3SiCH_2Mg relative to the center resonance of the methyl triplet of diethyl ether. *Anal.* Calcd. for $C_8H_{22}MgSi_2$: C, 48.4, H, 11.1. Found: C, 48.0; H, 10.2.

Properties

Base-free dialkylmagnesium is sensitive to oxygen and reacts violently with water. It melts at 200-205°, though it does not sublime until 180° in vacuum. It is insoluble in benzene and petroleum ether, although it is readily soluble in diethyl ether and tetrahydrofuran. The diethyl ether solutions are conveniently stored at 0-5°. The titer does not change over a period of 1 month. The compound can be further characterized as its N,N,N',N'-tetramethylethylene-diamine complex, $[Mg(Me_3Si)CH_2]_2 \cdot$tmed, mp 93-95°.

References

1. W. Schlenk, *Chem. Ber.*, **64**, 734 (1931).
2. E. C. Ashby and R. C. Arnott, *J. Organomet. Chem.*, **14**, 1 (1968).
3. C. W. Kamienski and J. F. Eastham, *J. Org. Chem.*, **34**, 116 (1969).
4. W. Schlenk and W. Schlink, *Chem. Ber.*, **62**, 920 (1929).
5. G. E. Coates and J. A. Heslop, *J. Chem. Soc., A*, **1968**, 514.
6. W. Strohmeier and F. Siefert, *Chem. Ber.*, **94**, 2356 (1961).
7. A. Jacot-Guillarmod, R. Tabacchi, and J. Porret, *Helv. Chim. Acta*, **53**, 1491 (1970); M. R. Collier, M. F. Lappert, and R. Pearce, *J. Chem. Soc., Dalton Trans.*, **1973**, 445; G. A. Razuvaev, V. N. Latyaeva, L. I. Vyshinskaja, A. N. Lingova, V. V. Droboteno, and V. K. Cherkasov, *J. Organomet. Chem.*, **93**, 113 (1975); E. Kohler, K. Jacob, and K.-H. Thiele, *Z. Anorg. Chem.*, **421**, 129 (1976); C. Santini-Scampucci, and J. G. Reiss, *J. Chem. Soc., Dalton Trans.*, **1976**, 195; and R. A. Andersen, E. Carmona-Guzman, J. F. Gibson, and G. Wilkinson, *J. Chem. Soc., Dalton Trans.*, **1976**, 2204.

8. F. C. Whitmore and L. H. Sommer, *J. Am. Chem. Soc.*, **68**, 481 (1946).

62. DODECAMETHYLCYCLOHEXASILANE

$$6(CH_3)_2SiCl_2 + 12M \rightarrow Si_6(CH_3)_{12} + 12MCl$$

Submitted by ROBERT WEST,* LAWRENCE BROUGH* and WIESLAW WOJNOWSKI[†]
Checked by DONALD A. VAN BEEK, JR.[‡] and A. LOUIS ALLRED[‡]

Dodecamethylcyclohexasilane (Si_6Me_{12}) is a starting material for the syntheses of many fully or partially methylated linear or cyclic polysilanes.[1] This compound was first prepared in 1949, when Burkhard found less than 1% in the product of the reaction of dichlorodimethylsilane (Me_2SiCl_2) with sodium at 115-200°C.[2] Later Gilman and Tomasi found that adding a small amount of triphenylsilyllithium in the reaction of Me_2SiCl_2 with lithium in tetrahydrofuran (THF) greatly improved the yield of Si_6Me_{12} to 60-70%.[3] Simultaneously, Stolberg reported a remarkably high 81% yield of Si_6Me_{12} from Me_2SiCl_2 and sodium-potassium (Na/K) alloy in refluxing THF.[4] The synthesis reported here is a modification of the method of Stolberg.[5]

Procedure

■**Caution.** *Na/K alloy is extremely reactive and may inflame spontaneously in air. Adequate measures should be taken so that a vigorous fire can be controlled. The reaction should be carried out in a hood and a suitable fire extinguisher such as Met-L-X[§] should be immediately available.*

The reaction is carried out in a 2-L three-necked flask set in a heating mantle and equipped with a Friedrich condenser with nitrogen inlet and a Tru-bore mechanical stirrer. The glassware is oven-dried overnight and purged with nitrogen, using the third neck as a nitrogen outlet. Once this has been completed a syringe septum is placed on the third neck and 400 mL of carefully predried

*Department of Chemistry, University of Wisconsin, Madison, WI 53706.
[†]Institute of Inorganic Chemistry & Technology, Technical University, Gdansk, Poland.
[‡]Department of Chemistry, Northwestern University, Evanston, IL 60201.
[§]Available from the Ansul Company, Marinette, WI 54143.

THF* is added by means of a syringe. Next, 92 mL (80 g, 2.4 moles) of Na/K alloy[†] (78% K by weight) is added with a syringe equipped with a metal stopcock. *Freshly distilled THF should be used, because THF rapidly absorbs both water and oxygen. Peroxides may be formed once the THF is distilled from the trace of antioxidant added to stabilize it. See Inorg. Synth., 12, 317 (1970).* The septum is removed and quickly replaced by a pressure-equalized addition funnel containing 122 mL of high purity Me_2SiCl_2[‡] (129 g, 1.0 mole).

(■**Caution.** *The following reaction is very exothermic. Excessive addition of Me_2SiCl_2 can result in very rapid refluxing that may exceed the capacity of the condenser.*) The stirrer is started and 25 mL of the Me_2SiCl_2 is added as rapidly as it will run in from the addition funnel. Within a few minutes the exothermic reaction brings the THF to vigorous boiling. The remaining Me_2SiCl_2 is then added over 10-15 minutes, to maintain a rapid reflux of solvent. The continued reaction of Me_2SiCl_2 with Na/K serves to keep the THF refluxing for about 30 minutes. When the reaction no longer evolves enough heat to maintain boiling, the heating mantle is used to keep the solution at reflux temperature.

From this point onward the amounts of Si_6Me_{12} and decamethylcyclopenta-silane (Si_5Me_{10}) in the mixture may be monitored by gas chromatography. Samples are removed from the reaction mixture with a syringe. However, the presence of salts and Na/K alloy make it necessary to prepare a glass wool filter to avoid plugging the needle. An approximately 12-cm long strip of glass wool is folded in half, and then one half is wound about the other half in a helical fashion, resulting in a fairly compact, elongated, glass wool ball. The syringe needle is then pushed through a septum and the end of the needle is inserted into the glass wool filter. The old septum is quickly removed from the neck of the reaction flask and is replaced by the syringe, and a sample is withdrawn. (A metal stopcock on the syringe is very useful.) The syringe needle is then pulled through the septum, leaving the glass wool filter in the reaction flask; the sample is placed in a vial and the insoluble material that comes through the filter is

---

*Aldrich Chemical Co. 940 W. St. Paul Ave., Milwaukee, WI 53233.

[†]Commercial grade, preformed Na/K alloy may be purchased from Callery Chemical Co., Callery, PA 16024, or it can be prepared from sodium and potassium as follows: Approximately 66 g of potassium and 19 g of sodium are scraped, cut, and weighed under an inert solvent (see Reference 5) and placed in a dry, nitrogen-purged, 500-mL three-necked flask equipped with a mechanical stirrer and nitrogen inlet. Rubber gloves are used to squeeze together a piece of sodium and a piece of potassium until a bead of Na/K alloy is formed and then the pieces are placed in the flask. The small amount of Na/K alloy prepared in this way serves to hasten the formation of alloy in the flask. The mechanical stirrer is then turned by hand to stir the sodium and potassium together. After the majority of the alloy is formed, mechnical stirring is started and the formation of the alloy is completed. The alloy is allowed to settle to permit any oxide impurities to separate and all but a few milliliters of the alloy is transferred to the reaction flask. 2-Propanol is used to destroy the remaining alloy.

[‡]Available from PCR, Inc., P.O. Box 1466, Gainesville, FL 32602.

separated by means of a centrifuge. For gas chromatographic analysis the authors used 225°, 50 mL/min He, 4.6 m × 6 mm, 20% QF-1 silicone on 60-80 mesh Chromosorb W. However, the gas chromatography conditions are not critical; other liquid phases such as SE-30 silicone can also be used. Initially the major product is $(Me_2Si)_n$ polymer and the ratio of Si_6M_{12} to Si_5Me_{10} is about 2:1. However, several hours after the addition of Me_2SiCl_2 the amount of Si_6Me_{12} present suddenly increases by a factor of about 3 in a period of less than 45 minutes, as the linear $(Me_2Si)_n$ polymer depolymerizes. The reflux time required before depolymerization takes place is 2-3 hours with pure Me_2SiCl_2. If the reaction is carried out on a larger scale, the addition of Me_2SiCl_2 must be more gradual and the reaction should be heated at reflux at least 15 hours. Also, reactions run with Me_2SiCl_2 of lower purity may require significantly longer refluxing before depolymerization and are likely to give lower yields. Depolymerization is accopanied by a characteristic color change from dark blue to black, which may also be used to monitor the reaction.

After depolymerization, the heating is continued at reflux for 15 hours and the mixture is cooled, at which time 400 mL of hexane is added. The excess Na/K alloy still present is destroyed by the gradual and *very slow* addition of water. (■Caution. *If water is added too rapidly the resulting heat may cause rapid refluxing that could exceed the capacity of the condenser. Under these conditions the possibility of fire is also increased.*) Initially 5 mL of water is added over a period of about 10 minutes with stirring, the rate of addition being determined by the rate of evolution of hydrogen gas. Once this has slowed, an extra 40 mL of water is added slowly over a period of 15 minutes. The rate of addition of water is then rapidly increased until 600 mL has been added. For efficient mixing the stirrer may have to be raised in the flask. Care should be taken to see that any Na/K alloy that may have splashed onto the sides of the flask or into one of the necks above the solvent level is decomposed.

The mixture is stirred for 10 minutes to dissolve the salts and then the hexane-THF layer is separated, filtered to remove $(Me_2Si)_n$ and other polymeric material, and evaporated by means of a vacuum rotary evaporator. A white crystalline solid that is almost entirely Si_6Me_{12} and 8-10% Si_5Me_{10} is obtained. The solid is recrystallized from boiling 95% ethanol-THF (7:1 v/v), which removes the Si_5Me_{10}, yielding about 49 g (84%) of Si_6Me_{12}, >99% pure by gas chromatographic analysis. About 70% of the material present in the mother liquor is Si_5Me_{10}, which may be isolated by reverse-phase high-pressure liquid chromatography on octadecylsilane-coated silica using methanol as a solvent.

Properties

The compound is generally soluble in organic liquids and is stable indefinitely in air. After two recrystallizations its melting point is 254-257° (sealed tube;

crystal transformations and sintering occur at lower temperatures).[3] The ^1H NMR spectrum of Si_6Me_{12} in CCl_4 consists of a singlet at $\delta = 0.13$ ppm.

References

1. M. Ishikawa and M. Kumada, *Syn. Inorg. Met.-Org. Chem.,* **1** (4), 229 (1971); M. Ishikawa and M. Kumada, *Syn. Inorg. Met.-Org. Chem.,* **1** (3), 191 (1971); W. Wojnowski, C. Hurt and R. West, *J. Organomet. Chem.,* **124**, 271 (1977).
2. C. A. Burkhard, *J. Am. Chem. Soc.,* **71**, 963 (1949).
3. H. Gilman and R. A. Tomasi, *J. Org. Chem.,* **28**, 1651 (1963).
4. U. G. Stolberg, *Angew. Chem. Int. Ed. Engl.,* **2**, 150 (1963).
5. W. S. Johnson and G. H. Daub, in *Organic Reactions,* Vol. 6, R. Adams (ed.), John Wiley & Sons, Inc., New York, 1951, p. 42.

63. IODOSILANE AND ITS METHYL DERIVATIVES

Submitted by JOHN E. DRAKE,* BORIS M. GLAVINČEVSKI,*
RAYMOND T. HEMMINGS,† and H. ERNEST HENDERSON*
Checked by C. G. NEWMAN,† JOHN DZARNOSKI, JR.,† and M. A. RING†

The position of iodosilane and iodotrimethylsilane in the heavy salt conversion series[1,2] is such that iodosilane and its methylated analogues may be used as sources for the $[(CH_3)_n SiH_{3-n}]^-$ – (n = 0-3) groups in the preparation of a wide variety of silicon halides,[3] pseudohalides,[4,6] chalcogenides,[5,7,8] and group V derivatives.[9]

Iodosilane has been prepared by the cleavage of phenylsilane[10] or chloro-phenylsilane[11,12] with hydrogen iodide, and by the reaction of monosilane with hydrogen iodide in the presence of a catalytic amount of aluminum triiodide.[4] The latter method has been extended to prepare the methylated iodosilanes.[5,13] Iodotrimethylsilane has also been prepared by the cleavage of trimethylphenyl-silane with iodine.[14]

We describe herein the synthesis of iodosilane by the reaction of phenylsilane with hydrogen iodide and the preparation of the methylated iodosilanes, namely $(CH_3)_n SiH_{3-n}I$ (n = 0-3), by the reaction of the parent hydride with hydrogen

*Department of Chemistry, University of Windsor, Windsor, Ontario, Canada, N9B 3P4.
†Department of Chemistry, Scarborough College, University of Toronto, Ontario, Canada 1A4.
‡ Department of Chemistry, San Diego State University, San Diego, CA 92182.

iodide in the presence of catalytic amounts of aluminum triiodide. These reactions give high yields in a relatively short time (about 4 hr) and may be modified to prepare polyiodinated silanes of the form $(CH_3)_n SiI_{4-n}$ (n = 1,2) and CH_3SiHI_2.

Apparatus

All manipulations are carried out on a previously cleaned and dried Pyrex-glass vacuum line of conventional design.[15] The authors' system consisted of two manifolds interconnected by four U-traps, and a central manifold leading to two liquid nitrogen backing-traps and mercury diffusion and rotary oil pumps. The vacuum in the system was monitored by a Pirani-type gauge fitted to the central manifold. Pressure readings in excess of 1 torr were registered by mercury manometers. Because the iodosilanes and hydrogen iodide undergo rapid decomposition in the presence of mercury, the manometers should be isolated when not in use. Kel-F oil over the mercury in the manometer may also be used to offset this difficulty. Any comparable system is suitable provided working pressures of <1 $\times$ 10^{-2} torr can be achieved. High-vacuum Teflon-in-glass valves and a silicone grease for ground-glass joints are preferred because of the marked solubility of the materials in hydrocarbon grease.

■**Caution.** *When working with silane and silyl compounds it is essential that the vacuum line is of particularly sound construction. Explosions can result if fractures occur during manipulation of these compounds.*

Starting Materials

The following starting materials are available from commercial sources: silane*, phenylsilane†, aluminum triiodide, chlorotrimethylsilane, chlorodimethylsilane, and dichloromethylsilane‡. Trimethylsilane, dimethylsilane, and methylsilane are prepared by the lithium tetrahydridoaluminate reduction of the corresponding chloromethylsilane in Bu_2O.[16a] Their purity is confirmed by ^{1}H NMR[20,22] spectroscopy. Hydrogen iodide*[17] is obtained by continued distillation of crude hydroiodic acid through traps at -78 (methanol-Dry Ice slush) and $-196°$ (liquid nitrogen); the pure product collecting in the latter trap is identified by IR analysis.[16b]

■**Caution.** *The silicon compounds in these preparations should be regarded as*

---

*Available from Matheson Gas Products, P.O. Box 85, 932 Paterson Plank Road, East Rutherford, NJ 07073.

†Available from Petrarch Systems, Levittown, PA 19059.

‡Available from Alfa Products, Ventron Corp., P.O. Box 299, Danvers, MA 01923.

toxic. They undergo rapid decomposition that may be explosive upon exposure to air and/or moisture. Manipulation should be carried out in a well-ventilated area.

A. IODOSILANE

$$C_6H_5SiH_3 + HI \longrightarrow SiH_3I + C_6H_6 \tag{1}$$

$$SiH_4 + HI \xrightarrow{AlI_3} SiH_3I + H_2 \tag{2}$$

Procedure (Eq. 1)

Phenylsilane ($C_6H_5SiH_3$; 0.3120 g, 2.88 mmole) and hydrogen iodide (about 6 mmole, measured in the gas phase using a mercury manometer) are condensed, *in vacuo*, into a 150-mL reaction flask (equipped with a 4-mm Teflon-in-glass valve and a MS19 ball joint (§ 18/9) for attachment to the vacuum line) held at $-196°$. The reactants are isolated in the flask by closing the valve and the temperature is allowed to rise to $-45°$ (chlorobenzene-liquid nitrogen slush). After the reaction has proceeded for 2 hours at $-45°$, the valve is opened and the volatile materials are allowed to distill through a series of U-traps at -78 (methanol-Dry Ice slush), -95 (toluene-liquid nitrogen slush), and $-196°$ with pumping. The $-78°$ fraction contains benzene and traces of diiodosilane (identified by their ^1H NMR parameters), whereas the $-196°$ fraction contains unreacted hydrogen iodide (about 3 mmole). The trap at $-95°$ retains pure iodosilane (SiH_3I; 0.3927 g, 2.49 mmole; identified by its ^1H NMR[18] and vibrational[19] spectra). The yield of iodosilane, based on the phenylsilane consumed, is 86%.

Procedure (Eq. 2)

The reaction flask is similar to that described in the preceding procedure.

Aluminum triiodide (about 2 g) is added to the flask (preferably in a dry box) and the flask is then attached to the vacuum line, surrounded by a $-78°$ slush bath and evacuated. Glass wool may be used above the vessel to minimize iodine contamination of the vacuum line. After evacuation, the bath is removed, and with the valve closed, the flask is gently heated to disperse the AlI_3 onto the sides. Silane (SiH_4; 0.1352 g, 4.2 mmole) and hydrogen iodide (about 6 mmole) are then condensed into the flask held at $-196°$. The valve is again closed and the contents are allowed to warm to $-45°$. The reaction occurs at $-45°$ for about 4 hours, at which time the volatile products are allowed to distill from the flask through a series of U-traps at -78, -95, and $-196°$ with pumping. Increased yields of SiH_3I occur if the reaction is allowed to proceed for a longer period.

Allowing the mixture to react overnight gave the following product distribution: SiH_3I (95%), SiH_2I_2 (3%), and SiH_4 (2%). The fraction at $-78°$ contains traces of diiodosilane; the fraction at $-196°$ contains unreacted hydrogen iodide and silane; and the fraction at $-95°$ retains pure iodosilane (SiH_3I; 0.5056 g, 3.2 mmole, identified by its ^1H NMR[18] and vibrational[19] spectra). Noncondensible hydrogen gas is also formed. The yield of iodosilane, based on SiH_4, is 76%.

Properties

Iodosilane is a clear, colorless liquid that is best stored at room temperature in break seal glass ampules. An analyzed sample of SiH_3I gives the following physical data[4]: mp $-57.0°$, bp $45.4°/760$ torr; vp 123.9 torr at $0°$. The ^1H NMR spectrum[18] gives a singlet due to δ_{SiH} at 3.44 ppm (shifted downfield of tetramethylsilane), whereas the infrared spectrum[19] gives prominent features above 300 cm^{-1} at 2201, 2192, 941, 903, 593, and 355 cm^{-1}.

B. IODOMETHYLSILANE AND IODODIMETHYLSILANE

$$(CH_3)_n SiH_{4-n} + HI \xrightarrow{AlI_3} (CH_3)_n SiH_{3-n}I + H_2$$

$$(n = 1, 2)$$

Procedure

The reaction flask is identical to the one described in Section A. The aluminum triiodide catalyst need not be discarded from the flask after each preparation, provided the vessel is properly evacuated prior to each usage. The methylsilane [(CH_3SiH_3, 4.0 mmole; $(CH_3)_2SiH_2$, 5.17 mmole] and hydrogen iodide (3.5 and 5.02 mmole, respectively) are condensed into the flask held at $-196°$. The valve is closed and the contents are allowed to react at $-45°$ for 45 minutes. After this time, the valve is opened and the volatile materials are allowed to distill out of the flask through U-traps held at $-23°$ (CCl_4 liquid N_2 slush) (■**Caution.** *Carbon tetrachloride is toxic and a suspected carcinogen. It should be handled in an efficient hood.*), $-78°$ and $-196°$ with pumping. In the case of CH_3SiH_2I, the $-23°$ fraction contains diiodomethylsilane and traces of iodomethylsilane, the $-78°$ fraction contains pure iodomethylsilane (CH_3SiH_2I, 0.5143 g, 2.99 mmole, 85%; identified by its ^1H NMR[20] and vibrational[21] spectra), and the $-196°$ fraction contains unreacted methylsilane (about 0.5 mmole). In the case of $(CH_3)_2SiHI$, the $-23°$ trap retains diiododimethylsilane (about 0.4 mmole) and traces of iododimethylsilane, the $-78°$ trap retains pure iododimethylsilane [$(CH_3)_2SiHI$, 0.7292 g, 3.92 mmole, 83%; identified by its ^1H NMR[22] and vibrational[23] spectra], and the −196° trap retains unreacted hydrogen iodide (about 0.3 mmole) and dimethylsilane (about 0.45 mmole). Noncondensible hydrogen gas is also formed. Increased yields of CH_3SiHI_2, CH_3SiI_3, or $(CH_3)_2SiI_2$ result if excess hydrogen iodide is used and/or the reaction is allowed to proceed for more than 45 minutes.

Properties

Iodomethylsilane (analyzed sample gives[5,24]: mp −109.5°, bp 71.8°/760 torr) and iododimethylsilane [analyzed sample gives[13]: mp −88°, bp (extrapolated) 92°; vp 32 torr at 0°] are clear, colorless liquids that may be stored at room temperature in break-seal glass ampules. They exhibit first-order[1]H NMR spectra as follows: CH_3SiH_2I[20], δ_{CH_3} at 0.93 ppm; δ_{SiH_2} at 4.08 ppm, $J_{^{29}SiH}$= 231.0 Hz; $(CH_3)_2SiHI$[22], δ_{CH_3} at 0.83 ppm; δ_{SiH} at 4.72 ppm, $J_{^{29}SiH}$ = 224.8 Hz (shifts measured in cyclohexane solution downfield from tetramethylsilane). The infrared spectra of iodomethylsilane and iododimethylsilane show strong absorptions above 300 cm^{-1} at: CH_3SiH_2I[21], 2985, 2874, 2190, 1418, 1264, 947, 880, 855, 735, 631, and 480 cm^{-1}; $(CH_3)_2SiHI$[23], 2952, 2890, 2160, 1412, 1243, 900, 856, 822, 763, 674, 645, 635, and 330 cm^{-1}.

C. IODOTRIMETHYLSILANE

$$(CH_3)_3SiH + HI \xrightarrow{\text{AlI}_3} (CH_3)_3SiI + H_2$$

Procedure

The reaction flask, catalyst, and experimental conditions are as described in Section A. Trimethylsilane [$(CH_3)_3SiH$; 0.7120 g, 9.61 mmole] and hydrogen iodide (about 13 mmole) are condensed into the flask held at −196°. The flask is isolated, the −196° bath is removed, and the contents are allowed to warm to room temperature. The reaction occurs at room temperature for about 2 hours; then the valve is opened and the contents are allowed to distill through U-traps at −78 and −196° with pumping. The −196° fraction retains unreacted hydrogen iodide (about 3.3 mmole) and traces of unreacted trimethylsilane. The −78° fraction retains pure iodotrimethylsilane [$(CH_3)_3SiI$; 1.8616 g, 9.31 mmole; identified by its ^{1}H NMR[22] and Raman spectra[25]]. The yield of iodotrimethylsilane, based on trimethylsilane, is 97%.

Properties

Iodotrimethylsilane (analyzed sample gives[1]: bp 106.8°/742 torr) is a clear, colorless liquid that is best stored in break-seal glass ampules at room temperature. The ^{1}H NMR spectrum[22] of $(CH_3)_3SiI$ gives a singlet due to δ_{CH_3} at 0.73 ppm (measured downfield of tetramethylsilane), and the Raman spectrum[25] gives sharp bands above 300 cm^{-1} at 2973, 2302, 1404, 1255, 845, 761, 704, 627, and 331 cm^{-1}.

References

1. C. Eaborn, *J. Chem. Soc.,* **1950**, 3077.
2. H. H. Anderson and H. Fischer, *J. Org. Chem.,* **19**, 1296 (1954); A. G. MacDiarmid, *Q. Rev.,* **10**, 208 (1956).
3. C. Newman, J. K. O'Loane, S. R. Polo, and M. K. Wilson, *J. Chem. Phys.,* **25**, 855 (1956); E. A. V. Ebsworth and H. J. Emeléus, *J. Chem. Soc.,* **1958**, 2150.
4. H. J. Emeléus, A. G. Maddock, and C. Reid, *J. Chem. Soc.,* **1941**, 353.
5. H. J. Emeléus, M. Onyszchuk, and W. Kuchen, *Z. Anorg. Allgem. Chem.,* **283**, 74 (1956).
6. E. A. V. Ebsworth and M. J. Mays, *J. Chem. Soc.,* **1961**, 4879; **1962**, 4844, **1963**, 3893; A. G. MacDiarmid, *J. Inorg. Nucl. Chem.,* **2**, 88 (1956); J. J. McBride, Jr. and H. C. Beachell, *J. Am. Chem. Soc.,* **74**, 5247 (1952); J. E. Drake, R. T. Hemming, and H. E. Henderson, *J. Chem. Soc., Dalton Trans.,* **1976**, 366.
7. J. E. Drake, B. M. Glavinčevski, R. T. Hemmings, and H. E. Henderson, *Can. J. Chem.,* **56**, 465 (1978).
8. A. J. Downs and E. A. V. Ebsworth, *J. Chem. Soc.,* **1960**, 3516; E. A. V. Ebsworth, H. J. Emeléus, and N. Welcman, *J. Chem. Soc.,* **1962**, 2290; H. J. Emeléus, A. G. MacDiarmid, and A. G. Maddock, *J. Inorg. Nucl. Chem.,* **1**, 194 (1955); H. J. Emeléus, and S. R. Robinson, *J. Chem. Soc.,* **1947**, 1592.
9. E. Wiberg and E. Amberger, *Hydrides of the Elements of Main Groups I-IV,* Elsevier Publishing Co., New York, 1971, p. 585.
10. G. Fritz and D. Kummer, *Z. Anorg. Allg. Chem.,* **304**, 322 (1960).
11. B. J. Aylett and I. A. Ellis, *J. Chem. Soc.,* **1960**, 3415.
12. L. G. L. Ward, *Inorg. Synth.,* **11**, 159 (1968).
13. H. J. Emeléus and L. E. Smythe, *J. Chem. Soc.,* **1958**, 609.
14. B. O. Pray, L. M. Sommer, G. M. Goldberg, G. T. Kerr, Ph. A. Digiorgio, and F. C. Withmore, *J. Am. Chem. Soc.,* **70**, 433 (1948).
15. D. F. Shriver, *The Manipulation of Air-Sensitive Compounds,* McGraw-Hill Book Co., New York, 1969.
16. (a) S. Tannenbaum, S. Kaye, and G. Lewenz, *J. Am. Chem. Soc.,* **75**, 3753 (1953); A. D. Norman, J. R. Webster, and W. L. Jolly, *Inorg. Synth.,* **11**, 170 (1968); (b) R. H. Pierson, A. N. Fletcher, and E. S. C. Gantz, *Anal. Chem.,* **28**, 1218 (1956).
17. A. I. Vogel, *A Textbook of Practical Organic Chemistry,* 3rd ed., Longmans, London, 1959, p. 182.
18. E. A. V. Ebsworth and J. J. Turner, *J. Phys. Chem.,* **67**, 805 (1963).
19. D. F. Ball, M. J. Buttler, and D. C. McKean, *Spectrochim. Acta,* **21**, 451 (1965).
20. E. A. V. Ebsworth and S. G. Frankiss, *Trans. Faraday Soc.,* **59**, 1518 (1963).
21. E. A. V. Ebsworth, M. Onyszchuk, and N. Sheppard, *J. Chem. Soc.,* **1958**, 1453.
22. E. A. V. Ebsworth and S. G. Frankiss, *Trans. Faraday Soc.,* **63**, 1574 (1967).
23. J. R. Durig and C. W. Hawley, *J. Chem. Phys.,* **58**, 237 (1973).
24. M. Abedini and A. G. MacDiarmid, *Inorg. Chem.,* **2**, 608 (1963).
25. J. Goubeau and H. Sommer, *Z. Anorg. Allg. Chem.,* **289**, 1 (1957).

64. SILYL SULFIDES

Submitted by JOHN E. DRAKE,* BORIS M. GLAVINČEVSKI,*
RAYMOND T. HEMMINGS,† and H. ERNEST HENDERSON*
Checked by E. A. V. EBSWORTH‡ and S. G. D. HENDERSON‡

Disilathiane has been the subject of several recent studies.[1-7] Interest has been centered on the importance of $(p\text{-}d)\pi$-bonding in determining both the structure and the base strength relative to the disiloxanes and carbon analogues.[2-6] The utility of disilathianes has been demonstrated by their conversion to a wide variety of silyl derivatives by exchange reactions or protolyses.[2,7]

Disilathianes, for example $(SiH_3)_2S$ and $[(CH_3)_3Si]_2S$, have been prepared by several routes,§ namely, the reaction of iodosilane with silver[8] and mercuric[2] sulfides; halosilanes with lithium sulfide,[7] $[NH_4]SH$,[9] and $[Me_3NH]SH$[7]; disilaselenane with H_2S[10]; and trisilylphosphine with sulfur.[10] Recently, the synthesis of hexamethyldisilathiane, $[(CH_3)_3Si]_2S$,# was described from the protolysis of 1-(trimethylsilyl)imidazole with H_2S and from the dehydrohalogenation of chlorotrimethylsilane and H_2S with a tertiary amine.[11] Both of these methods require about 18 hours.

The syntheses of disilathiane and its methylated analogues by the reaction of mercuric sulfide with gaseous iodosilanes is described here in detail. The procedures are convenient, take about 2 hours, and are well suited to small-scale vacuum-line techniques, using a minimum of special appartus. The yields are in the range of 90-95%.

Starting Materials

The manipulation of all volatile compounds is carried out in a Pyrex-glass vacuum system [consisting of four manifolds (volume about 150 mL) interconnected by a central manifold and pumping system and two sets of U-traps connecting adjacent manifolds], using conventional techniques.[12] Greaseless Teflon-

---

*Department of Chemistry, University of Windsor, Windsor, Ontario, Canada, N9B 3P4.

†Department of Chemistry, Scarborough College, University of Toronto, Ontario, Canada M1C A14.

‡Department of Chemistry, Edinburgh University, West Mains Road, Edinburgh, Scotland, EH9 3JJ.

§The checkers have prepared disilathiane and hexamethyldisilathiane in high yield by reacting the appropriate tertiary and secondary amine with H_2S for 2 hr and then treating the solid residue with an excess of the appropriate chlorosilane or bromosilane.

#Hexamethyldisilathiane may also be obtained commercially, (e.g., from Petrarch Systems, Levittown, PA 19059).

in-glass valves are used because the silanes show a marked solubility in hydrocarbon stopcock lubricants. Because mercury initiates decomposition of the iodosilanes, the manometers are isolated from the line when not in use.

Iodosilane,* iodomethylsilane, iododimethylsilane, and iodotrimethylsilane are prepared by the reaction of hydrogen iodide with the parent hydride,[†] namely, SiH_4, CH_3SiH_3, $(CH_3)_2SiH_2$, and $(CH_3)_3SiH$, in the presence of a catalytic amount of AlI_3.[13] The purity of the iodosilanes may be confirmed spectroscopically.[14] Commercial red mercuric sulfide is dried at 120° in a high vacuum before use.

■**Caution.** *The disilathianes should be regarded as toxic and are vile smelling. Their exposure to air and/or moisture is likely to promote rapid oxidation. Manipulations should be carried out in a sound vacuum system in a well-ventilated area. All preparations may be scaled-up to use 10 mmole of starting material.*

A. DISILATHIANE

$$2SiH_3I + HgS \rightarrow (SiH_3)_2S + HgI_2$$

Procedure

The reactor is a horizontal tube (about 25 mm id × 350 mm long) equipped with U-traps at either end (one end has a constriction and the other end is detachable by a B-24 (⊤ 24/40) joint to facilitate packing). The tube is packed alternately with glass wool and anhydrous red mercuric sulfide (about 20 g, 85 mmole), mixed with clean and dried powdered glass (about 10 g). The tube is attached to the vacuum line and is thoroughly evacuated for about 1 hour.[‡] (■**Caution.** *Always evacuate through the constricted end to avoid contamination of the vacuum line with the packing material.*) Iodosilane [SiH_3I; 0.4307 g, 2.73 mmole] is then allowed to pass through the column from one U-trap to the other. Typically after two double passes, the volatile products are fractionated on the vacuum line using cold traps at −45 (chlorobenzene-liquid nitrogen slush), −95 (toluene-liquid nitrogen slush) and −196° (liquid nitrogen). No volatile species should be present in the −45° trap unless disproportionation has occurred. The second trap retains disilathiane [$(SiH_3)_2S$; 0.1164 g, 1.24 mmole],

---

*The cleaveage of phenylsilane with hydrogen iodide also provides of iodosilane (based on details given in Reference 15).

[†]SiH_4, CH_3SiH_3, $(CH_3)_2SiH_2$, and $(CH_3)_3SiH$ are available commercially on some markets (Petrarch Systems). Alternatively, they may be synthesized by the reduction of the chlorides, $(CH_3)_nSiCl_{4-n}$, with $LiAlH_4$.[16]

[‡]The checkers found that by evacuating the tube for about 6 hr with occasional heating, the formation of siloxy impurities was eliminated.

and the final trap at $-196°$ should contain only traces of disiloxane (identified by its infrared spectrum[17]). No hydrogen is formed. The yield of disilathiane, based on the iodosilane consumed, is 91%. Optimum yields are achieved by minimizing the amount of moisture in the system.

Properties

Disilathiane is a colorless, volatile liquid (mp $-70°$, bp $58.8°$). Its vapor pressure relationship[2] is given in the range -40 to $+50°$ by log p (torr) = $7.977 - 1692/T$, which leads to ΔH_{vap} of 7743 cal/mole and a Trouton's constant of 23.3 (all physical data were obtained on analyzed samples). The vapor pressure at $0°$ is 61 torr. The ^{1}H NMR spectrum of disilathiane,[14] measured in cyclohexane, consists of a singlet (δ_{SiH}) at 4.35 ppm downfield of tetramethylsilane (J_{HH} = 0.70; $J_{^{29}SiH}$ = 224 Hz). The infrared spectrum[18] shows prominent band at 2180 (vs), 962 and 951 (vs), 907 (vs), 675 (ms), 635 (s), 610 (s), 517 (s), and 480 (ms) cm^{-1}. Disilathiane may be stored at room temperature in break-seal glass ampules.

B. 1,3-DIMETHYL-, 1,1,3,3-TETRAMETHYL-, AND HEXAMETHYLDISILATHIANE

$$2(CH_3)_n SiH_{3-n}I + HgS \rightarrow [(CH_3)_n SiH_{3-n}]_2 S + HgI_2$$

$$(n = 1, 2, 3)$$

Procedure

The reactor and conditions are identical to those described in Section A. The iodosilane [CH_3SiH_2I, 2.38 mmole; $(CH_3)_2SiHI$, 1.01 mmole; $(CH_3)_3SiI$, 1.94 mmole] is then passed through the tube four times. The volatile products are collected and fractionated, using cold traps at -45 and $-196°$. The first trap retains the appropriate methyl disilathiane, namely $(CH_3SiH_2)_2S$, 97%; $[(CH_3)_2SiH]_2S$, 92%; $[(CH_3)_3Si]_2S$, 95%. Any unreacted iodide or siloxane is collected in the trap held at $-196°$. The formation of siloxanes is again minimized by carefully removing moisture from the system by thorough evacuation.

Properties

1,3-Dimethyldisilathiane, 1,1,3,3-tetramethyldisilathiane (analyzed sample gives mp $-146°$, extrapolated bp $145°$, vapor pressure relationship: log p (torr) = $6.461 - 1498/T$ over the range $0-50°$; ΔH_{vap} = 6850 cal/mole; Trouton's constant, 16.4)³ and hexamethyldisilathiane (analyzed sample gives bp 162°)⁸ are all clear, colorless liquids that are stable at room temperature in break-seal glass ampules. These methyl disilathianes exhibit first-order ¹H NMR spectra[14,19,20] as follows: $(CH_3SiH_2)_2S$, δ_{CH_3} at 0.43, δ_{SiH_2} at 4.52 ppm, $J_{^{29}SiH}$ = 215.0 Hz; $[(CH_3)_2SiH]_2S$, δ_{CH_3} at 0.37, δ_{SiH} at 4.70 ppm, $J_{^{29}SiH}$ = 207.6 Hz; $[(CH_3)_3Si]_2S$, δ_{CH_3} at 0.33 ppm (shifts downfield from tetra-methylsilane in cyclohexane solution). The purity of these methyl disi-lathianes may be further checked by their infrared spectra.[21] $(CH_3SiH_2)_2S$: 2979, 2920, 2159 (vs), 1408, 1263, 949, 912, 875 (vs), 743, 701, and 510 cm⁻¹; $[(CH_3)_2SiH]_2S$, 2964, 2905, 2130 (s), 1425, 1255 (s), 897 (s), 832 (s), 769, 715, 663, 635, and 489 cm⁻¹; $[(CH_3)_3Si]_2S$[22], 2950 (s), 2890, 1452, 1406, 1252 (s), 865 (s), 845 (vs), 828 (vs), 755, 692, 628 (s), 493 (vs), and 441 cm⁻¹. Any likely impurities are unreacted iodosilane and disiloxane, which have prominent ν_{SiI} and ν_{SiOSi} absorptions at about 355 cm⁻¹ [14] and 1107 and 606 cm⁻¹ [17] in the infrared spectra.

References

1. A. Almenningen, K. Hedberg, and R. Seip, *Acta Chem. Scand.,* **17**, 2264 (1963).
2. H. J. Emeléus, A. G. MacDiarmid, and A. G. Maddock, *J. Inorg. Nucl. Chem.,* **1**, 194 (1955).
3. H. J. Emeléus and L. E. Smythe, *J. Chem. Soc.,* **1958**, 609.
4. E. A. V. Ebsworth, *Organometallic Compounds of the Grove IV Elements,* Vol. 1, Part I, A. G. MacDiarmid, (ed), Marcel Dekker, Inc., New York, 1968.
5. J. E. Drake and C. Riddle, *Q. Rev. Chem. Soc.,* **24**, 263 (1970).
6. H. Burger, *Angew. Chem. Int. Ed. Engl.,* **12**, 474 (1973).
7. C. Glidewell, *J. Inorg. Nucl. Chem.,* **31**, 1303 (1969).
8. C. Eaborn, *J. Chem. Soc.,* **1950**, 3077.
9. E. P. Lebedev, D. V. Fridland, and V. O. Reikhsfel'd, *Zh. Obshch. Khim.,* **44**, 2784 (1974).
10. B. J. Aylett, H. J. Emeléus and A. G. Maddock, *Research,* **6**, 30 S(1953).
11. D. A. Armitage, M. J. Clark, A. W. Sinden, J. N. Wingfield, E. W. Abel and E. J. Louis, *Inorg. Synth.,* **15**, 207 (1974).
12. D. F. Shriver, *The Manipulation of Air-Sensitive Compounds,* McBraw-Hill Book Co., New York, 1969.
13. H. J. Emeléus, A. G. Maddock, and C. Reid, *J. Chem. Soc.,* **1941**, 353; H. R. Linton and E. R. Nixon, *Spectrochim. Acta,* **10**, 229 (1958); E. A. V. Ebsworth, R. Mould, R. Taylor, G. R. Wilkinson, and L. A. Woodward, *Trans. Faraday Soc.,* **58**, 1069 (1962).
14. R. N. Nixon and N. Sheppard, *J. Chem. Phys.,* **23**, 215 (1955); *Trans. Faraday Soc.,* **53**, 282 (1957); H. R. Linton and E. R. Nixon, *Spectrochim. Acta,* **12**, 41 (1958); E. A. V. Ebsworth and J. J. Turner, *J. Chem. Phys.,* **36**, 2628 (1962); *J. Phys. Chem.,* **67**, 805 (1963).
15. L. G. L. Ward, *Inorg. Synth.,* **11**, 159 (1968); B. J. Aylett and I. A. Ellis, *J. Chem. Soc.,* **1960**, 3415.
16. A. D. Norman, J. R. Webster, and W. L. Jolly, *Inorg. Synth.,* **11**, 170 (1968).

17. R. C. Lord, D. W. Robinson, and W. C. Schumb, *J. Am. Chem. Soc.,* **78**, 1327 (1956).

18. E. A. V. Ebsworth, R. Taylor, and L. A. Woodward, *Trans. Faraday Soc.,* **55**, 211 (1959); H. R. Linton and E. R. Nixon, *J. Chem. Phys.,* **29**, 921 (1958).

19. E. A. V. Ebsworth and S. G. Frankiss, *Trans. Faraday Soc.,* **63**, (1967).

20. H. Schmidbaur and I. Ruidisch, *Inorg. Chem.,* **3**, 599 (1964).

21. H. Emeléus, M. Onyszchuk, and W. Kuchen, *Z. Anorg. Allgem. Chem.,* **283**, 74 (1956); E. A. V. Ebsworth, M. Onyszcuk, and N. Sheppard, *J. Chem. Soc.,* **1958**, 1453.

22. K. Kriegsman, *Z. Elektrochem,* **61**, 1088 (1957); K. A. Hooton and A. L. Allred, *Inorg. Chem.,* **4**, 671 (1965); H. Bürger, U. Goetze, and W. Sawodny, *Spectrochim. Acta,* **24A**, 2003 (1968).

65. CRYSTALLINE POLYAMMONIUM *CATENA*-POLYPHOSPHATE

$$P_4O_{10} + 2H_2O + 4NH_3 \rightarrow (NH_4PO_3)_4$$

$$\frac{n}{4} (NH_4PO_3)_4 + H_2O \rightarrow (NH_4)_n H_2 P_n O_{3n+1}$$

Submitted by RICHARD C. SHERIDAN* and JOHN F. McCULLOUGH*
Checked by N. E. STAHLHEBER† and C. Y. SHEN†

Crystalline polyammonium *catena*-polyphosphate, $(NH_4)_n H_2 P_n O_{3n+1}$, has been prepared by heating urea and monoammonium orthophosphate under ammonia vapor for 16 hours,[1] by ammoniation of superphosphoric acid,[2] by thermal condensation of urea phosphate,[3] and by heating various ammonium phosphates in a current of ammonia.[4,5] The procedure given below, in which crude ammonium tetrametaphosphate is reorganized and condensed to a long-chain polymer in a stream of ammonia, is straightforward and permits the use of common laboratory equipment and supplies.

■**Caution.** *The reaction of P_4O_{10} with aqueous ammonia is vigorous and strongly exothermic.*

Procedure

Phosphorus(V) oxide (28.4 g, 0.1 mole) is added in small portions with vigorous stirring to 100 mL of concentrated aqueous ammonia over a 15-minute period.

*Division of Chemical Development, Tennessee Valley Authority, Muscle Shoals, AL 35660.

†Monsanto Company, 800 N. Lindberg Blvd., St. Louis, MO 63166.

Phosphorus(V) oxide is very hygroscopic and the bottle must be closed except when an addition is made. The reactor is cooled with ice water so that the reaction temperature is maintained at 5-10°. Methanol (125 mL) is added dropwise to the stirred solution to precipitate the crude ammonium tetrametaphosphate, which is collected by suction filtration, washed with methanol, and dried in vacuum over anhydrous $CaSO_4$. The product weighs 33 g and contains about 65% of its phosphorus as tetrametaphosphate. The remainder is a mixture of ortho-, trimeta- and short-chain phosphates. Any lumps are crushed and the crude ammonium tetrametaphosphate is spread in a no. 1 porcelain evaporating dish and heated for 2 hours at 240° in a slow stream of anhydrous ammonia (about 100 mL/min or 3-5 g/hr). Care is taken to avoid absorption of moisture before or during heating, as this adversely affects the formation of the product.

■**Caution.** *Anhydrous ammonia does not generally constitute a fire or explostion hazard, but it is flammable in high concentrations, and contact with flames and electrical sparks should be avoided.*

The charge is conveniently heated in a 1-L resin reaction vessel* equipped with a heating mantle or in a small oven; a larger oven may be used if the ammonia is passed directly over the charge. The product should be cooled below 150° under ammonia to avoid deammoniation in handling. The product weighs 32.3 g and is a crystalline solid with the optical, X-ray powder diffraction pattern, and infrared absorption pattern reported for form I polyammonium *catena*-polyphosphate.[2,6] *Anal.* Calcd. for $(NH_4)_nH_2P_nO_{3n+1}$ with $n = 50$: N, 14.38; P, 31.81. Found: N, 14.4; P, 31.8.

Paper chromatographic analysis[7] shows that over 99% of the phosphate is present as a nonmoving high-molecular-weight species. However, the product contains a small amount of amorphous long-chain polyphosphates, and the purity of the crystalline form I polyammonium *catena*-polyphosphate is estimated to be 90%. The relatively soluble amorphous polyphosphates can be removed, if desired, by stirring in 100 mL of water for 15 minutes at room temperature. The product is collected by suction filtration, washed with methanol, and air dried. Yield 29.1 g (75% based on the P_4O_{10} charged).

Properties

Form I polyammonium *catena*-polyphosphate contains the shortest long-chain anions of several polymorphic forms having the same chemical composition.[1] It is slightly soluble in water and gives a cloudy, viscous solution. The solubility increases with the quantity of solid phase present; the apparent solubility of the pure compound at 25° has been estimated to be 0.15 g per 100 g of water.[1] The compound is more soluble in hot water or in the presence of other dissolved

*Available from Curtin Matheson Scientific, Inc., 1850 Greenleaf Ave., Elk Grove Village, IL 60007.

cations. It shows a slower hydrolytic degradation rate than Graham's salt, $Na_nH_2P_nO_{3n+1}$. The dry crystals are stable at room temperature but are converted to other crystalline modifications having the same chemical composition (forms II-V) by tempering at 200-400°. Paper chromatography, end-group titration, and other measurements indicate that the phosphate chain length of form I is above 50.[1]

References

1. C. Y. Shen, N. W. Stahlheber, and D. R. Dyroff, *J. Am. Chem. Soc.,* **91** (1), 62 (1969).
2. A. W. Frazier, J. P. Smith, and J. R. Lehr, *J. Agr. Food Chem.,* **13**, 316 (1965).
3. B. Dusek, F. Kutek, and P. Hegner, *Chem. Prum.,* **20**, No. 3, 106 (1970).
4. J. F. McCullough and R. C. Sheridan, U.S. Patent 3,912,802 (October 14, 1975).
5. S. I. Vol'fkovich, Z. G. Smirnova, V. V. Urusov, and L. V. Kubasova, *J. Appl. Chem. (USSR),* **45**, 483 (1972).
6. J. R. Lehr, E. H. Brown, A. W. Frazier, J. P. Smith, and R. D. Thrasher, *TVA Chem. Eng. Bull.,* 6 (1967).
7. T. C. Woodis, *Anal. Chem.,* **36**, 1682 (1964). The following modifications were employed: the sample was dissolved in 1% NaCl solution (to increase its solubility) and 30 μL of solution containing about 300 μg of phosphorus was spotted.

CORRECTION

PHOSPHORAMIDIC ACID AND ITS SALTS

Submitted by RICHARD C. SHERIDAN*

Ammonium hydrogen phosphoramidate releases ammonia upon heating and forms condensed phosphates.[1] The amount of polyphosphates formed depends on the length of time the sample has been heated or the rate of heating. Therefore the melting point, given in Reference 2, may vary and is not a suitable reference for judging the purity of this compound. X-Ray powder diffraction, infrared absorption, chromatography, chemical analysis, and chemical microscopy[3] are recommended as criteria of purity.

References

1. K. Dostal and L. Mezik, *Collect. Czech. Chem. Commun.,* **36**, 2834 (1971).
2. R. C. Sheridan, J. F. McCullough, and Z. T. Wakefield, *Inorg. Synth.,* **13**, 23 (1972).
3. M. L. Nielsen and W. W. Neilsen, *Microchem. J.,* **3**, 83 (1959).

*Division of Chemical Development, Tennessee Valley Authority, Muscle Shoals, AL 35660.

INDEX OF CONTRIBUTORS

SUBJECT INDEX

Names used in this Subject Index for Volume XIX, as well as in the text, are based for the most part upon the "Definitive Rules for Nomenclature of Inorganic Chemistry," 1957 Report of the Commission on the Nomenclature of Inorganic Chemistry or the International Union of Pure and Applied Chemistry, Butterworths Scientific Publications, London, 1959; American version, *J. Am. Chem. Soc.*, **82**, 5523-5544 (1960); and the latest revisions [Second Edition (1970) of the Definitive Rules for Nomenclature of Inorganic Chemistry]; also on the Tentative Rules of Organic Chemistry—Section D; and "The Nomenclature of Boron Compounds" [Committee on Inorganic Nomenclature, Division of Inorganic Chemistry, American Chemical Society, published in *Inorganic Chemistry, 7*, 1945 (1968) as tentative rules following approval by the Council of the ACS]. All of these rules have been approved by the ACS Committee on Nomenclature. Conformity with approved organic usage is also one of the aims of the nomenclature used here.

In line, to some extent, with *Chemical Abstracts* practice, more or less inverted forms are used for many entries, with the substituents or ligands given in alphabetical order (even though they may not be in the text); for example, derivatives of arsine, phosphine, silane, germane, and the like; organic compouncs; metal alkyls, aryls, 1,3-diketone and other derivatives and relatively simple specific coordination complexes: *Iron, cyclopentadienyl-(also as Ferrocene)*; *Cobalt(II), bis(2,4-pentanedionato)-* [instead of *Cobalt (II) acetylacetonate*]. In this way, or by the use of formulas, many entries beginning with numerical prefixes are avoided; thus *Vanadate (III), tetrachloro-*. Numerical and some other prefixes are also avoided by restricting entries to group headings where possible: *Sulfur imides*, with formulas; *Molybdenum carbonyl*, $Mo(CO)_6$; both *Perxenate*, $HXeO_6{}^{3-}$, and *Xenate(VIII)*, $HXeO_6{}^{3-}$. In cases where the cation (or anion) is of little or no significance in comparison with the emphasis given to the anion (or cation), one ion has been omitted; e.g., also with less well-known complex anions (or cations): $CsB_{10}H_{12}CH$ is entered only as *Carbaundecaborate(1-)*, *tridecahydro-* (and as $B_{10}CH_{13}$ – in the Formula Index).

Under general headings such as *Cobalt(III) complexes* and *Ammines*, used for grouping coordination complexes of similar types having names considered unsuitable for individual headings, formulas or names of specific compounds are not usually given. Hence it is imperative to consult the Formula Index for entries for specific complexes.

As in *Chemical Abstracts* indexes, headings that are phrases are alphabetized straight through, letter by letter, not word by word, whereas inverted headings are alphabetized first as far as the comma and then by the inverted part of the name. Stock Roman numerals and Ewens-Bassett Arabic numbers with charges are ignored in alphabetizing unless two or more names are otherwise the same. Footnotes are indicated by *n*, following the page number.

FORMULA INDEX

The Formula Index, as well as the Subject Index, is a cumulative index for Volumes XVI, XVII, XVIII, and XIX. The chief aim of this index, like that of other formula indexes, is to help in locating specific compounds or ions, or even groups of compounds, that might not be easily found in the Subject Index, or in the case of many coordination complexes are to be found only as general entries in the Subject Index. *All* specific compounds, or in some cases ions, with definite formulas (or even a few less definite) are entered in this index or noted under a related compound, whether entered specifically in the Subject Index or not.

Wherever it seemed best, formulas have been entered in their usual form (i.e., as used in the test) for easy recognition: Si_2H_6, XeO_3, NOBr. However, for the less simple compounds, including coordination complexes, the significant or central atom has been placed first in the formula in order to throw together as many related compounds as possible. This procedure often involves placing the cation last as being of relatively minor interest (e.g., $Co(C_5H_7O_2)_3Na$;$B_{12}H_{12}O$. Where they may be almost equal interest in two or more parts of a formula, two or more entries have been made: Fe_2O_4Ni and $NiFe_2O_4$; $NH(SO_2F)^{2-}$ $(SO_2F)_2NH$, and $(FSO_2)_2NH$ (halogens other than fluorine are entered only under the other elements or groups in most cases); $(B_{10}C_{11})_2Ni^{2-}$ and $Ni(B_{10}CH_{11})^{2-}$.

Formulas for organic compounds are structural or semistructural so far as feasible: $CH_3COCH(NHCH_3)CH_3$. Consideration has been given to probable interest for inorganic chemists, i.e., any element other than carbon, hydrogen, or oxygen in an organic molecule is given priority in the formula if only one entry is made, or equal rating is more than one entry: only $Co(C_5H_7O_2)_2$, but $AsO(+)-C_4H_4O_6Na$ and $(+)-C_4H_4O_6AsONa$. Names are given only where the formula for an organic compound, ligand, or radical may not be self-evident, but not for frequently occurring relatively simple ones like C_5H_5 (cyclopentadienyl), $C_5H_7O_2$ (2,4-pentanedionato), C_6H_{11} (cyclohexyl), C_5H_5N (pyridine). A few abbreviations for ligands used in the test, including macrocyclic ligands, are retained here for simplicity and are alphabetized as such, "bipy" for bipyridine, "en" for ethylenediamine or 1,2-ethanediamine, "diphos" for ethylenebis(diphenylphosphine) or 1.2-bis(diphenylphosphino)ethane or 1.2-ethanediylbis(diphenylphosphine), and "tmeda" for N,N,N',N'-tetramethylethylenediamine or N,N,N',N'-tetramethyl-1,2-ethanediamine.

Footnotes are indicated by *n*, following the page number.